Methods for Recovering Viruses from the Environment

Editor

Gerald Berg, Ph.D.
Department of Environmental Engineering
University of Cincinnati
Cincinnati, Ohio

CRC Press
Taylor & Francis Group
Boca Raton London New York

CRC Press is an imprint of the
Taylor & Francis Group, an **informa** business

First published 1987 by CRC Press
Taylor & Francis Group
6000 Broken Sound Parkway NW, Suite 300
Boca Raton, FL 33487-2742

Reissued 2018 by CRC Press

Library of Congress Cataloging-in-Publication Data

Methods for recovering viruses from the environment.

Bibliography: p.
Includes index.
1. Viruses--Isolation. 2. Sanitary microbiology--Technique. 3. Aquatic microbiology--Technique.
I. Berg, Gerald, 1928- .
QR385.5.M48 1987 616'.0194 87-10069
ISBN 0-8493-6246-6

A Library of Congress record exists under LC control number: 87010069

ISBN 13: 978-1-315-89536-9 (hbk)
ISBN 13: 978-1-351-07446-9 (ebk)

Visit the Taylor & Francis Web site at http://www.taylorandfrancis.com and the
CRC Press Web site at http://www.crcpress.com

FOREWORD

Without methods, there can be no research. Effective research requires effective methods, not always easy to come by.

The development of methods in environmental virology became a focus of growing interest about two decades ago. Progress has been significant since that time and in pure experimental systems, where there are no interferences, consistent high recoveries of viruses from environmental waters has been achievable for some time. In the natural environment, however, even in relatively clean waters, substances such as humic and fulvic acids interfere with viral recoveries and average recovery rates probably do not reach 20%. With sewage sludges and shellfish, recoveries are undoubtedly much lower.

Yet, even relatively low viral recovery rates have made possible the detection of viral hazards in drinking waters, recreational waters, sludges, and in shellfish and shellfish-growing waters. The hazards that exist are undoubtedly much greater than those demonstrated with the relatively inefficient methods developed thus far. Improving methods, as they are developed in the years to come, will undoubtedly bring the true extent of the hazards into better perspective.

CONTRIBUTORS

Gabriel Bitton, Ph.D.
Department of Environmental Engineering Sciences
University of Florida
Gainesville, Florida

G. Wolfgang Fuhs, Ph.D.
Division of Laboratories
California Department of Health Services
Berkeley, California

Charles P. Gerba, Ph.D.
Departments of Microbiology and Nutrition and Food Science
University of Arizona
Tucson, Arizona

Naomi Guttman-Bass, Ph.D.
Department of Environmental Health
Hebrew University
Jerusalem, Israel

John E. Herrmann, Ph.D.
Division of Infectious Diseases
University of Massachusetts Medical School
Worcester, Massachusetts

Christon J. Hurst, Ph.D.
Health Effects Research Laboratory
U.S. Environmental Protection Agency
Cincinnati, Ohio

Thomas W. Mix, Ph.D.
Merix Corporation
Babson Park
Wellesley, Massachusetts

V. Chalapati Rao, Ph.D.
Department of Virology and Epidemiology
Baylor College of Medicine
Houston, Texas

Mark D. Sobsey, Ph.D.
Department of Environmental Sciences and Engineering
University of North Carolina
Chapel Hill, North Carolina

Charles A. Sorber, Ph.D.
Department of Civil Engineering
University of Texas
Austin, Texas

TABLE OF CONTENTS

Chapter 1

RECOVERING VIRUSES FROM SEWAGE, EFFLUENTS, AND WATER

Charles P. Gerba

TABLE OF CONTENTS

I. INTRODUCTION

Waterborne outbreaks of disease produced by hepatitis A virus (enterovirus 72), rotaviruses, Norwalk and Norwalk-like agents, and other enteric viruses are now well documented.[1-5] It is now clear that the Norwalk agent is a major cause of waterborne nonbacterial gastroenteritis in the U.S.[5] While difficult to prove because of limitations of epidemiology and a lack of methods for detection, other groups of enteric viruses, such as the enteroviruses, most surely are responsible for undocumented waterborne disease. Recent studies with human volunteers have clearly shown that small numbers of enteric viruses delivered by the water route can cause disease.[6] The presence of even one virus, detectable in tissue culture, in a water supply poses a potential disease hazard. It has been calculated that if a water supply contains as little as one plaque-forming unit of virus per 50 gal of drinking water in a community utilizing 50 mgd (0.2% of which is ingested as drinking water), at a 30% infection rate, 600 individuals would daily contract a variety of clinical and subclinical infections.[7] Thus, most current and proposed standards for viruses in water (Table 1) suggest methods capable of detecting viruses in 10 to 1000 ℓ of water.

Although a great variety of methods for concentrating viruses from water have been evaluated, only a few have proven practical. The major limitations in methods have been the scale-up of systems to process large volumes, lack of ability to process samples on site (which is desirable when large volumes are processed), and cost.

Adsorption-elution of viruses from microporous filters is the most widely used method today, although other methods may be equally useful for some types of waters. This chapter will focus on the most promising and extensively used methods for virus concentration from water. Extensive literature reviews of this topic have been published elsewhere.[8,9]

II. VIRUSES AND THE WATER ENVIRONMENT

Concentration of viruses from water is not achieved simply by an application of methods; it is a strategy in which the types of viruses that must be isolated, the nature of those viruses, and their interaction with their environment are taken into consideration. The nature of the water from which a virus must be isolated is often a limiting factor in the choice of methods that may be employed. The types of water most commonly studied are potable water, freshwater, seawater, and sewage. These waters vary greatly in their physical-chemical composition and possess substances either dissolved or suspended in solution which may interfere with our ability to employ various concentration methods. The greatest limitation with the application of microporous filter adsorption-elution methods and ultrafiltration has been the turbidity of the water. While microporous methods do not depend upon the physical filtration of the virus, they do require pores of a small enough size to allow physical contact of the virus with adsorption sites on the filter. Although this problem can be overcome by increasing the surface area of the filter, where relatively high concentrations of viruses may be present (i.e., in sewage) methods involving flocculation or adsorption to particulate solids may be more practical.

Treated drinking water (i.e., with chemical flocculation and sand filtration) and most ground waters have little or no suspended matter, thus making processing of very large volumes by microporous filter optimal. Pleated microporous filters are capable of processing up to 5000 gal (19,000 ℓ) of finished drinking water[10] and even large diameter disc filters (142 to 293 mm) can easily process 100 to 400 ℓ of tapwater. Clogging problems with tapwater occur because of the retention by the filters of metal complexes that form when the water must be adjusted to low pH, and by the precipitation of humic acid.[11]

Concentration of viruses from sewage by microporous filters is often difficult because the high turbidities of these waters rapidly clog filters. Moreover, it may become difficult to

Table 1
STANDARDS FOR VIRUSES IN WATER

Water type	Concentration PFU/volume	Comment
Treated effluent	1/10 gal	Montgomery County, Md.
Effluent dominated stream		Arizona
Recreational use		
Full body	1/40 ℓ	
Partial body	125/40 ℓ	
Irrigation (used on feeds eaten raw)	1/40 ℓ	
Direct recycled water	0/10 ℓ	South Africa
Drinking water	0/10 ℓ	European Economic Community
Bathing water	0/10 ℓ	European Economic Community
World Health Organization	0/10—1000 ℓ	Tapwater

elute solids-associated viruses entrapped on filters, and soluble organics may interfere with virus adsorption. Thus, efficiency of virus concentration is often less from sewage compared with tapwater. Fortunately, large volumes of sewage do not normally have to be processed to detect viruses. Usually, only 100 mℓ to 2 ℓ need to be processed to detect viruses in raw sewage. Concentration of viruses from settled secondarily treated sewage (i.e., activated sludge) is more easily accomplished because of lower turbidities and lower concentrations of soluble organic matter. From this author's experience, processing 5 to 20 ℓ of sewage from a well-operated, activated sludge plant may be necessary to detect viruses, especially if chlorination is practiced. Volumes of 200 to 400 ℓ of secondarily treated sewage can be processed by microporous filter adsorption-elution methodologies.

Seawater also often contains high concentrations of suspended matter and dissolved solids. Although viruses may be readily detected in small volumes of highly polluted coastal waters, sample volumes of 20 to 100 gal (95 to 378 ℓ) are usually processed.[13] Microporous filter methods are capable of processing volumes of this size, but modifications of the procedure may be necessary.[13]

The nature of the virus may also place certain limitations on the method that may be employed or require modifications to improve chances of isolation. For example, many commonly used methods for the concentration of enteric viruses require alteration in the pH to either adsorb or elute a virus from a surface. Almost all enteric viruses are stable at pH 3 to 5 for long periods of time. Many enteroviruses are not rapidly inactivated between pH 10 and 11.5. However, there are strain differences in the inactivation of enteroviruses and probably other enteric viruses as well. For example, poliovirus 1 strain LSc is much more stable at pH 11.5 than the Chat strain.[14] Most other animal viruses are not as stable as the enteroviruses at pH levels above 10. Adenoviruses, reoviruses, and rotaviruses are rapidly inactivated above pH 10. Many coliphages are unstable at both high and low pH. Viruses also vary in their isoelectric points because of differences in the composition of the proteins which make up their outer capsids. This is especially significant for methods that depend upon adsorption where electrostatic phenomena are involved. In neutral solutions and in most natural waters, viruses are negatively charged, and there occur strong electrostatic repulsion forces between viruses and similarly charged surfaces. These electrostatic forces can, to a large measure, be controlled by altering environmental factors such as pH and electrolyte concentration. However, the variation in the isoelectric points can result in different optimal conditions for adsorption of different viruses to identical surfaces under a given set of conditions and can account to some extent for differences in efficiencies for adsorption of different viruses.

One of the greatest concerns in detection of viruses in water is the recovery of solids-associated viruses.[15-17] Enteric viruses in sewage discharges may be either adsorbed to the surface of the solids or embedded within.[15-17] Almost all sewage and water treatment methods reduce the amount of suspended material initially present, thereby reducing the amount of solids-associated virus in suspension. Studies on solids-associated viruses in discharges from activated sludge and trickling filter sewage treatment plants indicate that the percentage of solids-associated enteroviruses ranges from 3 to 100% of the total number of viruses in the discharge.[18] The association of viruses with solids is known to protect them from inactivation in natural waters[19] and from the action of disinfectants.[20] Thus, methods for virus concentration should be capable of detecting solids-associated viruses.

The presence of organic and proteinaceous substances may also interfere with virus detection. Organic matter may interfere with virus adsorption to surfaces[21,21a,21b] or with virus assay.[22] One may usually expect a lower efficiency of virus recovery from waters high in organic matter.

III. METHODS FOR CONCENTRATING VIRUSES FROM LARGE VOLUMES OF WATER

A. Filter Adsorption Methods

1. Negatively Charged Filters

The first attempts at recovering viruses from water employed adsorption to gauze pads.[23] In this method, a sterile gauze pad or sanitary napkin is suspended in flowing water for several days to allow virus particles to accumulate. The pH of pad and fluid content are raised to 8.0 to elute the viruses. The fluid is then expressed from the pad and assayed for viruses. Although convenient, the gauze pad method in only a qualitative method, and for this reason it is no longer extensively used.

Although others pointed out the potential for the use of microporous filters for virus concentration from water,[24,25] Wallis and Melnick[26] are credited with the development of the first practical microporous filter system for the concentration of viruses from water and wastewater. Early methods for virus concentration used nitrocellulose membrane filters that were clogged easily, and sample sizes were generally less than 5 ℓ.[26-29] Wallis et al.[30] found that adsorption of viruses to nitrocellulose filters could be greatly enhanced by addition of trivalent ($AlCl_3$) and divalent ($MgCl_2$) salts and adjustment of the pH of water to 3.5. They adsorbed enteroviruses to the filters and eluted them with 0.05 *M* glycine adjusted to pH 11.5 with NaOH. They also described a reconcentration procedure whereby viruses recovered from filters were readsorbed to smaller diameter filters and eluted with smaller volumes of the glycine, thus greatly reducing the amount of material to be assayed.

These developments provided the basis for the first portable virus concentrator for concentrating viruses in the field.[31,32] The original system was a continuous flow-through apparatus in which incoming water was passed through a series of five nonvirus-adsorbing clarifying textile filters followed by treatment with an anion-exchange resin to remove organics that clogged the adsorbent filters before adsorption of the viruses onto 293 mm diameter cellulose nitrate disc filters. Magnesium chloride was added via injection to enhance adsorption. Viruses were eluted from the filter with 1-ℓ volumes of pH 11.5 glycine buffer. The primary eluate was then reconcentrated on smaller diameter cellulose nitrate membranes from which viruses could be eluted with small volumes (approximately 20 mℓ) of pH 11.5 glycine. With this concentration-reconcentration procedure, the virus concentrator could recover 80% of exogenously added viruses during laboratory processing of 300 gal (1134 ℓ) of tapwater. Further refinement of the portable virus concentrator resulted in a unit in which the water was first clarified through a series of orlon or polyester depth spun fiber cartridge filters and the viruses adsorbed to a fiberglass or cellulose acetate spun fiber cartridge depth filter.[33] In addition to these changes, $AlCl_3$ (0.0005 *M*) was used instead of $MgCl_2$ to enhance virus adsorption since a 100-fold less concentration was needed. This system was used successfully in field studies for the isolation of naturally occurring viruses in wastewater and seawater.[33,34] Based on the prototype virus concentrator developed by Wallis and Melnick, the Carborundum Company (Niagara Falls, N.Y.) made commercially available a self-contained virus concentrator referred to as the "Aquella® Virus Concentrator".

These early systems efficiently concentrated viruses from large volumes of tapwater and small volumes of sewage and seawater, but several problems soon became evident. Suspended matter in wastewater and other turbid waters clogged filters and greatly reduced the flow rate. The use of the clarifying filters in front of the adsorbing filters reduced the magnitude of this problem, but often resulted in a decreased efficiency of virus recovery because of the loss of solids-associated viruses on the clarifying filters.[33,34]

Further study resulted in a modified version of the portable virus concentrator.[35] In this system, tapwater was acidified to pH 3.5 with 1 *N* HCl and then passed through a virus adsorber consisting of a fiberglass cartridge depth filter (K-27) and a 142 mm diameter, 0.65 μm pore size Cox® membrane filter in series. With this method, small quantities of poliovirus 1 in 100-gal (378-ℓ) volumes of tapwater were concentrated nearly 40,000-fold with average recoveries of 77%. This methodology was useful for concentrating viruses for 50- to 100-gal amounts of seawater[36] and well water beneath a wastewater land treatment site.[37]

Unfortunately, this system could not be effectively scaled up to process larger volumes at reasonable flow rates. Because of the limited surface area of the flat-disc adsorbent filters, maximum flow rates of only 3 gal (approximately 11 ℓ) per minute could be achieved with finished tapwater. Moreover, humic acid and other organic compounds were concentrated along with the virus.[38] These substances seriously interfered with reconcentration of the initial eluates on membrane filters[38] when greater than 100 gal of tapwater or smaller volumes of seawater and sewage were processed.

To overcome these limitations, Farrah et al.[39] tested a variety of membrane filters and found that Filterite® fiberglass membrane filters (Duo-Fine® series) were far less easily clogged than Cox® series AA, acrylonitrile polyvinyl chloride copolymer filters (Acropor series), and nitrocellulose (Millipore®) of approximately the same rated pore size. All filters adsorbed greater than 90% of poliovirus added to tapwater at pH 3.5. Filterite Duo-Fine® filters are manufactured as 10-inch (approximately 25.4 cm) long, pleated cartridges whose surface areas are 280 times that of a 47 mm diameter disc filter. Flow rates of up to 37.8 ℓ/min (10 gal/min) were obtained with the pleated membrane filter cartridges. Large volumes of tapwater would be processed without the need of a prefilter, but a prefilter (a 3 μm pore size Filterite® or a K-27 spun fiberglass depth filter) was used to spare the 0.25 or 0.45 μm final adsorbing filter when processing sewage and marine waters. The 0.45 μm nominal pore size filter was used with sewage and seawater to reduce any problems with clogging. With these filters, seeded poliovirus could be recovered from 472 to 1900 ℓ (125 to 500 gal) of tapwater, 378 ℓ (100 gal) of seawater and 19 to 190 ℓ (5 to 50 gal) of secondarily treated sewage with an average efficiency of 52, 53, and 50%, respectively. Because of clogging with humic acid and other organics eluted from the filters in the initial concentration step, reconcentration was accomplished by aluminum flocculation followed by hydroextraction.[40]

After autoclaving or soaking overnight in a concentrated solution of sodium hydroxide, the filters could be reused several times without loss of concentration efficiency, thus reducing operational costs.[41]

Tube filters (Balston®) have also been commonly used for virus concentration from tapwater. Studies with these epoxy-fiberglass filters were first described by Jakubowski et al.[42] Studies with this method have been described in detail in the 15th edition of *Standard*

Methods for the Examination of Water and Wastewater.[43] Virus adsorption to the filters was optimized by adjustment of pH to 3.5 and addition of $AlCl_3$ to the test water.

Elution of viruses was accomplished with pH 11.5 glycine buffer. The recovery of seeded poliovirus from 100-gal (378-ℓ) volumes of tapwater ranged from 42 to 57% with 8-m pore size tube filters.

Unfortunately, tube filters suffer from many of the drawbacks inherent in flat-disc membrane filters. The Balston® tube filters clog more readily than pleated membrane filters and cannot be used even with moderately turbid water. They cannot be used above an in-line pressure of 25 lb/in.^{2}, and thus high flow rates cannot be achieved. This limits their practical use to only finished tapwater of good quality. In contrast, pleated filters can operate with an in-line pressure of up to 100 lb/in.^{2} Also, systems utilizing flat and tube filters can only process tapwater at about 5 to 10 ℓ (2.5 gal) per minute[35,36,42] compared to almost 40 ℓ (approximately 10 gal) per minute for pleated cartridge filters.[39]

In an extensive comparative study, Sobsey et al.[44,45] evaluated the relative efficiencies of several microporous filters for concentrating viruses from large volumes of tapwater. Four primary virus adsorbent systems were evaluated (1) Balston® fiberglass-epoxy filter tubes, (2) Cox® fiberglass-asbestos-epoxy filter discs, (3) Millipore® cellulose nitrate membranes, and (4) commercial filters (fiberglass cartridge filters) and Cox® fiberglass-asbestos-epoxy filter discs in series. The efficiencies with these systems for recovering viruses from tapwater with a tentative standard method[44] were generally poor. For enteroviruses (poliovirus, coxsackievirus A9, coxsackievirus B3, echovirus 1), recoveries ranged from 0.7 to 32% and averaged 16% during 15 months of evaluation. Average recovery efficiency of natural enteroviruses mixtures obtained from raw sewage was only 3.9%. Further studies evaluating Zeta-plus, 1MDS Virozorb, and epoxyfiberglass Filterite® filters indicated that greatly improved, more consistent recoveries could be achieved with these filter systems. Optimal adsorption of viruses to the Filterite® filters required addition of magnesium and aluminum salts to the test water and reducing its pH to 3.5. Overall recoveries of poliovirus 1 were 34% for Filterite® and 48% for 1MDS from 1000 ℓ of tapwater. In these studies, temporal changes were observed for poliovirus adsorption and recovery efficiencies.[46] Further experimentation suggested that the observed variations in virus concentration efficiencies may have been related to changes in the concentrations of unidentified adsorption-interfering soluble or colloidal components in the tapwater used in the studies. Dahling and Wright[45a] recently reported better recoveries of poliovirus 1, in seeded experiments, when $MgCl_2$ was added to supply the cation than when $AlCl_3$ was used for this purpose.

2. *Positively Charged Filters*

Electronegative filters used for virus concentration require conditioning of the water by pH adjustment and/or addition of cations to optimize virus adsorption. This is necessary to reduce the electrostatic repulsion between the virions and the new filter surface. To reduce the need for preconditioning the water, Sobsey and co-workers[44-47] tested microporous filters which carry a less negative net charge in the pH range of most natural and tapwaters.

These filters are composed of cellulose-diatomaceous earth "charge modified" resin mixtures and are referred to as series S®, Zeta-plus filters (AMF/CUNO, Meriden, Conn.). These filters remain positively charged up to about pH 6 and may adsorb 99% of added poliovirus from water at pH 7.5 without the addition of salt. Small amounts of poliovirus in 380 ℓ of water were concentrated in a final volume of 25 mℓ with this two-stage procedure with an average efficiency of recovery of 22.5%.[47] It is important that the authors used monodispersed viruses for testing, which probably accounts for the overall apparent lower recovery of viruses because many elution procedures probably break up aggregates giving high apparent recovery of viruses. These studies suggested that positively charged filters might have major advantages over the negatively charged filter systems. Viruses could be adsorbed from most waters without prior conditioning of the water, and elution could proceed at lower pH levels thus reducing the risk of viral inactivation at high pH levels.

Utilization of Zeta-plus positively charged filter media for virus recovery is dependent upon the ambient pH of the water tested.[46,48] The efficiency of virus adsorption to the filters declines noticeably above pH 8.5.[46,48] Moreover, when large volume of water are processed, the ability of the filters to adsorb viruses from waters with a pH near neutral may be reduced.[49] This problem may be overcome by adjustment of the pH of the test water to 5.5 to 6.0 before filtration.[47] Because viruses possess different isoelectric points, the efficiency of virus concentration by Zeta-plus filters differs for different viruses.[50]

Because many viruses are inactivated at the low pH levels required for optimal adsorption to electronegative filters, Zeta-plus filters have been more useful for concentrating a wide variety of animal, plant, and insect viruses.[50-52]

Zeta-plus filters have also been useful for concentrating enteric viruses from sewage effluents.[53,54] Poliovirus 1 and coxsackievirus B3 could be recovered from 19 ℓ of activated sludge effluent with efficiencies of 40 to 56%.[54] In a comparative study of naturally occurring enteroviruses in primary sewage, no significant difference in virus recovery was observed between the electronegative Filterite® filters (which required effluent adjustment to pH 3.5 and the addition of $AlCl_3$) and Zeta-plus filters.[54] With the realization of the advantages of positively charged media for virus concentration, AMF/CUNO (Meriden, Conn.) developed a positively charged filter especially designed for concentrating viruses from water. This media, 1MDS Virozorb, was tested by Sobsey and Glass[46] for its efficiency in concentrating poliovirus from tapwater. One hundred gallon samples of tapwater were passed through double-layered, pleated-sheet cartridges of a design similar to the Filterite® negatively charged fiberglass media.[39] In a comparative study, adsorption of viruses from tapwater between pH 3.5 and pH 7.5 was more efficient with electropositive filters than with Filterite® filters. Beef extract-glycine at pH 9.5 eluted poliovirus effectively from 1MDS filters. In paired comparative studies, 1MDS filters, with adsorption at pH 7.5 and no polyvalent cations added, gave less variable results in virus concentration efficiencies than did Filterite® filters with adsorption at pH 3.5 in the presence of added $MgCl_2$. Recovery of poliovirus from 1000-ℓ (279-gal) tapwater volumes was approximately 30% efficient with both Virozorb 1MDS and Filterite® pleated cartridge filters, but the former were much simpler to use. The 1MDS Virozorb filters have been useful for concentrating enteroviruses from estuarine waters,[55] sewage,[54] swimming pool waters,[56] and ground waters.[57] Rotaviruses and hepatitis A virus can also be concentrated from water with these filters.[57,58]

Other positively charged filter cartridge material currently available include the Zeta-plus MK Filters and the spun fiber MW® filters. The MK® filter tubes of melamine resin bound fibers have been reported to effectively concentrate polioviruses from tapwater and are reusable and less expensive than other positively charged cartridge media.[59]

Asbestos fibers are also positively charged, and filters made of such fibers also exhibit a net positive charge.[47] Such filters effectively adsorb and concentrate viruses from water[47,60] but are not in common use in the U.S. because of the link between asbestos fibers and cancer in humans.

3. Elution of Viruses from Filters and Reconcentration

Many substances have been evaluated for their ability to elute adsorbed viruses from filters surfaces. Elution usually depends upon weakening the electrostatic and hydrophobic interactions that bind viruses to filter surfaces or is achieved by addition of substances that compete with viruses for adsorption sites. High pH buffers, which weaken electrostatic interactions, proved very useful for some groups of viruses but could not be used with other groups sensitive to inactivation at pH levels above 10. Moreover, elution with high pH buffers required rapid neutralization of the eluate to reduce virus inactivation. As a replace-

Table 2
SUBSTANCES USED FOR THE ELUTION OF VIRUSES FROM MICROPOROUS FILTERS

Eluent	Elution pH	Remarks
Glycine	10.5—11.5	If pH 11.5 is used, pH must be readjusted to 9.0 or below within 5 min of collection to prevent viral inactivation; not recommended for pH-sensitive viruses
Beef extract	9.0—10.0	May encounter problems with reconcentration because of poor flocculation with newer preparations of beef extract
Tryptose phosphate		May be better than beef extract for eluting rotaviruses
Nutrient broth, 5X	9.0	Has not been extensively used
Arginine, lysine		Has not been extensively used

Modified from Taylor, J. W., Gary, G. W., Jr., and Greenberg, H. B., *Am. J. Epidemiol.*, 114, 584, 1981.

ment for high pH, basic amino acids,[61] casein,[62] tryptose phosphate broth,[63] nutrient broth,[64] bovine serum albumin,[52] chaotropic agents,[65,66] beef extract,[67] urea,[68] etc. have been tested (Table 2).

Since many proteinaceous substances compete very effectively with viruses for adsorption, they have been the most often used. The most commonly used eluent in recent years has been a solution of 3% beef extract adjusted to pH 9.0 to 10.5.[67,69] Not only can beef extract be used at a lower pH, but reconcentration of the primary eluate is more easily achieved by a bioflocculation procedure originally described by Katzenelson and co-workers.[68a]

Farrah and co-workers[61,65,68,70,71] attempted to elucidate the mechanisms of virus interaction with filter surfaces as an aid to optimizing virus concentration by microporous filters. Utilizing chaotropic and antichaotropic salts, they studied the effects of poliovirus and bacteriophage MS-2 adsorbed to microporous filters. Chaotropic ions are large ions with a single charge, such as trichloroacetate, iodide, and thiocyanate, whereas antichaotropic ions are either small ions with a single charge, such as fluoride, or ions with multiple charges, such as sulfate and phosphate. Using both electronegative and electropositive filters, these authors found that solutions of chaotropic agents, such as sodium trichloroacetate (NaTCA) or sodium iodide (NaI) were similar to high pH buffers in their ability to elute viruses. Equimolar solutions of antichaotropic agents, such as sodium fluoride (NaF) or sodium phosphate ($NaHPO_4$), on the other hand, were less effective for viral elution. Since antichaotropic agents are believed to strengthen entropy-driven hydrophobic associations, and since this effect is antagonized by detergents (i.e., Tween® 80) and also by chaotropic salts which weaken hydrophobic associations, these authors hypothesized that adsorption of viruses to filters at high pH resulted from largely hydrophobic interactions. At low pH levels, they found that elution of viruses from filters was dependent on disruption of both electrostatic and hydrophobic interactions.[71] Thus, it was demonstrated that virtually 100% of the viruses adsorbed to filters at pH 4 could be eluted at the same pH with solutions containing both Tween® 80 and either $MgCl_2$ or NaCl.[71]

Utilizing these observations Moore et al.[66] developed a method using the chaotropic agent $NaNO_3$ for elution of poliovirus and echovirus from the electropositive 1MDS filters. Supplementation of beef extract solutions with $NaNO_3$ allowed for the effective elution of viruses from pH 5.5 to 9.5 at efficiencies of 70% and greater.

For detection of viruses in small amounts of fluids, a one-step adsorption-elution procedure is often used. With larger volumes, however, a second-step reconcentration method must be incorporated. The addition of second-step reconcentration to the virus concentration system may reduce virus recovery, but it reduces the amount of eluate to a manageable volume for subsequent assays.[72]

Techniques that have been used for reconcentration include: two-phase separation,[73] hydroextraction,[40,73] precipitation with inorganic salts,[73] continuous-flow ultracentrifugation,[74,74a] readsorption to and elution from successively smaller diameter filters,[32,75] and bioflocculation.[67]

Elution with glycine enables one to reconcentrate an eluate on another set of smaller diameter filters. Thus, the pH of a sample is adjusted to 3.5, an appropriate salt is added, and the sample is passed through filters. The adsorbed viruses are subsequently eluted with a high pH buffer such as 0.05 *M* glycine at pH 11.0 to 11.5. The pH of this eluate is readjusted to 3.5, appropriate salt is added, and viruses are adsorbed to smaller membrane filters that are subsequently eluted with a still smaller volume of the glycine.

These procedures do not work, however, when large volumes of tapwater[10] or turbid estuarine water[40] are sampled because organic compounds and metal ions present in water are adsorbed to membrane filters along with the viruses. Elution of viruses from these filters results in elution of the organic compounds as well.[76] Subsequent acidification of the eluate for sequential adsorption-elution results in the production of organic flocs that clog the smaller diameter filters used for the second concentration (reconcentration) step.

The organics and metal ions also form an insoluble precipitate when the eluate is neutralized and interfere with reconcentration on smaller filters when greater than 100 gal of tapwater or smaller volumes of seawater and sewage are processed.[10] By removing organic compounds with activated carbon and metal ions with ion-exchange resins, reconcentration can be performed on membrane filters.[76] However, to treat eluates with large amounts of organics, large amounts of carbon and resin are required, and a high percentage of virus adsorption to the columns occurs. Farrah and Bitton[68] reported an alternate reconcentration procedure in which they used a 3% lysine solution at pH 9 to elute viruses adsorbed to Filterite® filters. When this eluate (approximately pH 8.8) was passed through a 25-mm diameter Zeta-plus C-10 filter, 85 to 90% of the viruses adsorbed to the Zeta-plus filter, which was then eluted with 4 mℓ of 3% beef extract solution at pH 9. With this procedure, they were able to concentrate seeded poliovirus from 4 ℓ of tapwater to a final volume of 4 mℓ with an average recovery of 67%. Although this method seems to work well with small volumes, experiments need to be done with large volumes of tapwater to make it a feasible field procedure.

When a protein solution such as beef extract is used to elute viruses adsorbed to filters, the sequential membrane adsorption-elution scheme does not work for reconcentration of such eluates because proteins interfere with virus adsorption to filters.[33] To alleviate this problem, Katzenelson et al.[68a] described an organic flocculation method that does not involve membrane filters. In this method, viruses are efficiently eluted from filters by 3% beef extract solution at pH 9.[32,77,78] The pH of the eluate is then lowered to 3.5. This results in flocculation of proteins that are concentrated by centrifugation. This sediment obtained from flocculation of the beef extract is solubilized in 0.15 *M* Na_2HPO_4 (pH 9.0).

Two problems with the use of bioflocculation reconcentration procedures are virus-to-virus variation in concentration efficiency and variations in the size of the floc obtained with different preparations of beef extract. Morris and Waite[79] reported that efficiency of organic flocculation varied from a mean of about 40% for polioviruses to 7% for echovirus 1. Bass and Nasser,[79a] however, seeded tapwaters concurrently with four different enteroviruses, passed the waters through Balston® filters, and recovered almost all of the viruses with near equal efficiencies from beef extract eluates reconcentrated by organic flocculation. Rotaviruses appear to be less efficiently concentrated by beef extract.[80] Apparent changes in manufacturing processing have also greatly reduced the size of the floc obtained with beef extract, although efficiency of virus concentration does not appear to always be significantly affected.[81]

4. Portable Field Systems

Portable systems for concentration of viruses from large volumes of water have been developed and some have become commercially available. Carborundum (Niagara Falls, N.Y.) was the first company to manufacture a virus concentrator. The "Aquella® Virus Concentrator" was based on the filter system initially developed by Wallis and Melnick[31-33] and involved the passage of water through a series of spun fiber clarifying filters, acidification of the water by injection of 1 *N* HCl under positive pressure, and passage through a series of virus adsorbing filters, i.e., a fiberglass spun filter and a disc Cox® series filter. Hill et al.[72] described a portable system for use with Balston® filters that incorporated a fluid proportioner for injecting acid, salts, and, when necessary, sodium thiosulfate for neutralization of free chlorine. The device, when properly adjusted, automatically injects the correct amount of acid or other chemical additive and automatically adjusts the amount injected if changes in flow rate occur. Earlier systems required the injection of additives from a pressure tank and continuous monitoring with an in-line pH probe[10] or placement of the water sample in a large drum and adjustment of the pH before passage through the virus adsorbent. Unfortunately, hydraulic fluid proportioners are expensive, tend to break down frequently when used with turbid waters, and limit flow rate. To overcome these problems, Payment and Trudel[83,84] developed a simple, low cost injector that uses the Venturi effect to draw liquid by creating a partial vacuum and then injecting it in the main water stream.

The development of positively charged filters has greatly simplified potable water concentration systems.[46] The Microfiltration Products Division of AMF/CUNO (Meriden, Conn.) currently manufactures a virus concentration kit enclosed in an aluminum suitcase that contains a filter housing, flow meter, and adjustable chemical feeder for applications requiring the neutralization of chlorinated water. This system used the 1MDS Virozorb filters in a pleated cartridge housing. A portable virus concentrator system that uses disc Zeta-plus filters for adsorbing virus has been described by Logan et al.[51] Recently, Toranzos et al.[84a] adapted a pressure spray tank for supplying the positive pressure necessary for processing water through microporous filters in the field.

Currently, it is difficult to suggest any standard filter methodology for the concentration of viruses from water because of the rapid development of new techniques and filter media. A number of tentative schemes for a standard method for concentrating viruses from finished tapwaters are presented in the 15th edition of *Standard Methods for the Examination of Water and Wastewater.*[43] The USEPA has published step-by-step standard methods[43a] and the American Society for Testing Materials has round-robined one of the USEPA methods.[84b] A flow diagram of recommended strategies is shown in Figure 1. Differences in the strategies occur depending upon the type of adsorbent filter, type of eluent, and choice of reconcentration method.

Studies comparing various filter media for tapwater [42,44,45,85] have indicated that the Filterite® and 1MDS Virozorb give the best overall efficiencies of virus recovery when beef extract is used for the primary elution. Tables 2, 3, and 4 list the various media that have been used for virus concentration from water and methods that can be used for reconcentration of filter eluates.

Because of the complexity of the methods for concentrating viruses from large volumes of water, an evaluation of the efficiency of any method used with a vaccine strain of poliovirus should first be undertaken by those not familiar with the technique before attempts are made to isolate naturally occurring viruses. Moreover, because virus concentration methodology is so dependent on the nature of the water tested, it is beneficial to first evaluate recovery efficiency with artificially seeded viruses to obtain an idea of the efficiency of virus recovery from the water examined.

Because the nature and type of suspended matter vary from one type of water to another, one filter system may be more useful than another. The volume processed also influences

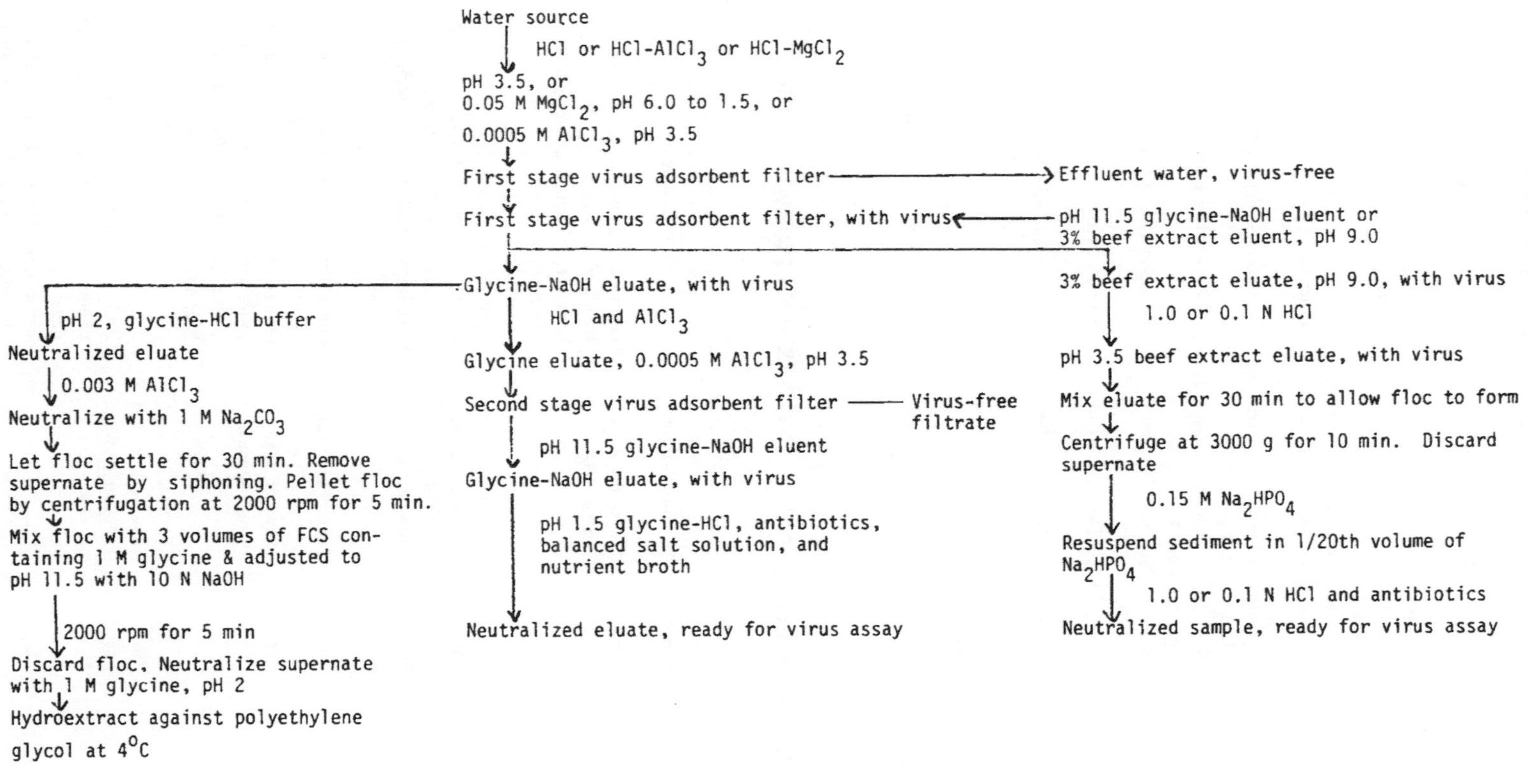

FIGURE 1. Method for concentrating enteroviruses from large volumes of water. (Modified from *Standard Methods for the Examination of Water and Wastewater*, 15th ed., American Public Health Association, Washington, D.C., 1980.)

Table 3
FILTERS COMMONLY USED FOR VIRUS CONCENTRATION

Composition	Manufacturer	Code
Nitrocellulose	Millipore Corp., Bedford, Mass.	HA
Epoxy-fiberglass-asbestos	Cox Instrument Corp., Detroit, Mich.	Series AA[a]
Epoxy-fiberglass	Filterite Corp., Timonium, Md.,	Duo-Fine series (cartridge filters)
Borosilicate glass microfiber-epoxy resin	Balston, Inc., Lexington, Mass.	Filter tubes
Fiberglass	Commercial Filter Div., Carborundum Co., Lebanon, Ind.	Wound fiber depth filter (K-27)
Fiberglass	Nucleopore, Pleasanton, Calif. (formerly Johns-Manville)	D39 (very fine prefilter)
Charge modified cellulose inorganic filter aid	AMF, CUNO Div., Meriden, Conn.	Zeta-plus (S and C grades)
Charge modified cellulose	AMF, CUNO Div., Meriden, Conn.	1-MDS (Virozorb)
Cellulose and glass fibers	AMF, CUNO Div., Meriden, Conn.	Micro-Klean tube filter

[a] No longer manufactured with asbestos.

Modified from Farrah, S. R., Goyal, S. M., Gerba, C. P., Wallis, C., and Shaffer, P. T. B., *Water Res.*, 10, 897, 1976.

Table 4
METHODS USED FOR CONCENTRATING VIRUSES FROM WATER

Method	Initial volume of water	Applications	Remarks
Filter Adsorption-Elution Negatively charged filters	Large	All but the most turbid waters	Only system shown useful for concentrating viruses from large volumes of tapwater, sewage, seawater, and other natural waters; cationic salt concentration and pH must be adjusted before processing
Positively charged filters	Large	Tapwater, sewage, seawater	No preconditioning of water necessary at neutral or acidic pH levels
Adsorption to metal salt precipitate, aluminum hydroxide, ferric hydroxide	Small	Tapwater, sewage	Have been useful as reconcentration methods
Charged filter aid	Small	Tapwater, sewage	40-ℓ Volumes tested, low cost; used as a sandwich between prefilters
Polyelectrolyte PE60	Large	Tapwater, lake water, sewage	Because of its unstable nature and lot-to-lot variation in efficiency for concentrating viruses, method has not been used in recent years
Bentonite	Small	Tapwater, sewage	
Iron oxide	Small	Tapwater, sewage	
Talcum powder	Large	Tapwater, sewage	Can be used to process up to 100-ℓ volumes as a sandwich between filter paper supports
Gauze pad	Large		First method developed for detection of viruses in water, but not quantitative or very reproducible

Table 4 (continued)
METHODS USED FOR CONCENTRATING VIRUSES FROM WATER

Method	Initial volume of water	Applications	Remarks
Glass powder	Large	Tapwater, seawater	Columns containing glass powder have been made that are capable of processing 400-ℓ volumes
Organic flocculation	Small	Reconcentration	Widely used method for reconcentrating viruses from primary filter eluates
Protamine sulfate	Small	Sewage	Very efficient method for concentrating reoviruses and adenoviruses from small volumes of sewage
Polymer two-phase	Small	Sewage	Processing is slow; method has been used to reconcentrate viruses from primary eluates
Hydroextraction	Small	Sewage	Often used as a method for reconcentrating viruses from primary eluates
Ultrafiltration			
Soluble filters	Small	Clean waters	Clogs rapidly even with low turbidity
Flat membranes	Small	Clean waters	Clogs rapidly even with low turbidity
Hollow fiber or capillary	Large	Tapwater, lake water	Up to 100 ℓ may be processed, but water must often be prefiltered
Reverse osmosis	Small	Clean waters	Also concentrates cytotoxic compounds that adversely affect assay methods

Modified from Gerba, C. P., *Viral Pollution of the Environment*, CRC Press, Boca Raton, Fla., 1983, 19.

the type of filter that must be used. For example, small diameter cellulose nitrate filters may be adequate for processing 1 or 2 ℓ of tapwater, but not the same volume of sewage or river water. The Filterite® fiberglass filters and the Nucleopore D39 (very fine fiberglass prefilter) are the most resistant to clogging of the negatively charged filter media.[39] The 1MDS Virosorb is also resistant to clogging, but is not as resistant as the fiberglass filters. The choice of filter combination is based on experience, cost, and availability. Certain modifications are often made in methodology dependent upon the type of virus that is being concentrated.

B. Ultrafiltration, Reverse Osmosis, and Hydroextraction

Ultrafiltration refers to the passage of solutions through membranes, usually of cellulosic material, with pore sizes that permit the passage of water and low molecular weight materials, but exclude viruses and macromolecules. The viruses and macromolecules are then concentrated on the membrane or in the liquid portion that does not pass through the membrane. In the past, the use of these systems for efficient virus concentration was limited by the rapid clogging of available membranes and the difficulty in recovering retained viruses from the membranes.[8] These problems have been greatly reduced by systems using tangential flow and hollow-fiber configurations.[9]

Three membrane filtration methods that have been used for virus concentration are reverse osmosis (RO), ultrafiltration, and hydroextraction. RO membranes are of a finer porosity than conventional ultrafilters and are characterized by their ability to retain macrosolutes and microsolutes that are within one order of magnitude of the size of the solvent molecules. They generally operate at high pressures and low fluxes. Most of the studies on the use of RO membranes for virus concentration were done in the early 1970s.[8,86,87] Although recoveries of viruses from distilled water were good, they were very poor from tapwater.[86] Poor recoveries from tapwater were attributed to virus aggregation or inactivation by substances co-concentrated from the water along with the viruses. Also, the excellent total solute rejection RO membranes concentrate materials toxic to cell cultures.

Ultrafilters consisting of aluminum alginate gel containing lanthanum ions that are soluble in sodium citrate have been used to concentrate viruses from 1- to 10-ℓ volumes of preclarified sewage and tapwater.[88] These filters clog easily, and most waters need to be prefiltered before testing resulting frequently in loss of viruses. Ultrafilters have been used more recently to further concentrate viruses recovered from treated wastewater by a filter adsorption-elution procedure.[89] A number of workers have reported the application of antisotropic membranes to the concentration of viruses from water.[90-92] Nupen and Stander[90] used polymeric membranes with a molecular weight "cut-off" of 30,000 (Amicon® Type PM30) to concentrate viruses from 10-ℓ effluent samples from the Windhoek wastewater reclamation plant. Viruses were recovered from the membrane with a salt solution containing 10% fetal calf serum and 0.5% lactalbumin hydrolysate. The average efficiency of recovery was 70%. With a battery of flat-sheet, cellulosic ultrafilter elements in a tangential flow system, Foliguet et al.[92] concentrated poliovirus 1 to a volume of 350 mℓ from 20 ℓ of tapwater and from 50-ℓ volumes of filtered and raw river water with recovery efficiencies of 50 to 100%.

Belfort et al.[91] used polysulphone asymmetric hollow fiber membranes to successfully concentrate poliovirus 1 from 50-ℓ volumes of tapwater to a final volume of 250 mℓ with an average recovery efficiency of 42%. They found that backwashing the membranes was necessary, and that polysulphone membranes were superior to cellulose acetate both for recovering viruses and in general chemical and biological stability. Further design improvements have lead to the development of tangential flow hollow fiber ultrafiltration (TFHFU) systems capable of processing lake water and tapwater.[9,93] With this system poliovirus 2 was concentrated from large volumes to 2 to 3 ℓ with an average efficiency of 69%.[85] For high turbidity feed waters (75 to 85 NTU), obtained by the addition of kaolin clay, virus recoveries varied from 30 to 49%.[94] Processing of a 100-ℓ volume required 79 min. Larger ultrafiltration units are available that concentrate viruses from 1900 ℓ of tapwater in about 3 hr. Although clogging of filters by turbidity appears to be less of a problem, high initial cost and lack of portability are still factors currently limiting TFHFU use.

Capillary (1.5 mm diameter capillaries) rather than hollow fibers have been used to concentrate viruses from 100-ℓ volumes of tapwater.[91] In this system, water is continuously recirculated through the unit until the volume is reduced to approximately 1 ℓ which requires about 86 min. During processing of the sample, the virus becomes adsorbed to the membrane. Viruses are eluted from the membrane by backwashing with 1 ℓ of 1% beef extract and further concentrated by organic flocculation.[67] With this procedure, poliovirus 1 was concentrated from 100 ℓ to approximately 35 mℓ with an average efficiency of 35%.

Hydroextraction is a simple method for concentrating viruses in which the sample is placed in a semipermeable membrane such as dialysis tubing and exposed to a macromolecular hydroscopic material such as polyethylene glycol, which results in the transfer of water but not macromolecules and viruses across the membrane to the hydroscopic material. This method has been used to concentrate viruses directly from water samples[95] and as a second-stage concentration step in recovery of viruses from filter eluates.[10,51,96] Hydroextraction is a simple method, but it is only useful for samples of 1 ℓ or less. It may also concentrate materials toxic to cell cultures.

C. Glass Beads

Small glass beads or glass powder have been used as an adsorbent by several French investigators for concentrating viruses from tapwater.[97,98] As with negatively charged microporous filters, virus adsorption to the glass beads is enhanced by adjustment of the pH of the water to 3.5 and by the addition of $AlCl_3$ to a final concentration of 0.0005 *M*. Viruses are eluted from the beads with 0.005 *M* glycine buffer, pH 11.5. Large volumes of water can be processed by passage through columns of glass beads.[99] With such a system, Schwartzbrod and Lucena-Gutierrez[99] recovered seeded poliovirus LSc 2ab from 50-ℓ volumes of tapwater with a mean efficiency of 55%. Processing 50 ℓ of tapwater requires 150 g of glass beads. The flow rate through the columns was 80 ℓ/hr. Systems using glass columns capable of processing 400-ℓ volumes of tapwater have been constructed.

Lucena et al.[100] recently used glass beads to detect naturally occurring enteroviruses at fresh and marine bathing beaches.

In this procedure, a glass column (50 × 340 mm high) is filled with glass beads (100 to 200 μm) and the conditioned water is passed through the column from the bottom. The system is easy to construct, the glass beads are reusable, and operational design eliminates clogging problems.[101]

D. Minerals, Clays, and Filter Aids

Talcum powder, a hydrous magnesium silicate, has been used for concentrating enteric viruses from water in a "sandwich" technique. A prefilter base is used to retain the talc.[102-105] This technique has been used extensively by Sattar and co-workers for enteric virus isolation from tapwater, river water, and sewage.[102-104] In practice, the pH of a water sample is adjusted to 6.0, Earle's balanced salt solution is added to a final concentration of 1:100, and the water is passed through a talc-celite layer. The layer is composed of three parts talc and one part celite filter aid sandwiched between two coarse filter papers. Viruses are eluted from the talc with 10% fetal calf serum or 3% beef extract adjusted to pH 9.0. The talc-celite concentration method can efficiently recover small numbers of viruses from up to 1000 ℓ of tapwater and from 5 ℓ of sewage effluent. In experiments with 100- to 1000-ℓ volumes, poliovirus 1 was concentrated 100,000-fold with an efficiency of 58 to 64%.[105] More recently, these investigators concentrated rotaviruses from 100-ℓ volumes of tapwater with a similar overall efficiency.[105]

Recently, positively charged filter aids similar to celite have become available. This material has a net positive charge similar to the Zeta-plus and 1MDS filters used for virus concentration. It is also manufactured by AMF/CUNO (Meriden, Conn.). The filter aid is suspended in a small volume of water and collected on a paper filter or fiberglass prefilter.[106,107] Because of the positive charge on the filter aids, the water processed does not have to be conditioned. Enteroviruses can be concentrated from 20-ℓ volumes of tapwater or sewage with 1.75 g of filter aid contained in a 47-mm filter housing with efficiencies ranging from 32 to 100%. Viruses were most efficiently eluted from the filter aid with a solution of 4% beef extract in 0.5 *M* NaCl, adjusted to pH 9.5. The method has several advantages over other methods for processing similar volumes including (1) it requires a small 47-mm filter housing for processing 20- to 40-ℓ volumes, (2) unit cost per filtration is less than 3 cents, (3) water can be processed at ambient pH.

The clay mineral, bentonite, has been used for concentrating viruses from small volumes (0.5 to 2 ℓ) of wastewater.[108,109] The method consists of adsorbing viruses to bentonite clay in the presence of 0.01 *M* $CaCl_2$ or adjustment of the pH to 4.0, and allowing a sufficient length of time for virus adsorption to the clay. The virus-clay complex is then removed by centrifugation and eluted with tryptose phosphate broth. Recovery efficiencies of 65 to 94% have been reported.[108,109]

Magnetic and nonmagnetic mineral iron oxides have also been used as adsorbents in the concentration of viruses from water. Iron oxides have been used in batch studies similar to those with bentonite[110,111] in a filter sandwich[112] or in packed columns.[103] In the batch procedure developed by Bitton and Mitchell,[113] $CaCl_2$ was added to obtain optimal adsorption of bacteriophage T7 to magnetic iron oxide (Fe_3O_4). The magnetic iron-virus complex was collected by passage of the mixture through stainless steel wool placed in a background magnetic field. The magnetite in batch systems can be retained at the bottom of a beaker by appropriate placement of a small magnet during decanting of the supernate. Viruses can be eluted from the matrix with either beef extract[67] or isoelectric casein[111] adjusted to pH 8 or 9. Bitton achieved a 90% recovery of poliovirus 1 from 2 to 4 ℓ of activated sludge effluents with a magnetite organic flocculation method.[110]

A filter system comprised of bituminous coal sandwiched between two prefilters recovered seeded poliovirus 1 as efficiently from 100 ℓ of tap water over a pH range of 3 to 7 as did Filterite® and Millipore® filters.[113a] Coal filters were as good as Millipore® filter discs for recovering viruses from sewage.

IV. METHODS FOR CONCENTRATING VIRUSES FROM SMALL VOLUMES OF WATER

A. Inorganic Precipitates

A number of procedures have been described for concentrating viruses from waters in which the viruses are either precipitated (coagulated) by or adsorbed to preformed precipitates of polyvalent metal salts. Efficient adsorption usually requires controlled pH and ionic conditions, suggesting that electrostatic forces are involved.[8] Wallis and Melnick[114] first concentrated viruses from water on precipitates of aluminum phosphate, aluminum hydroxide, and calcium phosphate. They found that aluminum hydroxide and calcium phosphate efficiently adsorbed enteroviruses and adenoviruses but not reoviruses, while aluminum phosphate did not adsorb any of these viruses. Wallis and Melnick concentrated viruses from 1-gal (3.8-ℓ) volumes of sewage by adding 1 g of preformed aluminum hydroxide to the sewage and collecting the floc by filtration through a Millipore® AP-20 prefilter. They are then resuspended in a 10% solution of fetal bovine serum. A similar procedure with ferric hydroxide flocs has also been described.[115]

Aluminum hydroxide flocs formed *in situ* have been used to concentrate viruses from eluates of filters used to concentrate viruses from large volumes of water.[116] In this procedure, aluminum chloride is added to neutralized glycine eluate (0.05 *M*) to a final concentration of 0.003 *M* which results in a pH drop to approximately 4. Neutralization of the samples with a sodium carbonate solution produces an aluminum hydroxide floc that is an efficient virus adsorbent. The floc is collected by centrifugation and mixed with 1 *M* glycine dissolved in fetal calf serum, pH 11.5. After mixing, the sample is centrifuged, and the supernate is neutralized.

None of the methods involving inorganic precipitates have been used to process more than a few liters of sample since none can be scaled-up to process large volumes of water.

B. Polyelectrolytes

Viruses have been concentrated successfully from sewage and water by adsorption onto and elution from insoluble polyelectrolytes. The polymer which has been used most extensively is PE60, produced by Monsanto. The compound PE60 is a cross-linked copolymer of isobutylene/maleic anhydride.[7] According to Cookson[117] the PE60 adsorption sites for viruses are carboxyl groups and ammonium radicals involved in hydrogen bonding and quaternary ammonium groups that electrostatically interact with the negatively charged virus surface. Virus elution from PE60 occurs at pH 8 to 9 in isotonic solutions. Elution is also enhanced in the presence of protein solutions such as fetal calf serum because proteins compete with viruses for adsorption sites. PE60 can be used both in batch process and in a sandwich between filter pads to concentrate viruses from water.[118,119] PE60 was effective for concentrating viruses from water and sewage,[8,117-119] but has not been used in recent years because of variability in its efficiency in concentrating viruses. PE60 is an unstable polymer, and with increasing storage time it becomes chemically altered and less efficient as a virus adsorbent. In addition, different production lots of PE60 varied in their chemical characteristics, including the optimum pH for virus adsorption. Moreover, the efficiency of PE60 is not identical for different enteric viruses.[8]

C. Protamine Sulfate

Protamines are basic proteins, rich in arginine, with an isoelectric point near 12 and are very useful precipitants near neutral pH for negatively charged macromolecules of high molecular weight such as viruses. England[120] developed and tested procedures utilizing the sulfate salt of protamine for concentrating enteric viruses from clarified sewage and treated effluent. In this method, bovine serum albumin is added to a sample as a precipitate enhancer along with protamine sulfate at pH 7.5 to 7.8, and the precipitate is collected by filtration. The precipitate is then dissolved in a small volume of NaCl and diluted to isotonicity before assay. Reoviruses and adenoviruses in experimentally contaminated wastewater were recovered with this technique at efficiencies of 80 to 100%, but enterovirus recoveries were variable and depended upon the specific virus tested.

Adams et al.[121] recently used a modification of this procedure for concentrating reoviruses from river water and sewage where the precipitates from 500- to 4000-mℓ volumes were collected on Millipore® AP20 fiberglass prefilters. Each precipitate was then dissolved in a small volume of 1 *M* NaCl. Reovirus recoveries of 80 to 100% were achieved by this procedure.

D. Two-Phase Separation

When two different organic polymers are dissolved in water, two liquid phases are often produced which are not compatible. Viruses and macromolecules can be partitioned between the two immiscible aqueous phases that are produced. Partitioning of viruses depends upon the type of polymers, their molecular weights, and the ionic composition and strength of the partitioning system. By selecting the appropriate polymers and controlling ionic composition, ionic strength, and pH, viruses can be partitioned largely into one of two phases. If the volume of the virus-containing phase is made small in relation to the original fluid volume, then a considerable degree of concentration can be achieved. The polymer mixtures used for concentrating viruses from water have usually been dextran-polyethylene glycol or dextran sulfate-polyethylene glycol.[122-125]

Two-phase separation has been used to concentrate viruses from sewage effluents[125] and tapwater[124] and in the reconcentration of eluates from filters.

Grindrod and Cliver[126] found that dextran sulfate was inhibitory to certain enteroviruses and interfered with their detectability in cell cultures. These authors recommended replacing dextran sulfate with dextran in the two-phase system.[127] Shuval et al.[124] used a two-step phase separation procedure to achieve concentration factors of 250 to 500 for viruses in 2 to 7 ℓ of raw sewage or tapwater. This is about the largest volume size that can be sampled by two-phase separation.

E. Adsorption to Living Membranes

Armon et al.[129] recently reported that adsorption, from small volumes of liquid, of bacteriophages ΦX174, MS2, f2, and poliovirus 1 to porcine erythrocyte membranes at pH 3.5 and elution of the adsorbed viruses from the membranes (after centrifugation of the erythrocytes) at pH 7.5 yielded virus recoveries in the 90 to 100% range.

V. SUMMARY AND CONCLUSIONS

Major advances have been made in the last 2 decades in the development of methods for concentrating viruses from large volumes of water. Development of microporous adsorption-elution techniques made possible the first practical field methods for concentrating enteroviruses from drinking water, sewage, and seawater. Although this method still has many inherent limitations, continued improvements have led to increased efficiencies of enterovirus recoveries and its use with other members of the enteric virus group such as rotaviruses and hepatitis A virus (enterovirus 72). These systems make possible the routine monitoring of water for enteric viruses. The flexibility of these systems to modification and increased knowledge of virus surface interactions should lead to continued improvement in this technology.

Although microporous filter systems have proved successful for concentrating viruses, this does not portend the abandonment of other technologies. Other methods may be more appropriate when smaller volumes need to be processed or where microporous filters are not readily available or are too costly. Although ultrafiltration is still a promising alternative, it suffers from lack of portability, high equipment costs, and it has not yet been evaluated with water of varying quality.

REFERENCES

1. **Haley, C. E., Gunn, R. A., Hughes, J. M., Lippy, E. C., and Craun, G. F.,** Outbreaks of waterborne disease in the United States, *J. Infect. Dis.*, 141, 794, 1980.
2. **Lippy, E. C.,** Waterborne disease: occurrence is on the upswing, *J. Am. Water Works Assoc.*, 73, 57, 1981.
3. **Lycke, E., Bloomberg, J., Berg, G., Eriksson, A., and Madsen, L.,** Epidemic acute diarrhoea in adults associated with infantile gastroenteritis, *Lancet*, 2, 1056, 1978.
4. **Sutmoller, F., Azerdo, R. S., Lacerda, M. D., Barth, O. M., Pereira, H. G., Hoffer, E., and Schatzmayer, H. G.,** An outbreak of gastroenteritis caused by both rotavirus ans *Shigella sonnei* in a private school in Rio de Janeiro, *J. Hyg. Camb.*, 88, 285, 1982.
5. **Taylor, J. W., Gary, G. W., Jr., and Greenberg, H. B.,** Norwalk-related viral gastroenteritis due to contaminated drinking water, *Am. J. Epidemiol.*, 114, 584, 1981.
6. **Schiff, G.M., Stefanovic, G. M., Young, B., and Pennekamp, J. K.,** Minimum human infectious dose of enteric virus (echovirus-12) in drinking water, in *Enteric Viruses in Water*, Vol. 15, Melnick, J. L., Ed., S. Karger, New York, 1984, 222.
7. **Gerba, C. P., Wallis, C., and Melnick, J. L.,** Viruses in water: the problem, some solutions, *Environ. Sci. Technol.*, 13, 1122, 1975.
8. **Sobsey, M. D.,** Methods for detecting enteric viruses in water and wastewater, in *Viruses in Water*, Berg, G., Bodily, H. L., Lennette, E. H., Melnick, J. L., and Metcalf, T. G., Eds., American Public Health Association, Washington, D. C., 1976, 89.
9. **Belfort, G. and Dziewulski, D. M.,** Concentration of virus from aqueous environments: a review of the methodology, in *Water Reuse*, Middlebrooks, E. J., Ed., Ann Arbor Science, Ann Arbor, Mich., 1982, 679.
10. **Farrah, S. R., Gerba, C. P., Wallis, C., and Melnick, J. L.,** Concentration of viruses from large volumes of tapwater using pleated membrane filters, *Appl. Environ. Microbiol.*, 31, 221, 1976.
11. **Farrah, S. R., Goyal, S. M., Gerba, C. P., Wallis, C., and Shaffer, P. T. B.,** Characteristics of humic acid and organic compounds concentrated from tapwater using the Aquella virus concentrator, *Water Res.*, 10, 897, 1976.
12. **Goyal, S. M., Gerba, C. P., and Melnick, J. L.,** Occurrence and distribution of bacterial indicators and pathogens in canal communities along the Texas coast, *Appl. Environ. Microbiol.*, 34, 139, 1977.
13. **Gerba, C. P., Farrah, S. R., Goyal, S. M., Wallis, C., and Melnick, J. L.,** Concentration of enteroviruses from large volumes of tapwater, treated sewage, and seawater, *Appl. Environ. Microbiol.*, 35, 540, 1978.
14. **Goyal, S. M. and Gerba, C. P.,** *Methods in Environmental Virology*, Marcel Dekker, New York, 1982, 59.
15. **Hejkal, T. W., Wellings, F. M., LaRock, P. A., and Lewis, A. L.,** Survival of poliovirus within organic solids during chlorination, *Appl. Environ. Microbiol.*, 38, 114, 1979.

16. **Wellings, F. M., Lewis, A. L., and Mountain, C. W.,** Demonstration of solids-associated virus in wastewater and sludge, *Appl. Environ. Microbiol.*, 31, 354, 1976.
17. **Stagg, C. H., Wallis, C., and Ward, C. H.,** Inactivation of clay-associated bacteriophage MS-2 by chlorine, *Appl. Environ. Microbiol.*, 33, 385, 1977.
18. **Gerba, C. P., Stagg, C. H., and Abadie, M. G.,** Characterization of sewage solids-associated viruses and behavior in natural waters, *Water Res.*, 12, 805, 1978.
19. **Gerba, C. P. and Schaiberger, G. E.,** Effect of particulates on the survival of virus in seawater, *J. Water Pollut. Control Fed.*, 47, 93, 1975.
20. **Stagg, C. H., Wallis, C., Ward, C. H., and Gerba, C. P.,** Chlorination of solids-associated coliphages, *Prog. Water Technol.*, 10, 381, 1978.
21. **Wallis, C. and Melnick, J. L.,** Concentration of enteroviruses on membrane filters, *J. Virol.*, 1, 472, 1967.
21a. **Sobsey, M. D. and Hickey, A. R.,** Effect of humic and fulvic acids on poliovirus concentration from water by microporous filtration, *Appl. Environ. Microbiol.*, 49, 259, 1985.
21b. **Guttman-Bass, N. and Cattalano-Sherman, J.,** Effect of humic materials on virus recovery from water, *Appl. Environ. Microbiol.*, 49, 1260, 1985.
22. **Bixby, R. L. and O'Brien, D. J.,** Influence of fulvic acid on bacteriophage adsorption and complexation in soil, *Appl. Environ. Microbiol.*, 38, 840, 1979.
23. **Melnick, J. L., Emmons, J., Opton, E. M., and Coffey, J. H.,** Coxsackie viruses from sewage, methodology including an evaluation of the grab sample and gauze pad collection procedures, *Am. J. Hyg.*, 59, 185, 1954.
24. **Metcalf, T. G.,** Use of membrane filters to facilitate the recovery of virus from aqueous suspensions, *Appl. Microbiol.*, 9, 376, 1961.
25. **Cliver, D. O.,** Factors in the membrane filtration of enteroviruses, *Appl. Microbiol.*, 13, 417, 1965.
26. **Wallis, C., and Melnick, J. L.,** Concentration of enteroviruses on membrane filters, *J. Virol.*, 1, 472, 1967.
27. **Cliver, D. O.,** Enterovirus detection by membrane chromatography, in *Transmission of Viruses by the Water Route*, Berg, G., Ed., Interscience, New York, 1967, 139.
28. **Wallis, C. and Melnick, J. L.,** Concentration of viruses from sewage by adsorption on Millipore membranes, *Bull. W.H.O.*, 36, 219, 1967.
29. **Berg, G., Dahling, D. R., and Berman, D.,** Recovery of small quantities of viruses from clean waters on cellulose nitrate membrane filters, *Appl. Microbiol.*, 22, 608, 1971.
30. **Wallis, C., Henderson, M., and Melnick, J. L.,** Enterovirus concentration on cellulose membranes, *Appl. Microbiol.*, 23, 46, 1972.
31. **Wallis, C. and Melnick, J. L.,** A portable virus concentrator for use in the field, in *Advances in Water Pollution Research*, Jenkins, S. J., Ed., Pergamon Press, New York, 1973, 119.
32. **Wallis, C., Homma, A., and Melnick, J. L.,** A portable virus concentrator for testing water in the field, *Water Res.*, 6, 1249, 1972.
33. **Homma, A., Sobsey, M. D., Wallis, C., and Melnick, J. L.,** Virus concentration from sewage, *Water Res.*, 7, 945, 1973.
34. **Metcalf, T. G., Wallis, C., and Melnick, J. L.,** Environmental factors influencing isolation of enteroviruses from polluted surface waters, *Appl. Microbiol.*, 27, 920, 1974.
35. **Sobsey, M. D., Wallis, C., Henderson, M., and Melnick, J. L.,** Concentration of enteroviruses from large volumes of water, *Appl. Microbiol.*, 26, 529, 1973.
36. **Sobsey, M. D., Gerba, C. P., Wallis, C., and Melnick, J. L.,** Concentration of enteroviruses from large volumes of turbid estuary water, *Can. J. Microbiol.*, 23, 770, 1977.
37. **Gilbert, R. G., Gerba, C. P., Rice, R. C., Bouwer, H., Wallis, C., and Melnick, J. L.,** Virus and bacteria removal from wastewater by land treatment, *Appl. Environ. Microbiol.*, 32, 333, 1976.
38. **Farrah, S. R., Goyal, S. M., Gerba, C. P., Wallis, C., and Shaffer, P. T. B.,** Characteristics of humic acid and organic compounds concentrated from tapwater using the Aquella virus concentrator, *Water Res.*, 10, 897, 1976.
39. **Farrah, S. R., Gerba, C. P., Wallis, S., and Melnick, J. L.,** Concentration of viruses from large volumes of tapwater using pleated membrane filters, *Appl. Environ. Microbiol.*, 31, 221, 1976.
40. **Farrah, S. R., Goyal, S. M., Gerba, C. P., Wallis, C., and Melnick, J. L.,** Concentration of enteroviruses from estuarine water, *Appl. Environ. Microbiol.*, 33, 1192, 1977.
41. **Farrah, S. R., Gerba, C. P., Goyal, S. M., Wallis, C., and Melnick, J. L.,** Regeneration of pleated filters used to concentrate enteroviruses from large volumes of tapwater, *Appl. Environ. Microbiol.*, 33, 308, 1977.
42. **Jakubowski, W., Hoff, J. C., Anthony, N. C., and Hill, W. F., Jr.,** Epoxy-fiberglass adsorbent for concentrating viruses from large volumes of potable water, *Appl. Microbiol.*, 28, 501, 1974.
43. *Standard Methods for the Examination of Water and Wastewater*, 15th ed. American Public Health Association, Washington, D.C., 1980.

43a. **Berg, G., Safferman, R. S., Dahling, D. R., Berman, D., and Hurst, C.,** Manual of Methods for Virology, EPA 600/4-84-013 Environmental Monitoring and Support Laboratory, U.S. Environmental Protection Agency, Cincinnati, 1984.

44. **Sobsey, M. D., Glass, J. S., Carrick, R. J., Jacobs, R. R., and Rutala, W. A.,** Evaluation of the tentative standard method for enteric virus concentration from large volumes of tapwater, *J. Am. Water Works Assoc.*, 72, 292, 1980.

45. **Sobsey, M. D., Glass, J. S., Jacobs, R. R., and Rutala, W. A.,** Modifications of the tentative standard method for improved virus recovery efficiency, *J. Am. Water Works Assoc.*, 72, 350, 1980.

45a. **Dahling, D. R. and Wright, B. A.,** Processing and transport of environmental samples, *Appl. Environ. Microbiol.*, 47, 1272, 1984.

46. **Sobsey, M. D. and Glass, J. S.,** Poliovirus concentration from tapwater with electropositive adsorbent filters, *Appl. Environ. Microbiol.*, 40, 201, 1980.

47. **Sobsey, M. D. and Jones, B. L.,** Concentration of poliovirus from tapwater using positively charged microporous filters, *Appl. Environ. Microbiol.*, 37, 588, 1979.

48. **Hou, K., Gerba, C. P., Goyal, S. M., and Zerda, K. S.,** Capture of latex beads, bacteria, endotoxin, and viruses by charge-modified filters, *Appl. Environ. Microbiol.*, 40, 892, 1980.

49. **Primrose, S. B., Seeley, N. D., and Logan, K.,** *Viruses and Wastewater Treatment*, Pergamon Press, New York, 1981, 211.

50. **Goyal, S. M., Zerda, K. S., and Gerba, C. P.,** Concentration of coliphages from large volumes of water and wastewater, *Appl. Environ. Microbiol.*, 39, 85, 1980.

51. **Logan, K. B., Scott, G. E., Seeley, N. D., and Primrose, S. B.,** A portable device for the rapid concentration of viruses from large volumes of natural freshwater, *J. Virol. Methods*, 3, 241, 1981.

52. **Goyal, S. M., Hanssen, H., and Gerba, C. P.,** Simple method for the concentration of influenza virus from allantoic fluid on microporous filters, *Appl. Environ. Microbiol.*, 39, 500, 1980.

53. **Chang, L. T., Farrah, S. R., and Bitton, G.,** Positively charged filters for virus recovery from wastewater plant effluents, *Appl. Environ. Microbiol.*, 42, 921, 1981.

54. **Rose, J. B., Singh, S. N., Gerba, C. P., and Kelley, L. M.,** Comparison of microporous filters for concentration of viruses from wastewater, *Appl. Environ. Microbiol.*, 47, 989, 1984.

55. **Kilgen, M. B. and Cole, M. T.,** Recovery of poliovirus type 1 from artificial seawater and natural estuarine water at varying salinities using electropositive filters, *Abstr. Annu. Meet. Am. Soc. Microbiol.*, 1983, 283.

56. **Keswick, B. H., Gerba, C. P., and Goyal, S. M.,** Occurrence of enteroviruses in community swimming pools, *Am. J. Public Health*, 71, 1026, 1981.

57. **Hejkal, T. W., Keswick, B., LaBelle, R. L., Gerba, C. P., Sanchez, Y., Dreesman, G., and Hafkin, B.,** Viruses in a community water supply associated with an outbreak of gastroenteritis and infectious hepatitis, *J. Am. Water Works Assoc.*, 74, 318, 1982.

58. **Toranzos, G. A. and Gerba, C. P.,** Concentration of rotavirus from water and sewage on microporous filters, *Abstr. Annu. Meet. Am. Soc. Microbiol.*, 1983, 269.

59. **Dizer, H., Tischer, E., and Lopez, J.,** Concentration of viruses from tapwater with positively charged filters which can be regenerated and reused repeatedly, in *Disinfection of Viruses in Water and Wastewater*, Butler, M., Medlen, A. R., and Morris, R., Eds., University of Surrey, Surrey, U. K., 1982, 96.

60. **Scutt, J. E. and Garelick, H.,** Concentration of virus from tapwater at ambient salt and pH levels using positively charged filter media, *Water Res.*, 15, 815, 1981.

61. **Farrah, S. R. and Bitton, G.,** Elution of poliovirus adsorbed to membrane filters, *Appl. Environ. Microbiol.*, 36, 982, 1978.

62. **Bitton, G., Feldberg, B. N., and Farrah, S. R.,** Concentration of enteroviruses from seawater and tapwater by organic flocculation using non-fat dry milk and casein, *Water, Air, Soil Pollut.*, 12, 187, 1979.

63. **Farrah, S. R., Goyal, S. M., Gerba, C. P., Conklin, R. H., Wallis, C., Melnick, J. L., and Dupont, H. L.,** A simple method for concentration of enteroviruses and rotaviruses from cell culture harvests using membrane filters, *Intervirology*, 9, 56, 1978.

64. **Hill, W. F., Jr., Akin, E. W., Benton, W. H., and Metcalf, T. G.,** Virus in water. II. Evaluation of membrane cartridge filters for recovering multiplicities of poliovirus in water, *Appl. Microbiol.*, 23, 880, 1972.

65. **Farrah, S. R., Shah, D. O., and Ingram, L. O.,** Effects of chaotropic and antichaotropic agents on the elution of poliovirus adsorbed to membrane filters, *Proc. Nat. Acad. Sci. U.S.A.*, 78, 1229, 1981.

66. **Moore, R. S., Wait, D. A., and Stokes, E. H.,** Improved methods for poliovirus recovery from water with electropositive adsorbent filters, *Annu. Meet. Am. Soc. Microbiol.*, 219, 1982.

67. **Berg, G., Dean, R. B., and Dahling, D. R.,** Removal of poliovirus 1 from secondary effluents by lime flocculation and rapid sand filtration, *J. Am. Water Works Assoc.*, 60, 193, 1968.

68. **Farrah, S. R. and Bitton, G.,** Low molecular weight substitutes for beef extract as eluents for poliovirus adsorbed to membrane filters, *Can. J. Microbiol.*, 25, 1945, 1979.

68a. **Katzenelson, E., Fattal, B., and Hostovesky, T.,** Organic flocculation: an efficient second-step concentration method for the detection of viruses in tapwater, *Appl. Environ. Microbiol.*, 32, 638, 1976.

69. **Landry, E. F., Vaughn, J. M., Thomas, M. Z., and Vicale, T. V.,** Efficiency of beef extract for the recovery of poliovirus from wastewater effluents, *Appl. Environ. Microbiol.*, 36, 544, 1978.

70. **Farrah, S. R.,** Chemical factors influencing adsorption of bacteriophage MS-2 to membrane filters, *Appl. Environ. Microbiol.*, 43, 659, 1982.

71. **Shields, P. A. and Farrah, S. R.,** Influence of salts on electrostatic interactions between poliovirus and membrane filters, *Appl. Environ. Microbiol.*, 45, 526, 1983.

72. **Hill, W. F., Jr., Akin, E. W., Benton, W. H., Mayhew, C. J., and Jakubowski, W.,** Apparatus for conditioning unlimited quantities of finished waters for enteric virus detection, *Appl. Microbiol.*, 27, 1177, 1974.

73. **Shuval, H. I., Cymbalista, S., Fattal, B., and Goldblum, N.,** Concentration of enteric viruses in water by hydro-extraction and two-phase separation, in *Transmission of Viruses by the Water Route*, Berg, G., Ed., Interscience, New York, 1967, 45.

74. **Anderson, N. G., Cline, G. B., Harris, W. W., and Green, J. G.,** Isolation of viral particles from large fluid volumes, in *Transmission of Viruses by the Water Route*, Berg, G., Ed., Interscience, New York, 1967, 75.

74a. **Mack, W. N., Yue-Shoung, L., and Coohon, D. B.,** Isolation of poliomyelitis virus from a contaminated well, *Public Health Rep.*, 87, 271, 1972.

75. **Shields, P. A., Berenfeld, S. A., and Farrah, S. R.,** Modified membrane filter procedure for concentration of enteroviruses from tap water, *Appl. Environ. Microbiol.*, 49, 453, 1985.

76. **Farrah, S. R., Wallis, C., Shaffer, P. T. B., and Melnick, J. L.,** Reconcentration of poliovirus from sewage, *Appl. Environ. Microbiol.*, 32, 653, 1976.

77. **Payment, P. and Trudel, M.,** Efficiency of several microfiber glass filters for recovery of poliovirus from tapwater, *Appl. Environ. Microbiol.*, 38, 365, 1979.

78. **Fields, H. A. and Metcalf, T. G.,** Concentration of adenovirus from seawater, *Water Res.*, 9, 357, 1975.

79. **Morris, R. and Waite, W. M.,** Evaluation of procedures for recovery of viruses from water. I. Concentration systems, *Water Res.*, 14, 791, 1980.

79a. **Bass, N. G. and Nasser, A.,** Simultaneous concentration of four enteroviruses from tap, waste, and natural waters, *Appl. Environ. Microbiol.*, 47, 1311, 1984.

80. **Smith, E. M. and Gerba, C. P.,** Development of a method for detection of human rotavirus in water and sewage, *Appl. Environ. Microbiol.*, 43, 1440, 1982.

81. **Rose, J., Vasel, R., and Gerba, C. P.,** personal observations, 1983.

83. **Payment, P. and Trudel, M.,** A simple low cost apparatus for conditioning large volumes of water for virological analysis, *Can. J. Microbiol.*, 26, 548, 1980.

84. **Payment, P. and Trudel, M.,** Improved method for the use of proportioning injectors to condition large volumes of water for virological analysis, *Can. J. Microbiol.*, 27, 455, 1981.

84a. **Toranzos, G. A., Gerba, C. P., and Hanssen, H.,** Simple field method for concentration of viruses from large volumes of water, *Appl. Environ. Microbiol.*, 48, 431, 1984.

84b. **Melnick, J. L., Safferman, R. S., Rao, V. C., Goyal, S., Berg, G., Dahling, D. R., Wright, B. A., Akin, E., Stetler, R., Sorber, C., Moore, B., Sobsey, M. D., Moore, R., Lewis, A. L., and Wellings, F. M.,** Round robin investigation of methods for the recovery of poliovirus from drinking water, *Appl. Environ. Microbiol.*, 47, 144, 1984.

85. **Jakubowski, W., Hill, W. F., and Clarke, N. A.,** Comparative study of four microporous filters for concentrating viruses from drinking water, *Appl. Environ. Microbiol.*, 31, 58, 1975.

86. **Sweet, B. A., Ellender, R. D., and Leong, J. K. L.,** Recovery and removal of viruses from water utilizing membrane techniques, *Dev. Ind. Microbiol.*, 15, 143, 1974.

87. **Sorber, C. A., Malina, J. F., and Sagik, B. F.,** Virus rejection by the reverse osmosis-ultrafiltration process, *Water Res.*, 6, 1377, 1972.

88. **Gartner, H. and Witt, G.,** Virologische Stuhl- und Abwasser-Untersuchungen, ein Beitrag zur praktischen Erfassung der Poliomyelitis-Situation, *Arch. Hyg. Bacteriol.*, 150, 189, 1966.

89. **Selna, M. W. and Miele, R. P.,** Virus sampling in wastewater-field experiences, *J. Environ. Eng. Div. Am. Soc. Civ. Eng.*, 103, 693, 1977.

90. **Nupen, E. M. and Stander, G. V.,** *Advances in Water Pollution Research*, Pergamon Press, New York, 1972, 133.

91. **Belfort, G., Rotem, Y., and Katzenelson, E.,** Virus concentration using hollow fiber membranes, *Water Res.*, 10, 279, 1976.

92. **Foliguet, J. M., Lavillaurex, J., and Schwartzbrod, L.,** Viruses in water. II. General review of the methods available to detect viruses in water, *Rev. Epidemiol. Med. Soc. Sante Publique*, 21, 185, 1973.

93. **Belfort, G., Baltutis, T. F., and Blatt, W. F.,** Automatic hollow fiber ultrafiltration: pyrogen removal and phage recovery from water, in *Ultrafiltraton Membranes and Applications*, Plenum Press, New York, 1980, 439.

94. **Dziewulski, D. M. and Belfort, G.,** Virus concentration from water using high-rate tangential-flow hollow fiber ultrafiltration, *Water Technol.*, 15, 75, 1983.

95. **Wellings, F. M., Lewis, A. L., and Mountain, C. W.,** Demonstration of solids-associated virus in wastewater and sludge, *Appl. Environ. Microbiol.*, 31, 354, 1976.
96. **Ramia, S. and Sattar, S. A.,** Second-step concentration of viruses in drinking and surface waters using polyethylene glycol extraction, *Can. J. Microbiol.*, 25, 587, 1979.
97. **Sarrette, B. A., Danglot, C. D., and Vilagines, R.,** A new and simple method for recuperation of enteroviruses from water, *Water Res.*, 11, 355, 1978.
98. **Hugues, B., Plisser, M., Andre, M., Pagliardini, D., and Laurent, A.,** Evaluation de la charge viral dans les eaux d'une station d'epuration biologique au moyen de deux methodes de concentration de virus par adsorption-elution sur poudre de verre ou sur microfibre de verre, *Water Res.*, 13, 1117, 1979.
99. **Schwartzbrod, L. and Lucena-Gutierrez, F.,** Concentration des enteroviruses dans les eaux par adsorption sur poudre de verre: proposition d'un appareillage simplifie, *Microbia*, 4, 55, 1978.
100. **Lucena, F., Finance, C., Jofre, J., Sancho, J., and Schwartzbrod, L.,** Viral pollution determination of superficial waters (river water and seawater) from the urban area of Barcelona (Spain), *Water Res.*, 16, 173, 1982.
101. **Block, J. C. and Schwartzbrod, L.,** *Analyse Virologique des Eaux*, Lavoiser, Paris, 1982.
102. **Sattar, S. A. and Westwood, J. C. N.,** Comparison of talc-celite and polyelectrolyte 60 in virus recovery from sewage: development of techniques and experiments with poliovirus (type 1, Sabin)-contaminated samples, *Can. J. Microbiol.*, 22, 1620, 1976.
103. **Sattar, S. A. and Westwood, J. C. N.,** Viral pollution of surface waters due to chlorinated primary effluents, *Appl. Environ. Microbiol.*, 36, 427, 1978.
104. **Sattar, S. A. and Ramia, S.,** Use of talc-celite layers in the concentration of enteroviruses from large volumes of potable waters, *Water Res.*, 13, 637, 1979.
105. **Ramia, S. and Sattar, S. A.,** Concentration of seeded simian rotavirus SA-11 from potable waters by using talc-celite layers and hydroextraction, *Appl. Environ. Microbiol.*, 39, 493, 1980.
106. **Singh, S. N. and Gerba, C. P.,** Concentration of coliphage from water and sewage with charge-modified filter aid, *Appl. Environ. Microbiol.*, 45, 232, 1983.
107. **Singh, S. N., Rose, J. B., and Gerba, C. P.,** Concentration of viruses from tapwater and sewage with a charge modified filter aid, *J. Virol. Methods*, 6, 329, 1983.
108. **Moore, B. E., Sagik, B. P., and Sorber, C. A.,** Procedure for the recovery of airborne human enteric viruses during spray irrigation of treated wastewater, *Appl. Environ. Microbiol.*, 38, 688, 1979.
109. **Simmonds, R. S., Lewis, G. D., Loutit, M. W., and Austin, F. J.,** A method for detecting and identifying enteroviruses in effluents, *N.Z. J. Sci.*, 25, 57, 1982.
110. **Farrah, S. R. and Bitton, G.,** Methods (other than microporous filters) for concentration of viruses from water, in *Methods in Environmental Virology*, Marcel Dekker, New York, 1983, 117.
111. **Bitton, G., Chang, L. T., Farrah, S. R., and Clifford, K.,** Recovery of coliphages from wastewater effluents and polluted lake water by the magnetite-organic flocculation method, *Appl. Environ. Microbiol.*, 41, 93, 1981.
112. **Rao, B. C., Sullivan, R., Read, R. B., and Clarke, N. A.,** A simple method for concentrating and detecting viruses, *J. Am. Water Works Assoc.*, 60, 1288, 1968.
113. **Bitton, G. and Mitchell, R.,** The removal of *Escherichia coli* bacteriophage T7 by magnetic filtration, *Water Res.*, 8, 549, 1974.
113a. **Dahling, D. R., Phirke, P. M., Wright, B. A., and Safferman, R. S.,** Use of bituminous coal as an alternate technique for field concentration of waterborne viruses, *Appl. Environ. Microbiol.*, 49, 1222, 1985.
114. **Wallis, C. and Melnick, J. L.,** Concentration of viruses on aluminum and calcium salts, *Am. J. Epidemiol.*, 85, 459, 1967.
115. **Schaefer, E.,** Experimental investigation of the quantitative analysis of poliovirus 2 in surface water (German), *GWF-Wasser/Abwasser*, 112, 109, 1971.
116. **Farrah, S. R., Goyal, S. M., Gerba, C. P., Wallis, C., and Melnick, J. L.,** Concentration of enteroviruses from estuarine water, *Appl. Environ. Microbiol.*, 33, 1192, 1977.
117. **Cookson, J. T.,** The chemistry of virus concentration by chemical methods, *Dev. Ind. Microbiol.*, 15, 160, 1974.
118. **Grinstein, S., Melnick, J. L., and Wallis, C.,** Virus isolation from sewage and from a stream receiving effluents of sewage treatment plants, *Appl. Microbiol.*, 42, 291, 1970.
119. **Wallis, C., Grinstein, S., Melnick, J. L., and Fields, J. E.,** Concentration of viruses from sewage and excreta on insoluble polyelectrolytes, *Appl. Microbiol.*, 18, 1007, 1969.
120. **England, B.,** Concentration of reovirus and adenovirus from sewage and effluents by protamine sulfate (salmine) treatment, *Appl. Microbiol.*, 24, 510, 1972.
121. **Adams, D. J., Ridinger, D. N., Spendlove, R. S., and Barnett, B. B.,** Protamine precipitation of two reovirus particle types from polluted waters, *Appl. Environ. Microbiol.*, 44, 589, 1982.

122. **Albertsson, P.,** Two-phase separation of viruses, in *Methods in Virology,* Maramorosch, K. and Koprowski, H., Eds., Academic Press, New York, 1967, 303.
123. **Lund, E. and Hedstrom, C. E.,** The use of an aqueous polymer phase system for enterovirus isolations from sewage, *Am. J. Epidemiol.*, 84, 287, 1966.
124. **Shuval, H. I., Fattal, B., Cymbalista, S., and Goldblum, N.,** The phase-separation method for the concentration and detection of viruses in water, *Water Res.*, 3, 225, 1969.
125. **Nupen, E. M.,** Virus studies on the Windhock wastewater reclamation plant (South-West Africa), *Water Res.*, 4, 661, 1970.
126. **Grindrod, J. and Cliver, D. O.,** Limitations of the polymer two-phase system for detection of viruses, *Arch. Gesamte Virusforsch.*, 28, 337, 1969.
127. **Grindrod, J. and Cliver, D. O.,** A polymer two-phase system adapted to virus detection, *Arch. Gesamte Virusforsch.*, 31, 365, 1970.
128. **Gerba, C. P.,** Methods for recovering viruses from the water environment, in *Viral Pollution of the Environment,* Berg, G., Ed., CRC Press, Boca Raton, Fla., 1983, 19.
129. **Armon, R., Kott, Y., and Neeman, I.,** Ghost cells as sorption matrix for virus concentration from water, *Appl. Environ. Microbiol.*, 47, 1337, 1984.

Chapter 2

RECOVERING VIRUSES FROM SEWAGE SLUDGES AND FROM SOLIDS IN WATER

Christon J. Hurst

TABLE OF CONTENTS

I. INTRODUCTION

Many viruses and some bacteria[1] adsorb onto suspended solids in wastewater[2-6] and natural water.[7] For at least one virus group, the enteroviruses, the degree of adsorption to these solids is type and strain dependent.[8] Solids-associated viruses are important because they are infective for cultured cells as well as living animals.[7,9] The association of viruses with suspended solids in aqueous systems seems not only to enhance the survival of the viruses,[10] but also to increase their resistance to both thermal inactivation[11] and chlorine disinfection.[12]

Because of the large numbers of viruses and other pathogens that concentrate in sewage sludges during settling processes,[13] the proper treatment and disposal of sludges merits caution and concern. Apart from incineration, the principal methods of ultimate sewage sludge disposal are discharge into bodies of water and land application. In general, discharge into bodies of water is seen as advantageous for disposal of many types of pollutants. The benefits seen are primarily those of dilution and transportation away from the discharge site. These same characteristics are, however, also the greatest hazards associated with water discharge. Rather than concentrating the contaminants in one place where they can be carefully monitored and where human exposure can be controlled, the viruses are broadly dispersed creating a widespread problem.

When sewage solids are discharged into bodies of water, the opportunity exists for rapid transportation of solids-associated viruses over great distances. Discharged sludges are a source of solids-borne viruses, but freely suspended virus particles in wastewater effluents and overland runoff that reach surface waters may well become adsorbed onto natural solids present in the receiving waters. There is also potential for transfer of viruses from sludge solids onto other natural solids in waters.[14] Because of the affinity of viruses for particulate solids, and the enhancement of survival that results, suspended solids in water probably deserve more attention than they currently receive in environmental monitoring.

Land application (disposal) of waste solids in the form of night soil has been practiced for thousands of years. In the past, this practice was conducted primarily for crop fertilization. It was also an inexpensive method of disposal that required no formal public works such as treatment facilities. More recently, land application of wastewater sludges has enjoyed a resurgence in popularity as an alternative to discharge into bodies of water. It has been suggested that by reverting to land disposal the potential epidemiological and ecological[15] consequences that may result from water discharge can be prevented. The more traditional benefits for the soil-crop system through recycling of nutrients in the sludge is now often viewed as a beneficial but secondary factor. In addition to reducing pollution of surface waters and some agricultural benefits, land application in some regions may result in lower overall treatment costs.[16] Land application is not, however, a practice entirely devoid of health hazards.[17] There is concern that viruses and other pathogens disposed of in this manner may contaminate crops[18] or may move with overland runoff and percolation into surface or ground waters subsequently used as sources of drinking water.[19,20]

Thus, viruses in sewage sludges are clearly subjects of public health concern. To fully assess the extent of these concerns, a need exists for the quantification of viruses associated with the solids involved. This chapter will address this need by summarizing the methods that have been proposed for recovering viruses from wastewater sludges and from suspended solids in water.

II. BACKGROUND

A. Persistence of Viruses in Sludges and on Solids

Enteric viruses, those which replicate in the digestive tract, are the major concern in waterborne virus disease. These viruses are excreted in concentrations as high as one million

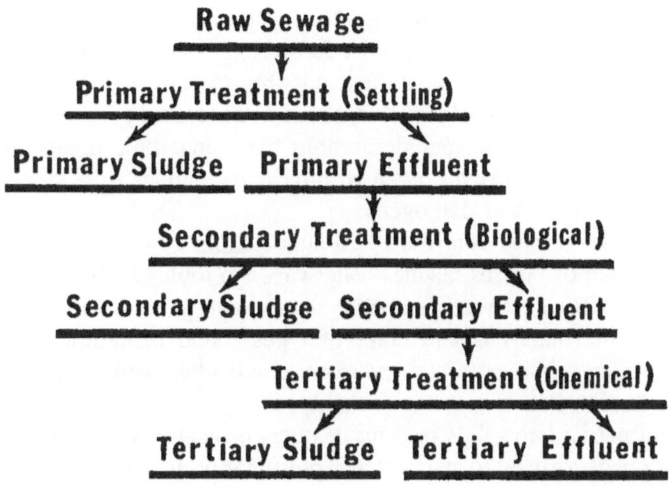

FIGURE 1. General outline of sludge-producing steps during sewage treatment process.

per gram of feces, and concentrations as high as 400,000 infective viruses per liter have been detected in raw sewage.[21] Among the families of human enteric viruses whose transmission through wastewater is of public health concern are the enteroviruses (a group that includes hepatitis A virus, polioviruses, echoviruses, and groups A and B coxsackieviruses), reoviruses, and adenoviruses. Of great concern also are other less well understood viruses that may be spread by the fecal-oral route. These include the Norwalk and Norwalk-like viruses and perhaps the astroviruses and other diarrhea-associated viruses.[22]

Because these viruses are excreted in fecal matter, a large part of the enteric virus content of raw sewage is associated with suspended solids and can be removed during primary settling. Primary sludges, the product of primary settling (see Figure 1), may contain numbers of enteric viruses as high as 1000 infective units per milliliter.[23] Of the enteric viruses that do not sediment during primary treatment because they are freely suspended in the effluent or attached to very small particulates that do not settle, 40 to 90% may subsequently adsorb onto and sediment with solids generated during secondary treatments.[24] Tertiary treatments, such as chemical coagulation and sedimentation, can also remove viruses from effluents. Chemicals commonly used to induce coagulation during tertiary treatment include aluminum sulfate (alum), ferric chloride, and calcium hydroxide (lime). As in primary and secondary settling processes, virus removal during chemical coagulation is generally due to specific association of virus particles with the precipitating solids. Although all three types of treatments (primary, biological, and chemical) remove viruses from effluents, the net result of each treatment is concentration of viruses into highly infective sludges.[23] These sludges must then be disposed of in a careful and controlled manner.[25]

The most commonly utilized methods for ultimate sludge disposal are incineration, discharge into bodies of water, and land application.[13] Of these three disposal means, incineration is the most effective destroyer of pathogens, but incineration is under increasing attack because of air pollution concerns. Discharge of sludges into receiving waters raises uncertainties of ensuing dispersal of pathogen-laden solids (and possible contamination of bathing and shellfishing facilities) and of potential ecological hazards associated with the formation of submarine sludge fields.[15] Because of the problems associated with incineration and water discharge, land application is frequently selected by communities as the method of choice for disposal of sludges. Nutrient recycling benefits may accrue to this means of disposal also. Unfortunately, land application is not completely free of health concern either.

Pathogens survive for long periods following land disposal. Viruses in land-disposed sludge solids and sludge-amended soils may persist for periods ranging from 2 to as many as 23 weeks depending primarily upon climate and possibly to some extent upon virus type.[26-33] To reduce the pathogen load in sludges, sludges are often treated prior to disposal. Processes commonly used for this purpose include aerobic[29] and anaerobic digestion,[30,34-36] lagooning,[31,37] and composting.[38] However, even with these techniques, processed sludges are not always free of viruses and other pathogens.

In addition to those viruses contributed to aqueous environments through discharge of sludges, a large burden of viruses reaches waterways via routes such as overland runoff and discharge of sewage effluents. Viruses readily adsorb onto particulate solids of many kinds, including both organic materials and clays of types found in natural waters, and native sediments.[7,9,14] It is therefore possible that many unadsorbed viruses that reach waterways will adsorb onto particulates.

The presence of particulate solids in natural waters protects viruses against inactivation.[39] Physical adsorption of viruses to particulates presumably accounts for this protection. In fact, of the variety of factors such as salinity, sunlight, sewage pollutants, and microbial antagonism[40-42] that influence the survival of viruses in water, adsorption to solids may well be the most important.[10,39] Because of the affinity between viruses and suspended solids, and because adsorption onto solids enhances survival of viruses in water, it is very likely that a large percentage of the infective viruses in natural waters is adsorbed to the solids fraction.[43,44]

Particulate solids in water can be classified as either suspended or deposited. Such a distinction is somewhat arbitrary, however, because the two are interchangeable with changes in water velocity and other factors. Nevertheless, it would seem to be those solids in suspension at any given time that are more likely to be rapidly transmitted over long distances. Moreover, within deposited sediments, it is in the uppermost thin layer, that layer in greatest equilibrium with suspended solids, where the largest numbers of viruses are probably to be found.[45] For these reasons, the viruses associated with suspended solids in natural waters are of first concern.

B. Need for and Categorization of Recovery Methods

In order to fully assess the viral hazards presented by sewage sludges, it is necessary to develop and standardize methods that can be used to monitor sludge virus content during treatment and following ultimate disposal. In addition to studies devoted to examining human viruses in sewage sludges, research has been done on the closely related topics of animal virus persistence in manure slurries and sewage sludges.[46,47]

Viral association with solids in natural waters has also been studied. Suspended solids may act as a reservoir of viruses in waters because of the protection against viral inactivation that solids provide. It is unfortunate that many of the studies conducted on viruses in water were concerned with only the freely suspended viruses and ignored those adsorbed to the waterborne solids.[44] As is the case with sewage sludges, full assessment of the viral hazards associated with waterborne solids requires the development of adequate detection methodology.

A large number of the studies on viruses in sludges and on solids have utilized bacteriophages, generally as models or indicators for animal viruses.[48-50] Bacteriophages are often used in virus studies because of the relatively small expense involved in their production and assay. There exist, however, strong doubts that bacteriophages can serve as adequate indicators for animal viruses.[51] One of the most important reasons for these doubts is the difference in disinfection rates for phages and animal viruses.[52,53] Moreover, bacteriophages may regrow under conditions that occur during sewage treatment[54] whereas animal viruses cannot. In addition, the occurrences of coliphages and enteric viruses often do not correlate well in field samples.[55] Bacteriophages may be useful in environmental studies, but the results obtained must be viewed with the knowledge that they may not be directly comparable to animal virus studies.

Techniques developed for quantifying viruses in sewage sludges and adsorbed onto waterborne solids can be divided into three general categories. These three categories are direct assay, elution of viruses physically adsorbed onto particulates, and procedures for virus concentration. Included in the first category is assay of whole sludges, slurries of water solids, and supernates or filtrates produced from suspensions of sludges and water solids. Included in the elution category is the assay of viruses eluted from solids on which they had been physically adsorbed. Included in the category of concentration are methods for processing whole sludge or solids suspensions and supernates and eluates derived from such suspensions. Most of the studies presented in the text of this chapter are listed in Tables 1, 2, and 3. The three cover, respectively, the categories of direct assay, elution, and concentration.

III. RECOVERING VIRUSES FROM SEWAGE SLUDGES

A. Direct Assay of Viruses in Sludges

The most direct method for enumerating viruses in sewage sludges is assay of either the whole intact sludge or diluted whole sludge. Ewert and Paynter[54] assayed samples of raw sewage, mixed-liquor activated sludge, and mixed-liquor effluent for indigenous bacteriophages by both electron microscopy and direct inoculation. Lund[23,56] assayed indigenous enteroviruses in primary and secondary sludges by direct inoculation. She recovered adenoviruses 1, 2, and 5, coxsackievirus B3, and echovirus 7. Derbyshire and Brown[46] assayed 10 mℓ samples of filtered manure slurry supernates by direct inoculation. They recovered indigenous porcine enteroviruses, adenoviruses, and coronaviruses. Brown et al.[18] assayed by direct inoculation indigenous coliphages in settled solids from aerobically digested sludges and sludge solids washed from foliage onto which the sludges had been sprayed. Gerba et al.[14] quantified indigenous solids-associated animal enteric viruses and coliphages in sewage effluents by collecting solids on membrane filters and assaying the solids by direct inoculation.

Lund and Ronne[20] assayed indigenous viruses in lime precipitates of toilet water, ferric chloride precipitates of raw sewage, aluminum sulfate precipitates of activated sludge supernate, and on whole activated sludge. The size of sewage samples evaluated was generally 1 to 2 ℓ. They recovered adenovirus 3 from lime precipitates, coxsackievirus B3 from both ferric chloride and aluminum sulfate precipitates, and coxsackievirus B5 and echovirus 19 from aluminum sulfate precipitates. Ward and Ashley[57] seeded rotavirus SA 11 into anaerobically digested sludge, sonicated the virus-sludge mixtures, serially diluted them in cell culture medium, and assayed by direct inoculation.

A variation on the approach of inoculating whole sludge into cell assay systems directly is to mix the sludge with buffer and thereby aid release of solids-associated virions. Although this method is similar to elution of viruses from collected sludge solids, particularly when it is the resulting sludge-buffer supernate that is assayed, the method does not distinguish between virions only mixed with sludges and those physically adsorbed onto the solids. Thus, Turk et al.[58] recovered seeded poliovirus 1 and indigenous enteric viruses from primary raw and digested liquid sludges by dispersing the sludges in equal volumes of distilled water, agitating the sludge-water mixtures by magnetic stirring, homogenization, or sonication, and assaying the viruses in the resulting supernates after concentrating them by bentonite adsorption-elution. The efficiency of virus recovery was 74% for the poliovirus. Sanders et al.[35] assayed aerobically digested sludge for seeded poliovirus 1 by direct inoculation of either sludge supernates or of sludge solids suspended in phosphate buffered saline. Damgaard-Larsen et al.[26] mixed 25 to 50 g of dewatered anaerobically digested sludge that contained seeded coxsackievirus B3 with 50 mℓ of Hanks' balanced salt solution, diluted the mixture in cell culture fluid, and assayed by direct inoculation.

Table 1
SUMMARY OF STUDIES ON DIRECT ASSAY OF VIRUSES FROM SLUDGES AND SOLIDS SUSPENSIONS

Method used	Sample type	Sample size	Viruses recovered[a]	Recovery efficiency	Ref.
Sludge assayed directly; viruses also counted by electron microscopy	Mixed-liquor activated sludge	Unknown	Indigenous bacteriophages	Undetermined	54
Whole sludge sonicated, then diluted and assayed	Anaerobically digested sludge	Unknown	Seeded simian rotavirus SA-11	Undetermined	57
Sludge solids assayed directly	Aerobically digested sludge	Unknown	Indigenous bacteriophages	Undetermined	18
Solids assayed directly	Solids collected from secondary sewage effluents	Solids from 4—19 ℓ of effluent	Indigenous bacteriophages	Undetermined	14
Sludges assayed directly	Lime precipitate from toilet water, ferric chloride precipitate of raw sewage, aluminum sulfate precipitate of activated sludge supernate, and whole activated sludge	1—2 ℓ of sewage	Indigenous adenovirus 3, coxsackieviruses B3, B5, and echovirus 19	Undetermined	20
Slurry supernate assayed directly	Supernate from manure slurry	10 mℓ	Indigenous, porcine adenoviruses, porcine coronaviruses, porcine enteroviruses	Undetermined	46
Sludge dispersed in distilled water by magnetic stirring, homogenization, or sonication, resulting supernates assayed after concentration by bentonite adsorption-elution	Raw and digested primary and lagooned sludges	1 ℓ	Indigenous, enteric viruses, seeded poliovirus 1	74% for poliovirus 1 (in combination with bentonite adsorption-elution concentration	58
Direct inoculation of sludge supernate and of sludge solids suspended in phosphate-buffered saline (PBS)	Anaerobically digested sludge	Unknown	Seeded poliovirus 1	Undetermined	35
Sludge solids mixed with Hanks' balanced salt solution, mixtures shaken, diluted, and assayed	Dewatered, anaerobically digested sludge	25—50 g of sludge solids	Seeded coxsackievirus B3	Undetermined	26

Sludge solids suspended in PBS containing 10% calf serum, suspensions sonicated and assayed	Mixed-liquor activated sludge	5 mℓ	Seeded poliovirus 1	Undetermined	4
Sludge homogenized with Earle's balanced salt solution containing 2% calf serum (pH 9.5), centrifuged, supernate assayed	Digested sludge	100 mℓ	Seeded coxsackieviruses, echoviruses 6, 9, and polioviruses 1, 3	32% for poliovirus 1	59
Sludge diluted in 0.1% bactopeptone plus 0.05% NaCl for bacteriophage assays, or mixed with equal volume of cell culture medium for enterovirus assays	Mixed-liquor activated sludge	0.5 mℓ	Indigenous bacteriophages, seeded coxsackievirus B1, and poliovirus 1	Undetermined	60
Sludge diluted in solution containing 5% fetal calf serum, 3% beef extract and 1% gelatin, sonicated, and assayed	Anaerobically digested sludge	5 mℓ of 1:10 diluted sludge	Seeded coxsackieviruses A9, B4, echovirus 11, and poliovirus 1	80—120% depending on virus type and strain	34
Sludge solids suspended in Tris buffer containing 0.1% peptone, 0.03% NaCl, 0.0005 *M* $MgSO_4$, and 10% calf serum pH 7.8, and suspension assayed directly	Mixed-liquor activated sludge	5 mℓ	Indigenous bacteriophages seeded bacteriophage f2	Undetermined	49
Sludge mixed with 0.05 *M* glycine buffer containing 0.1% Tween®, 0.1% sodium lauryl sulfate, or 0.01 *M* ethylenediamine tetraacetic acid, sonicated at various pH levels, centrifuged and supernate assayed	Anaerobically digested sludge	10 mℓ	Seeded poliovirus 1	0—57%	62
Sludge extracted with freon, freon phase re-extracted with PBS, water phase from both steps pooled and assayed	Primary sludge	100 mℓ	Indigenous enteric viruses	Unknown	63

Table 1 (continued)
SUMMARY OF STUDIES ON DIRECT ASSAY OF VIRUSES FROM SLUDGES AND SOLIDS SUSPENSIONS

Method used	Sample type	Sample size	Viruses recovered[a]	Recovery efficiency	Ref.
Clay suspension assayed directly	Clay solids	Varied depending on assay system (cell culture of mice)	Seeded mouse encephalomyocarditis virus	Undetermined	9
Solids suspensions assayed directly	Seawater containing various types of suspensed solids	1 mℓ	Seeded bacteriophage T2	Undetermined	10
Solids suspensions assayed directly	Seawater containing suspended clay and bacterial cells	Unknown	Seeded bacteriophage T7	Undetermined	72
Sewage effluent solids and clay suspensions assayed directly	Clay suspensions and sewage effluent solids	Unknown	Seeded bacteriophages f2, T2, T7 and poliovirus 1	Undetermined	7
Solids suspensions diluted in tris (hydroxymethyl) aminomethane (Tris) buffered saline containing gelatin and assayed	Waters containing various types of suspended solids	0.5 mℓ	Seeded bacteriophages T1, P1, Qβ, 80α, Ø11M15	Undetermined	39
Solids suspensions diluted in Tris-buffered saline, then assayed	Seawater containing suspended sediment	Unknown	Seeded echovirus 1 and poliovirus 1	Undetermined	73
Solids suspensions mixed with purified water, glycine buffer with 2% calf serum (pH 7 or 11), 2% trypticase soy broth (pH 7), 0.1% Tween® 80, or 0.01 *M* EDTA (pH 11), diluted, and assayed	Clay suspensions	1 mℓ	Seeded bacteriophage MS-2	Maximum of 18%	12

[a] All animal virus types listed are human serotypes unless specificially indicated to be otherwise.

Table 2
SUMMARY OF STUDIES ON ELUTION OF VIRUSES FROM SLUDGES AND SOLIDS SUSPENSIONS

Methods used	Sample type	Sample size	Viruses recovered[a]	Recovery efficiency	Ref.
Solids homogenized in phosphate buffered saline (PBS) with 10% calf serum, freeze-thawed, centrifuged, and supernates assayed	Waste and return activated and blended undigested sludges, also trickling filter stones	Unknown	Indigenous coxsackievirus B5, polioviruses 1, 2, 3, untyped enteroviruses, and reoviruses	Undetermined	3
Solids suspended in 1 *M* glycine (pH 11), 0.05 *M* glycine with and without ethylene diaminetetraacetic acid (EDTA) at several pH levels, or 3% beef extract (pH 9.5), agitated by shaking, magnetic stirring, or sonication, centrifuged and supernates assayed; eluate produced from elution with 0.05 *M* glycine at pH 11 subsequently processed by organic flocculation-membrane filter concentration	Mixed-liquor activated sludge	500, 600 mℓ	Indigenous coxsackievirus B1, echoviruses 7, 17, poliovirus 1 Seeded coxsackievirus B3, echovirus 7, poliovirus 1	32 to greater than 100% for poliovirus 1 depending on elutant and agitation technique; approximately 80% for coxsackievirus B3, echovirus 7, and poliovirus 1, in combination with organic flocculation membrane filter concentration	29
Solids stirred in 0.05 *M* glycine at pH 10.5—11, centrifuged, and viruses concentrated from supernates by organic flocculation-membrane filtration	Mixed-liquor activated, aerobically digested, anaerobically digested, and lagooned sludges	100—1000 mℓ	Seeded poliovirus 1	Average 14—72% depending on sludge type (in combination with organic flocculation-membrane filter concentration)	64

Table 2 (continued)
SUMMARY OF STUDIES ON ELUTION OF VIRUSES FROM SLUDGES AND SOLIDS SUSPENSIONS

Methods used	Sample type	Sample size	Viruses recovered[a]	Recovery efficiency	Ref.
Sludge solids shaken in 3% solutions of beef extract, casamino acid, glycine, lysine, or aspartic acid (pH 9), 0.5% isoelectric casein (pH 9), 0.1% solutions of sodium dodecylsulfate, Tween® 80, cetyltrimethylammonium bromide, or 1% lysine containing various concentrations of urea (adjusted to pH 9), centrifuged, and viruses in supernates concentrated by two-stage inorganic-organic flocculation. Viruses eluted from sludge solids with 4 *M* urea buffered at pH 9 with 0.05 *M* lysine	Mixed-liquor activated, aerobically digested, and wasted liquid sludges, and lagooned sludge applied to land	25—1000 mℓ for liquid sludges, 50—100 g for sludges applied to land	Indigenous coxsackievirus B4, echoviruses 1, 4, 7, poliovirus 1 Seeded coxsackievirus B3, echovirus 1, poliovirus 1	5—75% for poliovirus 1, depending on eluate (for elution evaluations only), approximately 40% coxsackievirus B3 and poliovirus 1, and 90% for echovirus 1 (in combination with two-stage inorganic-organic concentration)	65
Solids stirred in 0.05 *M* glycine (pH 11.5) or mixed with 4 *M* urea buffered with 0.05 *M* lysine (pH 9), centrifuged, and supernate from each elution procedure was concentrated by an organic flocculation-membrane filter process and by two-stage inorganic-organic flocculation	Aerobically digested, anaerobically digested, lagooned sludges, and lagooned sludge applied to land	Unknown	Indigenous coxsackievirus B4, echovirus 1, 4, 7, 15, polioviruses 1, 2, 3 Seeded poliovirus 1	25—55% for poliovirus 1 depending on sludge type (for glycine elution in combination with organic flocculation-membrane filter concentration)	37

Solids repeatedly sonicated in 0.1% sodium lauryl sulfate with 0.05 *M* glycine, pH 7.5, centrifuged, and supernate assayed	Anerobically digested sludge	10 and 100 mℓ	Seeded poliovirus 1	Average 98%	62
Solids collected by filtration and eluted *in situ* with 0.05 *M* glycine (pH 7, 10, or 11.5), 3% trypticase soy broth, sewage effluent, or by water at ambient pH levels	Solids from secondary treated sewage effluent	4—19 ℓ	Indigenous bacteriophages, animals viruses	Apparent recovery[b] 9—75% for indigenous bacteriophages, depending on eluting agent	14
Solids suspended in 3% solution of casein hydrolysate, lactalbumin hydrolysate, beef extract, or in 10% calf serum, all in physiologic saline (pH 7.2—7.4), centrifuged, and the supernate assayed	Presumably raw sludge	20 mℓ	Indigenous coxsackieviruses B3, B5, poliovirus 2, reovirus 2 Seeded poliovirus 1	Less than 0.4% for poliovirus 1	66
Solids repeatedly eluted with 2x tryptose phosphate broth, centrifuged, and the supernate assayed	Raw and digested primary and secondary sludges	1 ℓ	Indigenous enteric viruses	Undetermined	67
Solids suspended in 3% beef extract (pH 9.4—9.6), sonicated, stirred, or extracted with freon, centrifuged, and supernate hydroextracted, and assayed	Activated sludge and solids from sewage treatment plant influent and effluent	500 mℓ	Indigenous echovirus 22/23, polioviruses 1, 2, 3, untyped reoviruses	Undetermined	33
Solids homogenized in 3% beef extract, sonicated at pH 9, 10, or ambient sludge pH, centrifuged and supernate assayed	Dewatered and partially composted anaerobic digestor solids	10 g	Seeded poliovirus 1	Average 30—47% depending on elution conditions	

Table 2 (continued)
SUMMARY OF STUDIES ON ELUTION OF VIRUSES FROM SLUDGES AND SOLIDS SUSPENSIONS

Methods used	Sample type	Sample size	Viruses recovered[a]	Recovery efficiency	Ref.
Sludge solids mixed with 10% beef extract (pH 7), centrifuged, and supernate assayed; solids from alum-precipitated sludges resuspended in tris (hydroxymethyl) aminomethane buffer (pH 9) centrifuged, and supernate assayed	Raw and digested primary and secondary sludges, and alum precipitated sludges	50 mℓ	Indigenous adenovirus 2, coxsackieviruses B2, B5, poliovirus 3	Undetermined	30
Solids mixed with 1.0% beef extract in McIlvane's buffer (pH 7.0), centrifuged, and supernate assayed	Primary, mixed-liquor activated, and anerobic mesophilically digested sludge	100 mℓ	Indigenous enteroviruses	Undetermined	68
Solids eluted by sequential washing with sodium chloride buffer and several sodium phosphate buffers, centrifuged, and supernate assayed	Clay solids	5 mℓ	Seeded reovirus 3	Less than 1% of initial infectivity	74
Solids resuspended in borate saline (pH 9.0), centrifuged, and supernate assayed	Organic solids suspensions and clay suspensions	10 mℓ	Seeded bacteriophages f2, T2, T7, poliovirus 1	77—78% for poliovirus 1	7
Solids collected by filtration, eluted *in situ* with 0.1 *M* glycine buffer alone or supplemented with 10% calf serum or with 0.01 *M* EDTA with 1% calf serum (pH 11.0), or solids eluted with tryptose phosphate broth (pH 7.2)	Clay suspensions and solids from primary treated sewage effluent	Unknown	Seeded poliovirus 1	2—72% depending on eluting agent	44

Solids sonicated in tryptose phosphate broth (pH 7.3), centrifuged, and supernate assayed	Clay suspensions in sludge-water supernate	2 ℓ sludge-water supernate	Indigenous enteroviruses	74% (in combination with sludge-water elution process)	58
Solids mixed with beef extract solutions of various concentrations (with or without borate or Mc-Ilvane's buffer), centrifuged, and supernate assayed	Suspended river solids	15 g solids in 200 mℓ water	Seeded poliovirus 1	39—63%	75

[a] All animal virus types listed are human serotypes unless specificially indicated to be otherwise.
[b] Because experiments involved indigenous viruses, the exact initial title is not known, therefore, recoveries listed are apparent.

Table 3
SUMMARY OF STUDIES ON CONCENTRATION OF VIRUSES FROM SLUDGES AND SOLIDS SUSPENSIONS

Method used	Sample type	Sample size	Viruses recovered[a]	Recovery efficiency	Ref.
Calcium phosphate-gel adsorption	Anaerobically digested sludge	100 mℓ	Indigenous enteric viruses	Undetermined	36
Two-phase concentration	Activated sludge	5 ℓ	Coxsackievirus B5		
Concentration with PE-60 (polyelectrolyte) resin (pH 7.5), collection of the resin, and elution of the resin with cell culture medium (pH 8.0)	Manure slurry supernates	2 ℓ	Indigenous porcine adenoviruses, porcine enteroviruses	Undetermined	46
Hydroextraction against polyethylene glycol	Sludge-beef extract suspensions, beef extract eluates of sludge, and supernates from freon extraction of sludge	500 mℓ initial sewage	Indigenous echovirus 22/23, polioviruses 1, 2, 3, untyped reovirus	Undetermined	33

Table 3 (continued)
SUMMARY OF STUDIES ON CONCENTRATION OF VIRUSES FROM SLUDGES AND SOLIDS SUSPENSIONS

Method used	Sample type	Sample size	Viruses recovered[a]	Recovery efficiency	Ref.
Adsorption onto suspended marine sediment and filtration to collect the sediment	Marine water eluates of sewage solids	30 mℓ	Indigenous bacteriophages	Undetermined	14
Adsorption onto bentonite and elution of viruses from the bentonite solids by sonication in tryptose phosphate broth (pH 7)	Supernates from sludge-distilled water mixtures	Supernates made with 1 ℓ of sludge	Indigenous enteric viruses, seeded poliovirus 1	74% for poliovirus 1 (in combination with prior distilled water dispersion of sludge)	58
Organic flocculation of supernates	Supernates from sludge solids-beef extract mixtures	Supernates from 25—50 g sludge	Seeded poliovirus 1	60—65%	38
Viruses pelleted by ultracentrifugation, or viruses concentrated from 3% beef extract preparations by organic flocculation	Supernates from freon extraction of sludge, and phosphage buffered saline reextraction of freon phase	100—200 mℓ	Indigenous enteric viruses	Undetermined	63
Beef extract added to eluate (final concentration 3%), viruses concentrated by organic flocculation	Sodium lauryl sulfate-glycine buffer eluates of sludge solids	Eluates from 100 mℓ sludge-virus mixture	Seeded echovirus 7, poliovirus 1	Approximately 87% for seeded poliovirus 1, seeded echovirus 7 (in combination with prior SLS-glycine buffer elution of sludge solids)	62
Organic flocculation-membrane filter concentration, consisting of organic flocculation of sludge eluates utilizing only materials eluted from the sludge, with subsequent membrane filter concentration of the resulting organic flocculation supernate	Glycine buffer eluates of sludge solids	Eluates from 500—10,600 mℓ of sludge	Indigenous coxsackievirus B1, echoviruses 7, 17, poliovirus 1, seeded coxsackievirus B3, echovirus 7, poliovirus 1	91% for poliovirus 1	29

Organic flocculation-member filter concentration	Glycine buffer eluates of sludge	Eluates from 100—1,000 mℓ sludge	Seeded poliovirus 1	Average 14—72%, depending on sludge type (in combination with prior glycine buffer elution of sludge)	
Two-Stage inorganic-organic flocculation of sludge eluates, consisting of aluminum hydroxide ($AlOH_3$) flocculation of sludge eluate followed by elution of the $AlOH_3$ floc with 0.1 *M* ethylenediamine tetraacetic acid (EDTA) with 3% beef extract buffer (pH 9.0), dialysis of the $AlOH_3$ floc eluate to remove EDTA, then organic flocculation of the dialyzed $AlOH_3$ floc eluate	Urea-lysine buffer eluates of sludge solids	Eluates from 20—1,000 mℓ of sewage sludges, and 50—100 g of lagooned sludges applied to land	Indigenous coxsackievirus B4, echoviruses 1, 4, 7, poliovirus 1, seeded coxsackievirus B3, echovirus 1, poliovirus 1	Approximately 40% for seeded coxsackievirus B3 and poliovirus 1. Approximately 90% for seeded echovirus 1 (in combination with prior urea-lysine buffer elution of sludge solids)	65
Organic flocculation-membrane filter concentration of glycine buffer sludge eluates, 2-stage inorganic-organic flocculation concentration of urea-lysine buffer sludge eluates	Glycine or urea-lysine eluates of sludges, and sludge supernates	Unknown	Indigenous coxsackievirus B4, echoviruses 1, 4, 7, 15, polioviruses 1, 2, 3, untyped isolates	25—55% for poliovirus 1 (in combination with prior glycine or urea-lysine buffer elution of sludge)	
Organic flocculation	Beef extract eluates of river water solids	Unknown	Unknown	Unknown	75

[a] All animal virus types listed are human serotypes unless specifically indicated to be otherwise.

Balluz et al.[4] sonicated solids from 5 mℓ of mixed-liquor activated sludge in 1 mℓ of phosphate-buffered saline that contained 10% calf serum and assayed the resulting suspension for seeded poliovirus 1 by direct inoculation. Subrahmanyan[59] mixed 100 mℓ of seeded digested sludge with 50 mℓ of Earle's balanced salt solution that contained 2% calf serum, homogenized and centrifuged the mixture, and assayed the supernates for seeded viruses by direct inoculation. In this paper, Subrahmanyan evaluated recovery of polio virus type 1 from sewage sludge at different pH levels and determined that recovery was maximal at 32%, when the sludge mixture was eluted at pH 9.5. Serum concentrations higher than 2% produced no advantage in elution of viruses. With his method, Subrahmanyan recovered seeded coxsackieviruses A9, B2, B4, and B5, echoviruses 6 and 9, and polioviruses 1 and 3 from sludges. Glass and O'Brien[60] recovered seeded poliovirus 1 and coxsackievirus B1 and indigenous coliphages from activated sludge by direct inoculation. Prior to inoculation, the sludge was mixed with either an equal volume of cell culture medium for assay of enteroviruses, or diluted in a solution of 0.1% tryptone plus 0.05% NaCl for assay of coliphages. Balluz et al.[49] sonicated solids from 5 mℓ of mixed-liquor activated sludge at pH 7.8 in 5 mℓ of tris(hydroxymethyl) aminomethane (Tris) buffer that contained 0.1% peptone, 0.03% NaCl, 0.0005 *M* $MgSO_4$, and 10% calf serum and assayed the suspensions by direct inoculation for seeded bacteriophage f2 and indigenous bacteriophages.

Bertucci et al.[34] assayed anaerobically digested sludge for seeded viruses by direct inoculation after tenfold dilution of the sludge in a solution that contained 5% fetal calf serum, 3% beef extract, and 1% gelatin followed by subsequent sonication and freezing. From 80 to 120% of seeded coxsackieviruses A9 and B4, echovirus 11, and poliovirus 1 were recovered from sludge with this technique. Wellings et al.[33] recovered indigenous echovirus 7 and poliovirus 3 from liquid and dried activated sludges by adding beef extract (final concentration 3%) to the sludges, adjusting pH to 9.0, sonicating, and hydroextracting the mixtures against polyethylene glycol before assay. Glass et al.[38] used a beef extract method to recover viruses from liquid sludge, and dewatered, partially composted digested sludge. They mixed 12 g of beef extract with each 400 mℓ of liquid sludge and suspended each 25 g (dry weight) of dewatered sludge in 400 mℓ of 3% beef extract, sonicated and centrifuged, and concentrated the viruses in the supernates by organic flocculation. A similar method for elution of poliovirus 1 physically adsorbed to sludge solids yielded 47% recovery (see sludge elution section). Nielsen and Lydholm[30] inoculated raw and anaerobically digested sludges directly and also inoculated supernates from homogenates of sludge diluted in 10% beef extract. They recovered polioviruses 2 and 3 and coxsackievirus B5. Srivastava and Lund[47] recovered seeded bovine parvovirus 1 from liquid manure by direct assay of diluted manure and by assay of eluates from liquid manure diluted and shaken in 10% beef extract.

Ward and Ashley[61] recovered from 33 to 100% of seeded polioviruses 1 and 2 from raw and anaerobically digested primary sludges after sonication with 0.1% sodium dodecylsulfate (SDS) (final concentration). They assayed by direct inoculation both whole sludge-SDS mixtures and supernates produced from these mixtures. Abid et al.[62] recovered seeded poliovirus 1 from 10 mℓ quantities of anaerobically digested sludge in a factorial design experiment with three different elutants at four different pH levels. In this study, aliquots of sludge were diluted to twice their volumes with buffers that adjusted the mixtures to 0.05 *M* glycine plus either 0.1% Tween, 0.01 *M* ethylenediamine tetraacetic acid (EDTA), or 0.1% sodium lauryl sulfate (SLS). The mixtures were then adjusted to pH levels of 8.5, 9.5, 10.5, or 11.0 and sonicated, and the resulting supernates were assayed for viruses by direct inoculation. Maximal virus recoveries were achieved at pH 7.5 with SLS-glycine buffer. With three sequential processings of the solids, the authors recovered 121% of seeded echovirus 7 and an average of 98% of seeded poliovirus 1 from 100 mℓ amounts of sludge.

Brashear and Ward[63] extracted 100 mℓ volumes of raw primary sludge with equal volumes of freon, collected the aqueous phase, and reextracted the freon phase with an equal volume of phosphate buffered saline. In combination with various concentration techniques, according to the authors, almost all of the seeded viruses were recovered.

Ward[76] also recovered seeded poliovirus 1 from mixed-liquor suspended solids (mixed-liquor activated sludge) diluted in nutrient broth by direct inoculation. Brashear and Ward[77] used the freon-centrifugation-ether method[63] for recovering seeded poliovirus 1 from raw (primary) sludge. Hejkal et al.[86] suspended wastewater solids in tris(hydroxymethyl) aminomethane buffer that contained 0.1% casein (pH 9.0) and assayed for indigenous viruses by direct inoculation.

B. Elution of Viruses from Sludge Solids

Because most viruses in wastewater are adsorbed on particulate solids, many researchers have given specific attention to the elution of physically adsorbed virions from sewage solids. Usually, solids collected by sedimentation (including centrifugation) or by filtration are mixed with diluents under conditions designed to free adsorbed and imbedded virions, the solids are then re-collected, and the resulting elution supernate or filtrate (called eluates) is assayed for released viruses.

Cliver[3] recovered indigenous enteric viruses from grit solids, primary sludge, waste and return activated sludge, blended undigested and digested sludges, and from trickling filter stones and solids sloughed from them by elution with phosphate-buffered saline (pH 7.4) that contained 10% calf serum. The elution mixtures were homogenized and subjected to three or more freeze-thaw cycles before the eluates were assayed. Polioviruses 1, 2, and 3, coxsackievirus B5, and other unidentified enteroviruses and reoviruses were recovered.

Hurst et al.[29] eluted enteroviruses from mixed-liquor activated sludge solids with glycine buffers of various strengths, with and without EDTA, and at different pH levels. Recovery of viruses was most effective when sludge solids were suspended by magnetic stirring in five volumes of pH 11.5, 0.05 *M* glycine buffer. The resulting eluate contained 86% of seeded poliovirus 1 initially adsorbed onto the sludge solids. In combination with an organic flocculation-membrane filtration concentration procedure, recovery of three serotypes of seeded enteroviruses averaged 74%. With the combined technique, indigenous echovirus 7, poliovirus 1, and other enteroviruses were recovered from aerobically digested sludge solids and from 1000 to 10,600 mℓ volumes of activated sludge.

Pancorbo et al.[64] eluted seeded poliovirus 1 from sludge solids, mixed-liquor activated sludge, waste sludge, and from both aerobically and anaerobically digested sludges by suspending the centrifuged solids in 0.05 *M* glycine at pH 10.5 to 11.0, centrifuging the elution mixture, and processing the resulting supernate (eluate) by an organic flocculation-membrane filter procedure. The supernate that resulted from centrifugation of the virus-seeded sludge was also processed by this organic flocculation-membrane filter technique. The combined procedures yielded an average recovery of 14 to 72% depending upon sludge type.

Farrah et al.[65] seeded poliovirus 1 into sludges, collected the solids by centrifugation, and suspended them at pH 9 in either a 3% solution of beef extract, casamino acids, glycine, lysine, or aspartic acid, or in 0.5% isoelectric casein, or at ambient pH (6.5 to 7.0) in a solution of sodium dodecylsulfate, Tween 80®, or cetyltrimethylammonium bromide (CTAB) or at pH 9.0 in 1% lysine that contained either 2, 4, or 6 *M* urea. Virus recoveries ranged from 5 to 75% depending upon the eluting agent. Farrah et al.[37] eluted indigenous viruses from dried sludges and solids collected from liquid sludges with either 0.05 *M* glycine, pH 11.5 or 4 *M* urea in 0.05 *M* lysine, pH 9.0 and concentrated the eluates either by an organic flocculation-membrane filter process for glycine eluates, or by a two-stage inorganic-organic flocculation process for urea-lysine eluates. Sludges tested for indigenous enteric viruses included those treated by aerobic digestion, anaerobic digestion, lagooning, and solids applied to land. The viruses recovered included coxsackievirus B4, echoviruses 1, 4, 7, and 15, polioviruses 1, 2, and 3, and some untyped isolates.

Abid et al.[62] eluted seeded enteroviruses from 100 mℓ quantities of anaerobically digested sludges by sonicating the sludge solids at pH 7.5 in 0.05 *M* glycine buffer that contained 0.1% sodium lauryl sulfate. Three sequential elutions were used. The net efficiency of elution was 81% for poliovirus 1 and 122% for echovirus 7.

Gerba et al.[14] collected particulate solids from sewage effluents by filtration and eluted indigenous coliphages from the solids *in situ* on the filters with 3% trypticase soy broth, 0.05 *M* glycine at pH levels of 7, 10, and 11.5, distilled water, autoclaved sewage effluent, dechlorinated tapwater, freshwater, estuarine water, marine water, or artificial seawater. The waters used for elution were all evaluated at ambient pH levels. The eluates were assayed for viruses by direct inoculation.

Sattar and Westwood[66] eluted viruses from sludge solids with 3% solutions of casein hydrolysate, lactalbumin hydrolysate, and beef extract, and with a 10% solution of fetal calf serum, each in physiological saline (pH 7.2 to 7.5). Coxsackieviruses B3 and B5, poliovirus 2, and reovirus 2 were recovered from the eluates. The elution of seeded poliovirus 1 from sludge solids with 10% fetal calf serum yielded a maximum efficiency of 20%.[31] Moore et al.[67] tested primary and secondary sludges for indigenous coliphages and enteric viruses by eluting solids sequentially two to three times in 2x tryptose phosphate broth (TPB).

Wellings et al.[33] used 3% beef extract buffered to pH 9.4 to 9.6 to elute viruses from pelleted solids obtained from 500 mℓ volumes of sludges. The solids were suspended in the buffered beef extract and then either sonicated, agitated mechanically, or extracted with fluorocarbon. The eluates were hydroextracted and assayed for indigenous human enteric viruses. Maximal elution of viruses was obtained from sonicated solids-beef extract suspensions. Viruses isolated included echovirus 22/23, polioviruses, 1, 2, and 3, and untyped reoviruses. Glass et al.[38] recovered seeded poliovirus 1 from dewatered and partially composted digestor solids by homogenizing the sludge solids with 3% beef extract at pH 9, 10, and at ambient pH. The solids and beef extract were combined in a ratio of 10 g dry weight of solids per 200 mℓ of beef extract. Maximum recovery of seeded viruses averaged 47% with sonication at ambient pH. Nielsen and Lydholm[30] precipitated viruses with acid or alum from 50 mℓ amounts of raw and digested wastewater sludges and eluted the viruses from the acid precipitates by shaking them in four volumes of 10% beef extract at pH 7.0, and from the alum precipitates by suspending them in four volumes of Tris buffer at pH 9.0. Indigenous viruses recovered included adenovirus 2, coxsackieviruses B2 and B5, and polioviruses 1, 2, and 3. Koch[80] evaluated effectiveness of four procedures for recovering seeded poliovirus 1 from anaerobic mesophilically digested sludge that contained 0, 1.5, or 5.0 kg/m^3 CaO. In Method 1, sludge solids were collected by gravitational settling, 20 mℓ amounts of settled solids were centrifuged, the pelleted solids were washed with distilled water, and viruses were eluted from the washed solids in 20 mℓ of PBS-10% calf serum. The eluate, the sludge supernates (supernates from initial centrifugation of settle solids), and washes were all assayed for viruses. In Method 2, beef extract (final concentration 3%, final pH 9.0) was added to the sludge, the sludge was sonicated, the pH of the sludge was adjusted to 7.2, the sludge was centrifuged, and the eluate was assayed for viruses. In Method 3, sludges were homogenized, sonicated, and centrifuged and the supernates assayed for viruses. In addition, the collected solids were suspended in distilled water, mixed with 10% beef extract at a 1:4 ratio, shaken, and centrifuged to produce an eluate that was also assayed for viruses. In Method 4, beef extract (final concentration 1.5, 3.9, or 4.5%) was added to sludges, and the sludges were homogenized, sonicated, and centrifuged to yield eluates that were assayed for viruses. In this study, all materials assayed for viruses were first extracted with freon and diluted in PBS before inoculation. The author reported that Method 2 was the best of the four methods overall. With Method 4, no differences in effectiveness were observed with different concentrations of beef extract. Koch and Strauch[81] subsequently used Method 2 in conjunction with freon extraction and recovered seeded poliovirus 1 and bovine parvovirus BP from 40 mℓ amounts of raw and anaerobic mesophilically digested sludges that contained lime (final concentrations 1.5, 3.0, 5.0, 7.5, and 10.0 kg/m^3 CaO).

Goddard et al.[87] recovered indigenous enteroviruses from raw, thickened raw, anaerobic mesophilically digested, aerobic thermophilically digested, and thickened sludges by diluting sludges with a 1% solution of skim milk, shaking and centrifuging the mixtures, concentrating the viruses in the supernates by organic flocculation at pH 4.5 (controlling contamination with chloroform containing dithizone), and assaying for indigenous viruses. Polioviruses 1, 2, and 3 and coxsackievirus B1, B3, and B5 were recovered. These authors used the same procedure to recover indigenous viruses from mixed raw and humus sludge, surplus activated sludge, and sludges dewatered by centrifugation or filter plate pressing.[88] Some of the sludges examined had been conditioned before collection with lime [$Ca(OH)_2$], $FeSO_4$, or soluble cationic polyelectrolytes. Polioviruses 1, 2, and 3, coxsackieviruses B2, B3, B4, and B5, and echovirus 9 were recovered.

Lewis et al.[85] compared five methods for recovering seeded poliovirus 2 and indigenous enteric viruses from raw and anaerobically digested primary sludges. In one method, sludge was blended with distilled water, and a mixture of Tris-HCl, pH 7.5 (final concentration 0.1 *M*) NaCl (final concentration 0.2 *M*), glycerol (final concentrations 5% v/v), and calf serum (final concentration 10% v/v) was added. The sample was shaken, centrifuged, and the supernate collected. The pelleted solids were twice again extracted with the mixture. The three supernates were combined, centrifuged to pellet viruses, and the pellets were sonicated in cell culture medium and assayed. In a second method 5 mℓ amounts of sludge were treated as described in the first method, but with deionized water in place of the mixture. In a third method, 5 mℓ amounts of sludge were mixed with 1.0 *M* glycine-NaOH buffer (pH 11.5), the mixtures were sonicated, vortexed, and centrifuged, and the supernates were neutralized and assayed. In a fourth method, sludges were extracted with freon and centrifuged and the aqueous phase was assayed. In a fifth method, samples were adjusted to pH 7.5 with Tris-HCl (final concentration 0.1 *M*), mixed with beef extract (final concentration 1 to 10%), the samples were shaken and centrifuged, and the supernates were assayed. Extraction with the mixture was more efficient than any of the other procedures. Addition of nonionic or weakly ionic surfactants (NP-40, Tween® 80, or Triton® X-100) in a final concentration of 0.1% suppressed recovery of viruses. Addition of SDS in a final concentration of 0.1% increased recovery of viruses, but resulted in handling difficulties because SDS-salt precipitates formed at temperatures below 30°C. Elution of viruses with the above mixture depended on sludge solids content. Virus elution did not occur at solids levels above 0.5 to 0.75%. Efficiency of virus elution was lower with digested sludge than with raw sludge. Hamparian et al.[93] centrifuged aerobically and anaerobically digested sludges for 30 min at 800 $\times g$ and eluted viruses from the solids by mixing 5 mℓ of 3% beef extract that contained 0.1% SDS with each 6 g of wet solids and maintaining the elution pH at 7.5 during 1 hr of mixing at room temperature. The mixtures were centrifuged at 4°C for 30 min at 800 xg, the supernates were disinfected with chloroform, these mixtures were centrifuged to separate the chloroform, and the supernates were frozen at −20°C until assayed for viruses in BGM, RD, or HeLa cells. Dewatered sludges were similarly eluted and assayed. Polioviruses, echoviruses, coxsackieviruses, and reoviruses were recovered.

Berman et al.[68] and Berg and Berman[69a] concentrated viruses in primary, activated, and in anaerobically digested sludges by salting and acidifying the sludges and eluting the viruses thus adsorbed on the centrifuged solids with 10% beef extract buffered at pH 7.0. This method has been adopted by the U.S. EPA as a standard method[69b] and has been recommended as a standard method by an ASTM task group.[69c]

Berg et al.[69b] also evaluated organic flocculation as a means for concentrating viruses from buffered beef extract eluates obtained from 100 mℓ volumes of primary, activated, and anaerobic mesophilically digested sludges. The weighted average efficiency for concentrating indigenous viruses from the beef extract eluates was 58%.

C. Concentration of Viruses from Sludge

The numbers of viruses in sludge samples are often small, particularly so in sludges that have digested extensively or have undergone long periods of lagooning. Because of the expense in assaying large quantities of inoculum, it is generally helpful to concentrate samples before inoculating them into cell cultures.

Palfi[36] used a calcium phosphate gel adsorbent to detect indigenous viruses in 100 mℓ amounts of whole anaerobic digestor sludge. Lund[23,56] recovered indigenous enteroviruses from primary and secondary sludges by concentrating the viruses with a two-phase polymer extraction system. She recovered adenoviruses 1 and 6, coxsackieviruses B3 and B5, and an untyped echovirus. Lund and Ronne[20] used the same aqueous two-phase polymer system for concentrating viruses from 5ℓ samples of activated sludge. They recovered indigenous coxsackievirus B5 with this procedure. Derbyshire and Brown[46] concentrated indigenous porcine adenoviruses and enteroviruses from filtered 1ℓ samples of manure slurry supernates by mixing the supernates (pH adjusted to 4.5) with polyelectrolyte PE-60 resin, collecting the resin by filtration, and eluting the viruses adsorbed to the resin with 5 mℓ of cell culture medium at pH 8.0. Wellings et al.[33] recovered indigenous enteric viruses from 3% beef extract eluates of sludges. They concentrated viruses from whole sludge-beef extract mixtures by hydroextracting the eluates and mixtures against polyethylene glycol.

Gerba et al.[14] found that indigenous coliphages that had been eluted from sewage effluent solids could be concentrated by adsorption onto marine sediment. Turk et al.[58] concentrated indigenous enteric viruses in liquid sludge-distilled water mixtures by adding 200 mg of bentonite to each liter of sludge-water supernate, adding $CaCl_2$ (final concentration 0.02 *M*) to the clay-supernate suspensions, adjusting the pH of each to 6.0, centrifuging, and eluting the collected solids with TPB (pH 7.3).

Epp and Metz[78] recovered seeded poliovirus 1, echovirus 6, coxsackieviruses A9 and B5, and adenovirus 1 from irradiated fresh and anaerobically digested sewage sludges by homogenizing and centrifuging the sludge samples, passing the supernates through alginate filters, dissolving the filters in 5% sodium citrate, and assaying the resulting solutions for viruses. Epp and Metz also used this method to recover wild-type polioviruses, echoviruses, coxsackieviruses, and adenoviruses from aerobically digested and activated sludges.

Of the methods used for concentrating viruses from sludge supernates and eluates of sludge solids, organic flocculation has received the greatest attention. This procedure consists of reducing the pH of liquid samples to 3.5, which results in a denaturation and flocculation of much of the sample's organic molecule content, collecting resulting floc by centrifugation, and dissolving the floc in a small volume of 0.15 *M* Na_2HPO_4. During the course of this process, a large percentage of the viruses in the initial sample adsorb onto the organic precipitate and are recovered in the buffer. Glass et al[38] used organic flocculation to concentrate viruses from beef extract eluates of seeded sludges and dewatered composted solids. From 60 to 65% of seeded poliovirus 1 were recovered. Brashear and Ward[63] concentrated viruses from freon extracts of raw primary sludge supernates either by pelleting the viruses by ultracentrifugation or by adding beef extract (final concentration 3%) to the freon extract supernates and concentrating the viruses by organic flocculation. Abid et al.[62] concentrated enteroviruses from SLS-glycine buffer eluates of seeded anaerobically digested sludge solids by adding beef extract (final concentration 3%) to the eluates, three times flocculating the beef extract-eluate solutions at pH 3.5, and dissolving the centrifuged solids from all three flocculations in 0.15 *M* Na_2HPO_4-10% calf serum, pH 9.0, buffer. They recovered 125% of the poliovirus 1 and 89% of the echovirus 7 seeded.

Hurst and Goyke [83] used buffered beef extract elution[68] coupled with organic flocculation[70,82] and hydroextraction[33] for recovering indigenous enteric viruses from primary and mixed liquor activated sludges. They evaluated three procedures for reducing cytotoxicity in the concentrated eluates: freon extraction, cationic polyelectrolyte precipitation, and washing infected cell monolayers with saline that contained calf serum before adding overlay medium. The washing procedure was most effective for reducing toxicity. Organic flocculation was better than hydroextraction for concentrating viruses in the sludge eluates because beef extract contains nondialyzable, low pH-soluble cytotoxic components. Lydholm and Nielsen[84] compared direct inoculation with two other procedures for enumerating indigenous viruses in raw and anaerobic mesophilically digested sludges. In one of these procedures, 100 mℓ of 10% beef extract was added to 50 mℓ of sludge, the mixture agitated and centrifuged, and eluate assayed. The other method was Zetag precipitation in which 2 mℓ of a 0.2% stock solution of Zetag 94 (soluble cationic polyelectrolyte) was added to 20 mℓ of sludge, the pH of the mixture was adjusted to 5.5, and the resulting floc was allowed to settle. The floc was collected, mixed with 20 mℓ of 10% beef extract (pH 7.0), stirred, and centrifuged. The pH of the eluate was adjusted to 7.2, and the eluate was assayed for viruses. All samples were decontaminated with ether before assay. With the Zetag precipitation method, 12 of 22 sludge samples evaluated were determined virus positive. By combining Zetag precipitation with hydroextraction, the number of virus-positive samples was increased to 17. With the beef extract elution procedure, 8 out of 22 sludge samples were positive. Of the 22 samples, 9 were positive by direct inoculation. During the course of the study, indigenous adenovirus 1, coxsackieviruses A18, B3, B4, and B5, and echoviruses 3, 7, 11, 15, 17, 25, and 27 were recovered.

An alternative approach to organic flocculation has been the formation at low pH of virus-adsorbing precipitates from those organic materials normally solubilized during elution of sludges with high pH eluting agents. This approach was developed separately by two groups, Hurst et al.[29] for recovering viruses form sludge, and Sato and Ose[71] for coagulating suspensions of clay solids and bacterial cells. Hurst et al.[29] used a combined organic flocculation-membrane filter adsorption-elution procedure for concentrating indigenous and seeded enteroviruses from eluates of mixed-liquor activated sludges and from aerobically digested sludges. The pH of glycine eluates of sludges were adjusted to 3.5, the floc that formed was collected by centrifugation, and the viruses in the floc were eluted with calf serum. The viruses remaining in the supernate were concentrated by the membrane filter adsorption-elution technique. With this technique, 94% of seeded poliovirus 1 was recovered. Pancorbo et al.[64] recovered 14 to 72% of seeded poliovirus 1 from a variety of sludges by organic flocculation-membrane filter concentration of both the sludge supernates and sludge solids eluates.

Farrah et al.[65] used a two-stage inorganic-organic flocculation procedure to recover seeded indigenous enteric viruses from supernates of sludges and from urea-lysine buffer eluates of sludge solids. The supernates were flocced with aluminum hydroxide and the viruses adsorbed onto the aluminum floc were eluted with a solution of 0.1 *M* EDTA and 3% beef extract, pH 9. The floc-supernate was dialyzed to remove the EDTA, and the viruses in the dialyzed supernate were recovered by organic flocculation. Altogether, about 40% of poliovirus 1 and coxsackievirus B3 seeded into 500 to 1000 mℓ of sludge were recovered. Indigenous viruses recovered included coxsackieviruses, echoviruses, and polioviruses. Farrah et al.[37] recovered indigenous enteric viruses from glycine buffer eluates of sludge solids by organic flocculation-membrane filter concentration. Indigenous enteric viruses were also concentrated by two-stage inorganic-organic flocculation of supernates of the sludge and from urea-lysine eluates of sludge solids. This study included aerobically digested, anaerobically digested, and lagooned sludges and sludge solids applied to land. Indigenous viruses recovered included coxsackievirus B4, echoviruses 1, 4, 7, and 15, and polioviruses 1, 2, and 3.

Wullenweber and Joret[90] eluted viruses from sludges with glycine (pH 11.5) and then concentrated the viruses on Balston® filters. They found this method superior to several others tested.

IV. RECOVERING VIRUSES FROM SOLIDS IN WATER

A. Direct Assay of Solids-Associated Viruses

Viruses in natural waters tend to associate with suspended particulate solids.[43,44] Association with suspended solids appears to afford the viruses some degree of protection against natural inactivation processes.[41] Because of these findings, studies have been directed at determining infectivity of solids adsorbed virions, and the effects of adsorption to solids upon susceptibility of viruses to disinfection.

The most common method for recovering viruses associated with solids in water has been direct inoculation. Schaub and Sagik[9] assayed encephalomyocarditis virus adsorbed onto clay (montmorillonite) solids by inoculating the clay into cell cultures and mice. Suckling mice were infected via the oral route and weanling mice via the intracerebral route. Gerba and Schaiberger[10] assayed clay suspensions for seeded bacteriophage T2 by direct inoculation. Bitton and Mitchell[72] recovered, by direct inoculation, seeded bacteriophage T7 from clay suspensions diluted in nutrient broth. Moore et al.[7] studied the infectivity of bentonite-adsorbed poliovirus 1, bacteriophages f2, T2, and T7, and of sewage effluent solids-adsorbed poliovirus 1 by direct inoculation. Babich and Stotzky[39] assayed by direct inoculation suspensions of clays seeded with bacteriophages T1, P1, Qβ, 80α, and ϕ11M15 after diluting the clays in Tris-buffered saline (pH 7.0) that contained 0.1% gelatin. LaBelle and Gerba[73] assayed by direct inoculation suspensions of natural sediments for seeded poliovirus 1 and echovirus 1 after diluting the sediments in Tris-buffered saline that contained 2% calf serum.

Stagg et al.[12] evaluated the effect of diluents upon the recovery of clay solids-adsorbed seeded bacteriophage MS2. They mixed 1 mℓ of a virus-seeded, solids suspension in 2 mℓ of purified water, glycine buffer at pH 7 or 11, 0.1% Tween® 80 at ambient pH, 2% trypticase soy broth at pH 7, 2% calf serum at pH 7 or 10, or 0.01 *M* EDTA at pH 11, diluted each suspension in purified water, and assayed for viruses by direct inoculation.

B. Elution of Viruses from Solids

To examine the nature of virus adsorption to solids and to provide cleaner preparations for inoculation into assay systems, methods have been developed for eluting viruses from clays and river solids. Taylor et al.[74] sequentially washed clay solids (allophane) with sodium chloride buffer (pH 6.0) and various sodium phosphate buffers (all pH 7.0) to elute adsorbed reovirus 3. Less than 1% of the initial input virus was recovered. Moore et al.[7] adsorbed poliovirus 1 and bacteriophages f2, T2, and T7 onto organic and inorganic (clay) solids and eluted the viruses form the solids with borate saline at pH 9.0. The efficiency of virus elution ranged from zero to greater than 99% depending on the type of virus and solids. Schaub and Sorber[44] seeded bentonite clay and sewage solids with poliovirus 1, collected the solids by filtration, and eluted the viruses with tryptose phosphate broth (pH 7.2), glycine buffer (pH 11.0), or with glycine buffer (pH 11.0) that contained either 10% calf serum or 0.01 *M* EDTA and 1% calf serum. Virus recovery ranged from 2 to 71%, depending on the eluting agent used. Turk et al.[58] evaluated elution of viruses adsorbed to bentonite clay as part of a sludge concentration procedure. They mixed clay solids with TPB (pH 7.3), sonicated the mixture, and assayed the supernate for viruses. Both seeded and indigenous viruses were recovered. Berg and Dahling[75] adsorbed poliovirus 1 onto river water solids and eluted the viruses with unbuffered or with buffered beef extract solutions. The unbuffered beef extract solutions tested ranged in concentration from 3 to 50%. The buffered solutions tested consisted of 10% beef extract in either borate buffer (pH 9.8) or McIlvane's buffer (pH 7.0). Maximal recoveries of poliovirions adsorbed onto river water solids was 39 to 63%, obtained by eluting 15 g of wet solids with 40 mℓ of 10% beef extract in the McIlvane's buffer. Simmonds et al.[79] adsorbed poliovirus 1 onto clay particles (bentonite) and eluted the viruses by suspending the clay particles in 6% beef extract at pH 9.0. The suspensions were then centrifuged, and the supernates were neutralized and assayed. The seeded poliovirus 1 was recovered with efficiencies ranging from 58 to 94%. With this method, indigenous polioviruses 1, 2, and 3, coxsackievirus B4, and echovirus 21 were recovered from wastewaters and effluents.

C. Concentration of Viruses from Solids

It could be anticipated that in the examination of viruses associated with waterborne solids, as with the examination of viruses in many other environmental systems, a need might exist for concentrating viruses in samples before assay. Unfortunately, few data are available on such concentrations.

Berg and Dahling[75] used organic flocculation to concentrate viruses from beef extract eluates of river water solids.

Metcalf and Melnick[89] recovered seeded poliovirus 1 and rotavirus SA11 from artificial water-suspended solids (a silt loam-kaolin mixture) by eluting *in situ* 3.9 to 5.8 g of the material, retained on a membrane filter, with 800 mℓ of 3% beef extract (pH 10.5). The eluate was concentrated by organic flocculation and the resulting product suspended in 0.05 *M* glycine, freeze-thawed, clarified by centrifugation, and assayed. Recovery of polioviruses and rotaviruses averaged, respectively, 55 and 48% for elution alone, and 33 and 24% for combined elution-concentration.

Goyal et al.[91] suspended 300 to 400 g of ocean sediment in 1.5 to 2.0 ℓ of 6% beef extract (pH 10.5), shook the mixture vigorously for 5 min, centrifuged it at low speed, and concentrated the viruses eluted into the supernate by organic flocculation with elution in 0.05*M* glycine (pH 11.0). Enteroviruses were recovered from several sediments. Johnson et al.[92] recovered enteroviruses from estuarine sediments of various compositions with isoelectric casein supplemented with lecithin. The best recoveries were obtained with isoelectric casein supplemented with 3% semipurified soybean lecithin. Crude soybean lecithin produced lesser recoveries. Isoelectric casein without lecithin was not as effective as isoelectric casein with lecithin. The authors also noted that an increased clay content of sediments was correlated with lesser recoveries of viruses.

V. SUMMARY AND CONCLUSIONS

In general, the methods used for quantifying viruses in sludges and solids in water may be classified in two major groups. The first group consists of directly assaying samples collected in the field, samples mixed with buffer, or supernates and filtrates produced from samples mixed with buffer. The second group entails recognition of the great extent to which many types of viruses adsorb onto solids. Within this group are approaches in which collected solids are mixed with an eluting agent and a subsequently produced eluate is assayed for viruses. Some researchers have combined the two approaches. Here, whole sludge is mixed with an eluting buffer, the solids are collected and resuspended in fresh buffer, and the supernates from both the buffer-sludge mixture (representing direct assay) and the resuspended solids (representing elution) are assayed for viruses.

Several types of diluents and elution buffers have been used for recovering viruses from sludges and solids. These include distilled water, saline, detergents, high pH amino acid buffers, and solutions of dissolved organic materials that may or may not contain buffering salts. Combinations of these elutants have also been evaluated.

A variety of methods have been tested for concentrating sewage sludges and waterborne solids that contain viruses. Concentration is helpful when it increases the ability to detect viruses present in small numbers, and when it reduces the high cost of assaying large quantities of material. The approaches that have been evaluated for this purpose include: high speed centrifugation to pellet the virus, filter adsorption-elution, two-phase polymer treatment, chemical precipitation, hydroextraction, organic flocculation, and adsorption onto clay solids. Of these methods, those with the greatest recent interest are hydroextraction and organic flocculation.

REFERENCES

1. **Sayler, G. S., Nelson, J. D., Jr., Justice, A., and Colwell, R. R.,** Distribution and significance of fecal indicator organisms in the Upper Chesapeake Bay, *Appl. Microbiol.*, 30, 625, 1975.
2. **Berg, G.,** Removal of viruses from sewage, effluents, and waters, *Bull. W.H.O.*, 49, 451, 1973.
3. **Cliver, D. O.,** Virus asssociation with wastewater solids, *Environ. Lett.*, 10, 215, 1975.
4. **Balluz, S. A., Jones, H. H., and Butler, M.,** The persistence of poliovirus in activated sludge treatment, *J. Hyg.*, 78, 165, 1977.
5. **Rao, V. C., Lakhe, S. B., Waghmare, S. V., and Raman, V.,** Virus removal in primary settling of raw sewage, *J. Environ. Eng. Div. Am. Soc. Civ. Eng.*, 107, 57, 1981.
6. **Farrah, S. R., Goyal, S. M., Gerba, C. P., Conklin, R. H., and Smith, E. M.,** Comparison between adsorption of poliovirus and rotavirus by aluminum hydroxide and activated sludge flocs, *Appl. Environ. Microbiol.*, 35, 360, 1978.
7. **Moore, B. E., Sagik, B. P., and Malina, J. F., Jr.,** Viral association with suspended solids, *Water Res.*, 9, 197, 1975.
8. **Gerba, C. P., Goyal, S. M., Hurst, C. J., and LaBelle, R. L.,** Type and strain dependence of enterovirus adsorption to activated sludge, solids and estuarine sediments, *Water Res.*, 14, 1197, 1980.
9. **Schaub, S. A. and Sagik, B. P.,** Association of enteroviruses with natural and artificially introduced colloidal solids in water and infectivity of solids-associated virions, *Appl. Microbiol.*, 30, 212, 1975.
10. **Gerba, C. P. and Schaiberger, G. E.,** Effect of particulates on virus survival in seawater, *J. Water Pollut. Control Fed.*, 47, 93, 1975.
11. **Liew, P. F., and Gerba, C. P.,** Thermostabilization of enteroviruses by estuarine sediment, *Appl. Environ. Microbiol.*, 40, 305, 1980.
12. **Stagg, C. H., Wallis, C., and Ward, C. H.,** Inactivation of clay-associated bacteriophage MS-2 by chlorine, *Appl. Environ. Microbiol.*, 33, 385, 1977.
13. **Bitton, G. and Farrah, S. R.,** Viral aspects of sludge application to land, *ASM News*, 46, 622, 1980.
14. **Gerba, C. P., Stagg, C. H., and Abadie, M. G.,** Characterization of sewage solids-associated viruses and behavior in natural waters, *Water Res.*, 12, 805, 1978.
15. **Griggs, G. B. and Hopkins, T. S.,** The delineation and growth of a sludge field, *Water Res.*, 10, 501, 1976.
16. **Dick, R. I.,** Sludge treatment, utilization, and disposal, *J. Water Pollut. Control Fed.*, 50, 1096, 1978.
17. **McLean, D.,** Sludge irrigation: health benefit or hazard?, *Ann. Intern. Med.*, 82, 112, 1975.
18. **Brown, K. W., Jones, S. G., and Donnelly, K. C.,** The influence of simulated rainfall on residual bacteria and virus on grass treated with sewage sludge, *J. Environ. Qual.*, 9, 261, 1980.
19. **Foster, D. H. and Engelbrecht, R. S.,** Microbial hazards in disposing of wastewater on soil, in *Recycling Treated Municipal Wastewater and Sludge Through Forest and Cropland*, Sopper W. E. and Kardos, L. T., Eds., Pennsylvania State University, University Park, 1973, chap. 15.
20. **Lund, E. and Ronne, V.,** On the isolation of virus from sewage treatment plant sludges, *Water Res.*, 7, 863, 1973.
21. **Buras, N.,** Recovery of viruses from waste-water and effluent by the direct inoculation method, *Water Res.*, 8, 19, 1974.
22. **Middleton, P. J., Szymanski, M. T., and Petric, M.,** Viruses associated with acute gastroenteritis in young children, *Am. J. Dis. Child.*, 131, 733, 1977.
23. **Lund, E.,** Disposal of sludges, in *Viruses in Water*, Berg, G., Bodily, H. L., Lennette, E. H., Melnick, J. L., and Metcalf, T. G., Eds., American Public Health Association, Washington, D.C., 1976, chap. 15.
24. **Gerba, C. P., Wallis, C., and Melnick, J. L.,** Viruses in water: the problem, some solutions, *Environ. Sci. Technol.*, 9, 1122, 1975.

25. **Hudson, J. A. and Fennel, H.,** Disposal of sewage sludge to land: chemical and microbiological aspects of sludge to land policy, *Water Pollut. Control*, 79, 370, 1980.
26. **Damgaard-Larsen, S., Jensen, K. O., Lund, E., and Nissen, B.,** Survival and movement of enterovirus in connection with land disposal of sludges, *Water Res.*, 11, 503, 1977.
27. **Duboise, S. M., Moore, B. E., Sorber, C. A., and Sagik, B. P.,** Viruses in soil systems, *Crit. Rev. Microbiol.*, 7(3), 245, 1979.
28. **Grigor'eva, L. V., Korchak, G. I., Bondarenko, V. I., and Bei, T. V.,** Sanitary characteristics (virological and bacteriological) of sewage, sludge, and soil in suburbs of Kiev, *Hyg. Sanit.*, 33, 360, 1968.
29. **Hurst, C. J., Farrah, S. R., Gerba, C. P., and Melnick, J. L.,** Development of quantitative methods for the detection of enteroviruses in sewage sludges during activation and following land disposal, *Appl. Environ. Microbiol.*, 36, 81, 1978.
30. **Nielsen, A. L. and Lydholm, B.,** Methods for the isolation of virus from raw and digested wastewater sludge, *Water Res.*, 14, 175, 1980.
31. **Sattar, S. A. and Westwood, J. C. N.,** Recovery of viruses from field samples of raw, digested, and lagoon-dried sludges, *Bull. W.H.O.*, 57, 105, 1979.
32. **Tierney, J. T., Sullivan, R., and Larkin, E. P.,** Persistence of poliovirus 1 in soil and on vegetables grown in soil previously flooded with inoculated sewage sludge or effluent, *Appl. Environ. Microbiol.*, 33, 109, 1977.
33. **Wellings, F. M., Lewis, A. L., and Mountain, C. W.,** Demonstration of solids-associated virus in wastewater and sludge, *Appl. Environ. Microbiol.*, 31, 354, 1976.
34. **Bertucci, J. J., Lue-Hing, C., Zenz, D., and Sedita, S. J.,** Inactivation of viruses during anaerobic sludge digestion, *J. Water Pollut. Control Fed.*, 49, 1642, 1977.
35. **Sanders, D. A., Malina, J. F., Jr., Moore, B. E., Sagik, B. P., and Sorber, C. A.,** Fate of poliovirus during anaerobic digestion, *J. Water Pollut. Control Fed.*, 51, 333, 1979.
36. **Palfi, A.,** Survival of enteroviruses during anaerobic sludge digestion, in *Proc. 6th Int. Conf. Advances in Water Pollution Research*, Jenkins, S. H., Ed., Pergamon Press, Oxford, 1973, A/13/26/1—6.
37. **Farrah, S. R., Bitton, G., Hoffman, E. M., Lanni, O., Pancorbo, O. C., Lutrick, M. C., and Bertrand, J. E.,** Survival of enteroviruses and coliform bacteria in a sludge lagoon, *Appl. Environ. Microbiol.*, 41, 459, 1981.
38. **Glass, J. S., Van Sluis, R. J., and Yanko, W. A.,** Practical method for detection poliovirus in anaerobic digestor sludge, *Appl. Environ. Microbiol.*, 35, 983, 1978.
39. **Babich, H. and Stotzky, G.,** Reductions in inactivation rates of bacteriophages by clay minerals in lake water, *Water Res.*, 14, 185, 1980.
40. **Hurst, C. J. and Gerba, C. P.,** Stability of simian rotavirus in fresh and estuarine water, *Appl. Environ. Microbiol.*, 39, 1, 1980.
41. **Kapuscinski, R. B. and Mitchell, R.,** Processes controlling virus inactivation in coastal waters, *Water Res.*, 14, 363, 1980.
42. **Cubbage, C. P., Gannon, J. J., Cochran, K. W., and Williams, G. W.,** Loss of infectivity of poliovirus 1 in river water under simulated field conditions, *Water Res.*, 13, 1091, 1979.
43. **Berg, G.,** Reassessment of the virus problem in sewage and in surface and renovated waters, *Prog. Water Technol.*, 3, 87, 1973.
44. **Schaub, S. A. and Sorber, C. A.,** Viruses on solids in water, in *Viruses in Water*, Berg, G., Bodily, H. L., Lennette, E. H., Melnick, J. L., and Metcalf, T. G., Eds., American Public Health Association, Washington, D. C., 1976, chap. 9.
45. **LaBelle, R. L.,** The Role of Sediment in the Ecology of Enteric Viruses in the Estuarine Environment, Ph.D. thesis, Baylor College of Medicine, Houston, Tex., 1979.
46. **Derbyshire, J. B. and Brown, E. G.,** Isolation of animal viruses from farm livestock waste, soil and water, *J. Hyg.*, 81, 295, 1978.
47. **Srivastava, R. N. and Lund, E.,** The stability of bovine parvovirus and its possible use as an indicator for the persistence of enteric viruses, *Water Res.*, 14, 1017, 1980.
48. **Balluz, S. A. and Butler, M.,** The influence of operating conditions of activated-sludge treatment on the behavior of f2 coliphage, *J. Hyg.*, 82, 285, 1979.
49. **Balluz, S. A., Butler, M., and Jones, H. H.,** The behaviour of f2 coliphage in activated sludge treatment, *J. Hyg.*, 80, 237, 1978.
50. **Scarpino, P. V.,** Human enteric viruses and bacteriophages as indicators of sewage pollution, in *Discharge of Sewage from Sea Outfalls*, Gameson, A. L. H., Eds., Pergamon Press, Oxford, 1975, 49.
51. **Berg, G.,** The indicator system, in *Indicators of Viruses in Water and Food*, Berg, G., Ed., Ann Arbor Publishers, Ann Arbor, Mich., 1978, chap. 1.
52. **Evison, L. M.,** Inactivation of enteroviruses and coliphages with ozone in waters and waste waters, *Prog. Water Technol.*, 10, 365, 1978.
53. **Kott, Y., Roze, N., Sperber, S., and Betzer, N.,** Bacteriophages as viral pollution indicators, *Water Res.*, 8, 165, 1974.

54. **Ewert, D. L. and Paynter, M. J. B.,** Enumeration of bacteriophages and host bacteria in sewage and the activated-sludge treatment process, *Appl. Environ. Microbiol.*, 39, 576, 1980.
55. **Vaughn, J. M. and Metcalf, T. G.,** Coliphages as indicators of enteric viruses in shellfish and shellfish raising estuarine waters, *Water Res.*, 9, 613, 1975.
56. **Lund, E.,** The effect of pretreatments on the virus contents of sewage samples, *Water Res.*, 7, 873, 1973.
57. **Ward, R. L. and Ashley, C. S.,** Effects of wastewater sludge and its detergents on the stability of rotavirus, *Appl. Environ. Microbiol.*, 39, 1154, 1980.
58. **Turk, C. A., Moore, B. E., Sagik, B. P., and Sorber, C. A.,** Recovery of indigenous viruses from wastewater sludges, using a bentonite concentration procedure, *Appl. Environ. Microbiol.*, 40, 423, 1980.
59. **Subrahmanyan, T. P.,** Persistence of enteroviruses in sewage sludge, *Bull. W.H.O.*, 55, 431, 1977.
60. **Glass, J. S. and O'Brien, R. T.,** Enterovirus and coliphage inactivation during activated sludge treatment, *Water Res.*, 14, 877, 1980.
61. **Ward, R. L. and Ashley, C. S.,** Inactivation of poliovirus in digested sludge, *Appl. Environ. Microbiol.*, 31, 921, 1976.
62. **Abid, S. H., Lue-Hing, C., and Sedita, S.,** Development of a method for concentrating enteroviruses in anaerobically digested sludge, Report 78-13, The Metropolitan Sanitary District of Greater Chicago, Department of Research and Development, Chicago, 1978.
63. **Brashear, D. A. and Ward, R. L.,** Comparison of methods for recovering indigenous viruses from raw wastewater sludge, *Appl. Environ. Microbiol.*, 43, 1413, 1982.
64. **Pancorbo, O. C., Scheuerman, P. R., Farrah, S. F., and Bitton, G.,** Effect of sludge type on poliovirus association with and recovery from sludge solids, *Can. J. Microbiol.*, 27, 279, 1981.
65. **Farrah, S. R., Scheuerman, P. R., and Bitton, G.,** Urea-lysine method for recovery of enteroviruses from sludge, *Appl. Environ. Microbiol.*, 41, 455, 1981.
66. **Sattar, S. A. and Westwood, J. C. N.,** Comparison of four eluents in the recovery of indigenous viruses from raw sewage, *Can. J. Microbiol.*, 22, 1586, 1976.
67. **Moore, B. E., Sagik, B. P., and Sorber, C. A.,** Land application of sludges: minimizing the impact of viruses on water resources, in *Proc. Conf. Risk Assessment and Health Effects of Land Application of Municipal Wastewater and Sludges*, Sagik, B. P. and Sorber, C. A., Eds., University of Texas at San Antonio, 1978, 154.
68. **Berman, D., Berg, G., and Safferman, R. S.,** A method for recovering viruses from sludges, *J. Virol. Methods*, 3, 283, 1981.
69a. **Berg, G. and Berman, D.,** Destruction by anaerobic mesophilic and thermophilic digestion of viruses and indicator bacteria indigenous to domestic sludges, *Appl. Environ. Microbiol.*, 39, 361, 1980.
69b. **Berg, G., Safferman, R. S., Dahling, D. R., Berman, D., and Hurst, C.,** *Manual of Methods for Virology*, EPA 600/4-84-013, Environmental Monitoring and Support Laboratory, Cincinnati, 1984.
69c. **Goyal, S. M., Schaub, S. A., Wellings, F. M., Berman, D., Glass, J. S., Hurst, C. J., Brashear, D. A., Sorber, C. A., Moore, B. E., Bitton, G., Gibbs, P. H., and Farrah, S. A.,** Round robin investigation of methods for recovering human enteric viruses from sludge, *Appl. Environ. Microbiol.*, 48, 531, 1984.
70. **Katzenelson, E., Fattal, B., and Hostovesky, T.,** Organic flocculation: an efficient second-step concentration method for the detection of viruses in tap water, *Appl. Environ. Microbiol.*, 32, 638, 1976.
71. **Sato, G. and Ose, Y.,** Floc-forming substances extracted from activated sludge by sodium hydroxide solution, *Water Res.*, 14, 333, 1980.
72. **Bitton, G. and Mitchell, R.,** Effect of colloids on the survival of bacteriophages in seawater, *Water Res.*, 8, 227, 1974.
73. **LaBelle, R. L. and Gerba, C. P.,** Influence of estuarine sediment on virus survival under field conditions, *Appl. Environ. Microbiol.*, 39, 749, 1980.
74. **Taylor, D. H., Bellamy, A. R., and Wilson, A. T.,** Interaction of bacteriophage R17 and reovirus type III with the clay mineral allophane, *Water Res.*, 14, 339, 1980.
75. **Berg, G. and Dahling, D. R.,** Method for recovering viruses from river water solids, *Appl. Environ. Microbiol.*, 39, 850, 1980.
76. **Ward, R. L.,** Evidence that microorganisms cause inactivation of viruses in activated sludge, *Appl. Environ. Microbiol.*, 43, 1221, 1982.
77. **Brashear, D. A. and Ward, R. L.,** Inactivation of indigenous viruses in raw sludge by air drying, *Appl. Environ. Microbiol.*, 45, 1943, 1983.
78. **Epp, C. and Metz, H.,** Virological analyses of irradiated sewer sludge, *Zentralbl. Bakteriol. Parasitenk. Infektionskr. Hyg. Abt. 1: Orig. Reihe B*, 171, 86, 1980.
79. **Simmonds, R. S., Lewis, G. D., Loutit, M. W., and Austin, F. J.,** A method for detecting and identifying enteroviruses in effluents, *N.Z. J. Sci.*, 25, 57, 1982.
80. **Koch, K.,** Report on the recovery of poliomyelitis virus from digested sewage sludge dependent upon the pH value, *Zentralbl. Bakteriol. Parasitenk. Infectionskr. Hyg. Abt. 1: Orig. Reihe B*, 174, 325, 1981.

81. **Koch, K. and Strauch, D.,** Removal of polio- and parvovirus in sewage sludge by lime treatment, *Zentralb. Bakteriol. Parasitenk. Infektionskr. Hyg. Abt. 1: Orig. Reihe B*, 174, 335, 1981.
82. **Berg, G., Berman, D., and Safferman, R. S.,** A method for concentrating viruses recovered from sewage sludges, *Can. J. Microbiol.*, 28, 553, 1982.
83. **Hurst, C. J. and Goyke, T.,** Reduction of interfering cytotoxicity associated with wastewater sludge concentrates assayed for indigenous enteric viruses, *Appl. Environ. Microbiol.*, 46, 133, 1983.
84. **Lydholm, B. and Nielsen, A. L.,** The use of a soluble polyelectrolyte for the isolation of virus from sludge, in *Viruses and Wastewater Treatment*, Goddard, M. and Butler, M., Eds., Pergamon Press, Oxford, 1981, 85.
85. **Lewis, M. A., Nath, M. W., and Johnson, J. C.,** A multiple extraction-centrifugation method for the recovery of viruses from wastewater treatment plant effluents and sludges, *Can. J. Microbiol.*, 29, 1661, 1983.
86. **Hejkal, T. W., Wellings, F. M., Lewis, A. L., and LaRock, P. A.,** Distribution of viruses associated with particles in wastewater, *Appl. Environ. Microbiol.*, 41, 628, 1981.
87. **Goddard, M. R., Bates, J., and Butler, M.,** Recovery of indigenous enteroviruses from raw and digested sewage sludges, *Appl. Environ. Microbiol.*, 42, 1023, 1981.
88. **Goddard, M. R., Bates, J., and Butler, M.,** Isolation of indigenous enteroviruses from chemically treated and dewatered sludge samples, *Appl. Environ. Microbiol.*, 44, 1042, 1982.
89. **Metcalf, T. G. and Melnick, J. L.,** Simple apparatus for collecting estuarine sediments and suspended solids to detect solids-associated virus, *Appl. Environ. Microbiol.*, 45, 323, 1983.
90. **Wullenweber, M. and Joret, J. C.,** Comparative study on method to recover indigenous viruses from samples of activated sewage sludge, *Zentralbl. Baketeriol. Mikrobiol. Hyg.*, 177, 482, 1983.
91. **Goyal, S. M., Adams, W. N., O'Malley, M. L., and Lear, D. W.,** Human pathogenic viruses at sewage sludge disposal sites in the middle Atlantic region, *Appl. Environ. Microbiol.*, 48, 758, 1984.
92. **Johnson, R. A., Ellender, R. D., and Tsai, S. C.,** Elution of enteric viruses from Mississippi estuarine sediments with lecithin supplemented eluents, *Appl. Environ. Microbiol.*, 48, 581, 1984.
93. **Hamparian, V. V., Ottolenghi, A. C., and Hughes, J. H.,** Enteroviruses in sludge. Multiyear experience with four wastewater treatment plants, *Appl. Environ. Microbiol.*, 50, 281, 1985.

Chapter 3

RECOVERING VIRUSES FROM AEROSOLS

Charles A. Sorber

TABLE OF CONTENTS

I. INTRODUCTION

Aerosolized viruses have been of public health interest for many years. Of particular note are applications related to biological warfare, hospital infections, and, more recently, a variety of wastewater treatment unit processes. The latter are the focus of this chapter.

The four most studied wastewater treatment processes are activated sludge aeration basins, trickling filters, and spray land application of both wastewaters and sludges. The challenges associated with measuring concentrations of indigenous aerosolized viruses from each of these sources are essentially the same. The primary considerations in the selection of appropriate methods are the expected concentration of viruses in the aerosol and, to some extent, the type(s) of viruses to be quantified.

Commonly, airborne viruses are quantified in terms of the number of infectious units per unit volume of air. In environmental work, the units of expression are usually plaque forming units per cubic meter of air (PFU/m^3).

A. Aerosol Sources

Aerosols are any solid or liquid materials that remain suspended in air for any appreciable time. That is, aerosols act like gases in air. Typically, aerosols are in the size range of 0.01 to 20 μm. Anytime a significant interaction occurs between water and an air stream, an aerosol is generated. Thus aerosols are generated through a number of wastewater treatment unit processes.

In the case of an aeration basin, the action of the diffused air or the mechanical aerator generates wastewater aerosols at the air-water interface. The simple passage of air (wind) over the surface of a trickling filter results in aerosol formation as wastewater falls between the distribution arm and the medium or as it splashes on the medium. Similarly, when wastewater or sludge is sprayed, as in some land application practices, aerosols are formed. This aerosol formation is the direct result of the relatively violent interaction between the water stream and the atmosphere as imparted by the spray machinery.

Once viruses are aerosolized, the concentration of these viruses in air is reduced by atmospheric diffusion and, possibly, by biological decay. It is this reduction, coupled with the relatively small numbers of viruses aerosolized from wastewater, that makes their quantification in air difficult.

B. Factors Affecting Aerosolized Viruses

The concentrations of viruses that have been found in wastewater aerosols vary considerably but are relatively small, usually less than 10^{-2} PFU/m^3. Furthermore, the specific concentrations of viruses in wastewater aerosols depend on a number of variables including

- Concentration of viruses in the wastewater or sludge at the point of aerosolization
- Amount of wastewater or sludge aerosolized per unit time
- Impact of aerosolization (i.e., the immediate negative impact on organisms upon release to the atmosphere from a water medium
- Atmospheric diffusion of the aerosol
- Biological decay with distance (time) traveled

Once viruses have been aerosolized and have survived the initial shock or impact of aerosolization, there are a number of factors which will affect the concentration of those viruses at any point downwind from the source. All of these factors are related to the meteorological conditions that exist at the time of aerosolization and have an effect on atmospheric diffusion and biological decay. They include

- Atmospheric stability
- Wind speed
- Relative humidity
- Temperature
- Sunlight

A number of papers have reviewed the effect of these factors;[1-5a] therefore, this information will not be repeated here in detail. However, it is important to understand that aerosolized viruses act as nonconservative air pollutants. As such, concentrations downwind of an aerosol source will be smaller, perhaps significantly smaller, than when aerosolized as a result of the above factors. In general, airborne viruses will be more concentrated under stable atmospheric conditions including moderate wind speeds, moderate temperatures, moderate to high relative humidities, and extensive cloud cover or darkness.

C. Why Measure Wastewater Aerosols?

Although some early investigators found the evaluation of wastewater aerosols of largely academic interest,[6] the topic received extensive attention when widespread use of spray irrigation as an alternative wastewater treatment scheme in the U.S. came under consideration. Investigations conducted in the early 1970s indicated that little information was available on the survival and transport of microbiological aerosols generated as a result of wastewater treatment practices.[1,7] As a result, a number of field studies were initiated in an attempt to describe the risk to workers and individuals living near sources of wastewater aerosol generation.[2,8-18] These efforts necessitated the development of methods appropriate for the measurement of the low levels of viruses aerosolized from wastewater.

Despite these extensive efforts, the more sensitive methods described herein are not normally appropriate for routine monitoring of wastewater aerosols. In fact, there is some question as to the requirement for this type of aerosol monitoring. Fundamentally, these techniques are research or special study tools used to estimate aerosol concentration, validate models, or for some other very specific purpose. Nevertheless, there can be other occasions when these measurements are required.

II. SAMPLING AEROSOLS

The first step in any analytical effort is to collect a representative sample that lends itself to the procedures that follow. The collected sample must be representative of the medium sampled. Further, the sampling procedures must be sufficiently sensitive to permit capture of several times the smallest concentration of viruses expected in the sample. Generally, this is translated into air volume sampled. Thus, if a virus concentration of 10^{-3} PFU/m^3 of air is anticipated, use of a sampler that handles 10 to 20 ℓ of air per minute is of no value. This fact has led to the use of "high volume" samplers.

In any case, sampling of airborne viruses is usually accomplished by collection of the viruses in a sampling fluid that is compatible with storage, shipping, and assay methods. Further, reconcentration of the viruses in the sampling fluid is usually required to obtain maximal sampling and assay sensitivity.

A. Air Sampling Equipment

Although there is a wide spectrum of sampling procedures and equipment available for microbial aerosols, only a limited number of sampler types are of value for sampling viruses aerosolized from wastewaters and sludges. This is primarily due to the low levels of viruses found in air and, thus, the sampling and assay sensitivity required for their enumeration.

Table 1 lists selected common sampling equipment suitable for sampling viruses. A more extensive discussion of each of these samplers has been presented elsewhere.[5]

Table 1
SELECTED COMMON SAMPLING EQUIPMENT FOR THE RECOVERY OF AIRBORNE VIRUSES

Sampler	Sampling mechanism	Air sampling capacity
All-glass impinger	Impingement	Low
Multistage impinger	Impingement	Low
Litton Model M; LEAP	Electrostatic precipitation	High
Cyclone scrubber	Centrifugal separation	High

1. Impingement

Impingement involves striking the viruses into a suitable collection medium which facilitates viral assay. The all-glass impinger (AGI) is a simple, inexpensive device designed to operate at 0.0125 m^3/min. In fact, the AGI-30, which has a jet-to-base distance of 30 mm, has been referred to as a "standard" sampler by many aerobiologists. Its major liability lies in its low air sampling capacity when compared to that normally required for environmental samples of airborne viruses. Obviously, sampling capacity may be increased by operating multiple AGIs in parallel. Multistage liquid impingers are used most frequently for particle size studies.

2. Electrostatic Precipitation

The development of large-volume electrostatic precipitator samples has made possible the recovery of airborne viruses from environmental sources where use of low-volume sampling equipment had not been successful. These samplers are designed for air flow rates of either 1 m^3/min or 10 m^3/min. Consequently, they have been used in a number of environmental studies.[2,8,13-17]

In electrostatic precipitator samplers, air drawn into the sampler through a converging nozzle passes through the center of a high-voltage plate. The air flows radially between the plate and a rotating collection disk with collection medium added at the center of the disk. Because of the centrifugal force of the disk, a thin moving film of medium moves over the entire disk surface. An electric potential of 12,000 to 15,000 V is maintained across the spacing between the plate and disk creating an electrical charge on the particles in the air and providing an electrical field that induces migration of the charged particles to the rotating disk. Particles collected on the film of collection medium on the rotating disk are collected and pumped back to the collection medium reservoir for recirculation through the sampler.

Electrostatic precipitator samplers have some limitations. They lose significant efficiency under conditions of high relative humidity, and they require special care and attention during operation and for sterilization. Moreover, sampler fluid may foam, and evaporative losses may have to be replaced.

3. Cyclone Scrubbing

Cyclone scrubber aerosol samplers collect particles by tangential impingement into a continous fine sampling fluid film on the sampler wall. Operating at air sampling rates of up to 1 m^3/min, these devices appear to be fairly efficient for a number of test organisms when compared to the AGI-30.[5]

As in the case of the electrostatic precipitator samplers, foaming of sampler liquid may present some problems, and collection medium losses may have to be replaced. On the other hand, these devices are relatively easy to sterilize and to operate in the field.[5]

Table 2
CORRECTION FACTOR FOR LARGE VOLUME SAMPLER OPERATING VOLTAGE (REFERENCED BASIS OF 12 kV)[15]

Operating voltage (kV)	Correction factor (F)
6	0.33
8	0.36
9	0.38
10	0.42
11	0.57
11.5	0.80
12	1.00
12.5	1.15
13	1.25
13.5	1.32
14	1.33
14.5	1.32
15	1.29
16	1.24
17	1.22
18	1.21

B. Sampler Data/Calibration

All large volume samplers used for the quantification of viruses in air require calibration. This is necessary because large-volume electrostatic precipitator sampler efficiency varies with operating voltage. Moreoever, these samplers are less efficient for collecting microorganisms than are the AGI-30 samplers.

Typically, sampler calibration is tedious and expensive. In one study, extensive calibration efforts were expended to improve the quality of the data to be collected by Litton Model M samplers.[15] It involved rigorous experimentation in a controlled environment to establish correction factors for a number of operating conditions.

The calibration experiments were conducted by the Naval Biosciences Laboratory, Oakland, Calif., and data were obtained for collection efficiencies for the large-volume sampler relative to AGI samplers. The test environment included an atomizer that created a predetermined amount of aerosol in an enclosed wind tunnel.[15] The test organism was *Bacillus subtilis* var *Niger*, operating voltages were varied from 8000 to 15,000 V, and relative humidity was varied from 51 to 81%.[15]

Evaluation of the data by personnel of the Southwest Research Institute, San Antonio, Tex., indicated that the only significant correction justified from the test data related to operating field voltages.[15] They further concluded that the experimental protocol did not permit the evaluation of such factors as different microorganisms, concentration of microorganisms, particle size, or wind gusts. Moreover, inadequate data were generated to evaluate the wide temperature ranges in the field.[15]

Consequently, a correction factor table was developed for various operating voltages (Table 2). These data were used in the calculation of microorganism densities sampled in the air as determined through assays of the large volume sampler collection fluid. For an individual sampler the relationship used was

$$C = \frac{A \times V}{F \times R \times D}$$

where C = concentration of detectable microorganism units/m^3 of air (e.g., colony-forming units (CFU/m^3); A = concentration of detectable microorganism units per milliliter of collection fluid assayed; V = final volume of collection fluid, mℓ; R = air sampling rate (usually 1 m^3/min); D = sampling duration, minutes; F = correction factor for operating voltage, from Table 2.

These data suggest a fourfold difference in sampler collection efficiency for the test organism over a range of 6 to 18 kV. These differences may not be much greater than normal laboratory errors found in the concentration and assay of viruses.

C. Sampling Strategies

A typical aerosol sampling strategy involves the location of samplers at different distances downwind of an aerosol source. An upwind sampler or samplers not under the influence of the aerosol source to be evaluated are also required.

Unfortunately, it is rare that samplers and mean prevailing wind direction can be aligned perpendicular to the aerosol source at all times. Thus, care must be taken that the samplers are located within the source's aerosol plume. Figure 1 provides an example of a sampler layout associated with a wastewater spray irrigation line source. This and other sampling arrangements have been used in various studies.[2,13,15-18]

Proper meteorological conditions must prevail for an aerosol run to be attempted. These include suitable and consistent wind speeds and wind direction and no precipitation. Once the prevailing wind direction is determined, sampler positions are assigned. Before initiating an aerosol run, meteorological conditions must again be evaluated for changes in prevailing wind direction and direction variability.

At the end of the run, wind direction data from the meteorological stations and other run parameters must be evaluated according to the following criteria which have been suggested as guidelines for a good aerosol run.[19]

- Average wind direction for the segments of the run should be no more than 30° from the wind direction angle used for sampler set up.
- The extremes of wind direction (excluding outlier spikes) should be no more than 60° from the wind direction average.
- At least one sampler at each sampling location should complete the run.
- No significant precipitation should occur during the run.

Under most field conditions, estimates of concentrations of viruses in the aerosol may be below the sensitivity of even large-volume samplers. This situation may be mitigated by using all available samplers in tandem operation and concentrating the collection fluid. This approach has been used to sample air volumes of more than 1000 m^3 in several studies.[2,15,17] Unfortunately, unless large numbers of samplers are available, the opportunity for evaluating multiple distances is lost with this approach.

Clearly, under any sampling conditions the availability of real-time, local meteorological data is essential. The sampling decisions described above and the interpretation of the results obtained depend upon these data. Thus, it is common to have an on-site meteorological station capable of recording wind speed, wind direction (both on a 2 m tower), temperature, relative humidity, and solar radiation. The significance of solar radiation is discussed below.

III. QUANTIFYING AIRBORNE VIRUSES

A. Expected Levels in Air

A number of factors determine the concentration of viruses in wastewater aerosols. Perhaps most important of these is the amount of wastewater aerosolized by the specific process

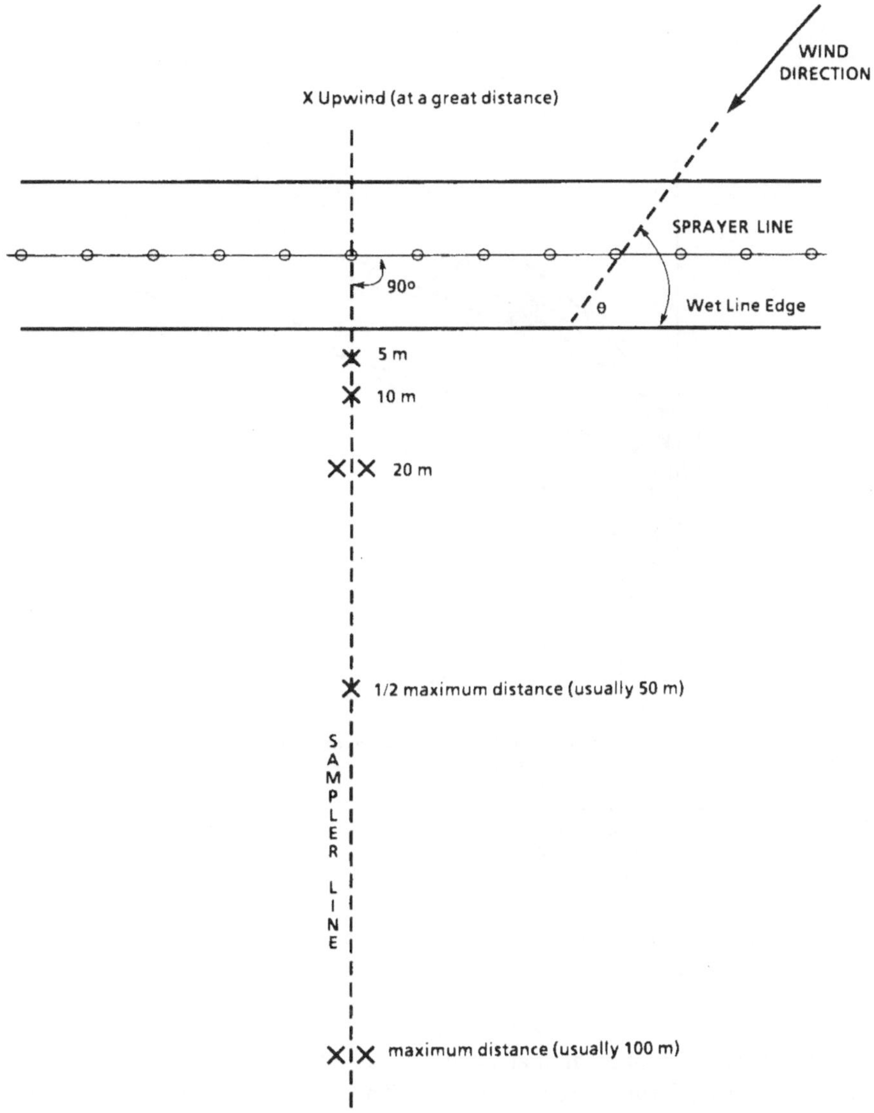

FIGURE 1. Typical sampler array for a spray line source.[2]

under consideration. A number of studies have attempted to quantify this information at wastewater and sludge spray application sites.

Early studies found the amount of aerosolized wastewater leaving the wetted zone to be dependent upon spray nozzle type, spray arc height, and spray pressure.[20] These factors are different at each site. Detailed site specific studies have found the median value to range from 0.3 to 0.5% for rotating impact sprinklers.[2,8,15-18] A regression equation of data from one of these studies suggested that the amount of wastewater aerosolized is primarily influenced by wind speed (shear forces), air temperature, and solar radiation (evaporation rate).[2]

Further, airborne microorganism concentrations can be reduced by a number of factors related to the aerosolization process (rapid water evaporation, turbulence, etc.).[4] However, this type of reduction does not appear to be significant for certain viruses, such as enteroviruses.

Since microorganisms (including viruses) are not conservative materials, viability will be lost with time. Again, within the time of downwind travel of interest related to wastewater aerosols, enteroviruses do not seem to be significantly affected by biological decay.[2,15]

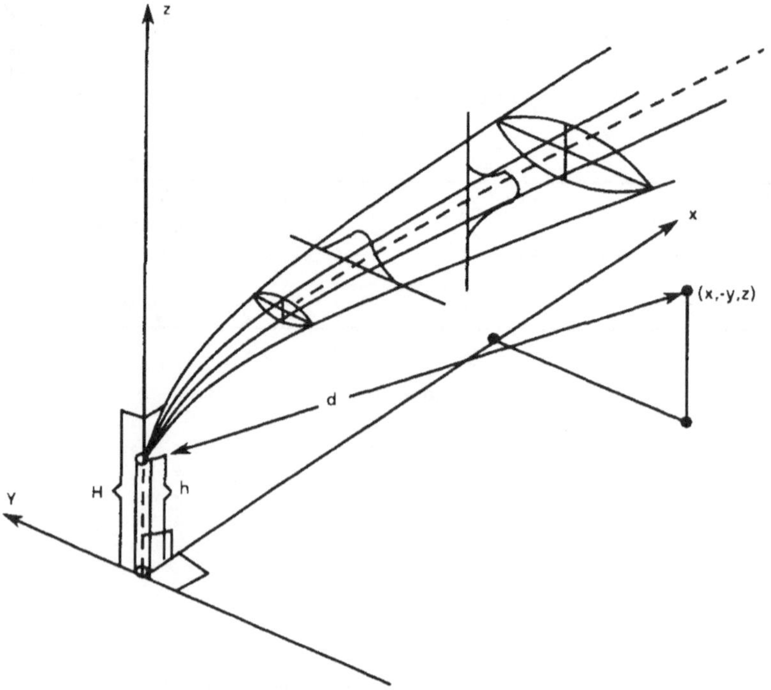

FIGURE 2. Gaussian distribution of aerosol plume in crosswind and vertical direction downwind from a point source.[21]

Lastly, concentrations in air of any aerosolized material are affected by atmospheric diffusion. Although a number of mathematical models are available for estimating this function, the Gaussian form is used most often. The general diffusion pattern of this type model is illustrated in Figure 2. This model assumes that pollutants (including airborne viruses) diffuse in a random manner, resulting in a normal distribution of pollutants at any specific downwind location.

This fundamental model has been modified to yield a microbiological dispersion model for use with point or line sources of aerosolized microorganisms.[4] The general expression for this model is

$$C_d = D_d \times Q_a \times M_d + B$$

where C_d = airborne concentration of a specific microorganism at distance d (units/m^3); D_d = atmospheric diffusion factor at d (sec/m^3); Q_a = aerosol source strength adjusted for loss of specific microorganism viability during the aerosolization process; M_d = fraction of the specific microorganism that remains viable due to die-off at distance d; B = natural background concentration of the specific microorganism.

As may be expected from earlier discussions, Q_a is a compound factor. For spray irrigation systems it has been estimated to be the product of the microorganism density in the wastewater

Table 3
ENTEROVIRUS CONCENTRATIONS OBSERVED IN AEROSOL FROM INDICATED SOURCE

Source	Location	Distance from source (m)	Concentration: Source (PFU/ℓ)	Concentration: Air (PFU/m³)	Ref.
Spray irrigation (wastewater)	Pleasanton, Calif.	63	110[a]	0.006[a]	2,4
	Wilson, Tex.	44	2200	16	15
		46	100	0.01	15
		49	66	0.01	15
		60	160	0.003	15
Spray application (sludge)	Shreveport, La.	50	0.7[b]	<0.002	17
Aeration basin	Chicago, Ill.	30	200	<0.02	3
	Portland, Ore.	30	500	<0.002	13

[a] Mean of two values.
[b] PFU/g TSS.

(W), the wastewater flow rate during irrigation (F), the fraction of the sprayed wastewater that becomes aerosolized (E), and the fraction of the viable microorganisms in the aerosolized wastewater not affected by all other processes during aerosol formation (I). Mathematically, this relationship may be represented by

$$Q_a = W \times F \times E \times I$$

where for a specific microorganism, the units of W and F are microorganisms per liter of wastewater and liters per second, respectively, and both E and I are unitless. This approach may be modified for any type of source.[4] The factors I and M_d are variable and a wide range of values has been observed for different microorganisms. However, the limited data available suggest that enteroviruses are not particularly affected by either factor.

In fact, the three factors that have the most bearing on the concentrations of viruses to be expected in air are the concentration in the source (W), the amount of the source material aerosolized (E), and the atmospheric dispersion factor (D_d). Although D_d can be estimated by relationships such as the Turner modification of the Paguell-Gifford model,[21] it is dependent on a variety of meteorological conditions including intensity of incoming solar radiation, wind speed, and, obviously, distance from the source. A more comprehensive discussion of these factors can be found elsewhere.[1,2,4,5,8]

Examples of concentrations of viruses observed in aerosolized wastewater and sludge at land application sites are shown in Table 3. With one exception, observed virus concentrations are quite low. The single anomaly defies explanation.

B. Analytical Methods

Clearly, because of the very low levels of viruses expected in environmental aerosol samples, extraordinary efforts must be made if quantitative recoveries are to be realized. These efforts generally involve pooling of the collection medium from a number of samplers and subjecting the medium to a concentration procedure prior to assay. In selecting concentration procedures, consideration must be given to the possibility of the collection medium containing high concentrations of both particulates (including microorganisms) and soluble organic compounds.

These problems led investigators to consider methods used for the recovery of viruses from cell culture fluids rather than methods commonly used for recovering viruses from natural waters.[22] Other parameters that must be considered in the selection of a methodology include ease of manipulation, time required for the process, volume reduction, efficiency of virus recovery, and reproducibility of results. Additional considerations should include the requirement for field concentration, shipping and storage of concentrates, and assay methods.

1. Concentration Procedures

A number of methodologies have been considered for concentrating aerosol sampler collection fluids. Published information includes evaluation of several two-phase polymer systems as described by Philipson and co-workers,[23] Shuval and colleagues,[24] and Grindrod and Cliver[25] as well as polyethylene glycol (PEG) hydroextraction as described by Shuval and co-workers.[26]

In one study, the researchers concluded that of the three, a workable field concentration procedure for a brain heart infusion broth (BHI) sampler collection fluid containing 0.1% Tween® 80 was a classical two-phase polymer system.[22] Appropriate volumes of BHI with a 0.1% Tween® 80 were mixed with sodium dextran sulfate, polyethylene glycol (PEG 6000), and sodium chloride to final concentrations of 0.2%, 6.45%, and 0.15 *M*, respectively.

The total volume was shaken vigorously in a separatory funnel and held at 4°C for 18 hr to permit adequate phase separation. The lower phase and interphase volumes were collected and diluted (1:3) with Hanks balanced salt solution with appropriate antibiotics. The concentrates were shipped at 4°C to the laboratory for assay.

In another extensive study, organic flocculation was evaluated as an alternate concentration procedure.[15] Three enteric viruses (poliovirus 1, coxsackievirus B3, and echovirus 6) were used in comparative testing. An extensive sampling/assay scheme was developed to permit the detection of all three viruses in one sample.[15]

Organic flocculation is performed by adding organics (beef extract) to a sample. These organics are precipitated out of solution when the pH is reduced to approximately 3.5. Virions are entrapped in the organic floc and removed by centrifugation. The amount of organics present in a solution frequently affects viral recovery rates; therefore, experiments were performed to determine the optimal amount of beef extract that should be added to the sampler fluid BHI + 0.1% Tween® 80.

Poliovirus 1, coxsackievirus B3, and echovirus 6 were added to 3 ℓ of BHI + 0.1% Tween® 80 to give a final concentration of approximately 10 to 100 PFU/mℓ and mixed for 15 min, 10 mℓ of the sample were removed to establish actual input titers and the remaining sample was aliquoted into 500-mℓ test volumes. Beef extract was added, resulting in final concentrations of 0, 1, 2, and 3%. The pH of each aliquot was adjusted to 3.5 by the dropwise addition of 1 *N* HCl. The samples then were mixed for 30 min and centrifuged for 10 min at 8000 × g. After the supernate was decanted, each pellet was resuspended in 10 mℓ of 0.15 *M* Na_2HPO_4 (pH 9.0), and the pH was adjusted to 7.0. The final volume was measured and the sample assayed. For comparative testing, a 500-mℓ aliquot of seeded sampler fluid was concentrated by two-phase separation as described by Moore et al.[22]

Results shown in Table 4 demonstrate that 2% beef extract provided optimal recovery when compared to the other beef extract concentrations evaluated. Organic flocculation with 2% beef extract also consistently outperformed two-phase separation, especially in the recovery of echovirus 6. Therefore, the following protocol was adopted for the detection of viruses in aerosols for this study.[15]

The total volume of BHI + 0.1% Tween® 80 from an aerosol run was measured, and 100 mℓ of the sample were removed for routine organism determinations. The amount of beef extract added to the sample was calculated on the basis of total volume minus 100 mℓ. The beef extract was added to a final concentration of 2% and mixed until the beef extract

Table 4
CONCENTRATION EFFICIENCY OF ORGANIC FLOCCULATION AND TWO-PHASE SEPARATION

Concentration procedure	Poliovirus 1[a] recovered (%)	Coxsackievirus B3[a] recovered (%)	Echovirus 6[b] recovered (%)
Organic flocculation			
0% beef extract	33	53	60
1% beef extract	41	61	79
2% beef extract	55	77	84
3% beef extract	33	62	81
Two-phase separation	50	61	43

[a] Results are an average of four experiments.
[b] Results are an average of two experiments.

went into solution. The pH of the sample was then reduced to 3.5 with 1 *N* HCl. After 30 min of mixing, the organic floc was recovered by centrifugation at 8000 ×g for 10 min. The pellet was resuspended in 140 mℓ of 0.15 *M* Na_2HPO_4 (pH 9.0). The pH of the final eluate was adjusted to 7 and assayed on two different cell lines. Prior to being assayed, the sample was treated with chloroform to reduce bacterial and fungal contamination.

2. *Assay and Identification Procedures*

Any viral assay procedure compatible with the concentration procedure selected may be used. Consideration should be given to those methods described elsewhere in this book and in the published literature.[16,17,22] However, the assay system chosen should provide a broad sensitivity for the viruses expected in the aerosol.

It is imperative that all virus isolates obtained from environmental aerosol samples be confirmed and, preferably, indentified. Appropriate methodologies for these functions are contained elsewhere in this book.

IV. SUMMARY AND CONCLUSIONS

Quantification of airborne viruses from environmental sources is a tedious process. Although sampling equipment is available and methodologies have been developed to accomplish this task, they are not readily adaptable to routine monitoring efforts. Thus, determining virus concentrations of environmental aerosols must be relegated to the category of special studies or research.

In fact, there is no standard approach to this problem. Each study design must be tailored to the specific site, and it must consider the physical, meteorological, and microbiological conditions of the site. Nevertheless, the methodologies presented herein have been useful in attempting to answer some of the questions about exposure to human viruses from a number of sources. Future developments will undoubtedly result in the improvement of these methodologies.

REFERENCES

1. **Sorber, C. A., Bausam, H. T., and Schaub, S. A.,** An assessment of potential virus hazard associated with land application of wastewaters, in *Virus Survival in Water and Wastewater Systems,* Malina, J. F. and Sagik, B. P., Eds., Center for Research in Water Resources, University of Texas, Austin, 1974, 241.
2. **Johnson, D. E., Camann, D. E., Sorber, C. A., Sagik, B. P., and Glennon, J. P.,** Aerosol monitoring for microbial organisms near a spray irrigation site, in *Risk Assessment and Health Effects of Land Application of Municipal Wastewater and Sludges,* Sagik, B. P. and Sorber, C. A., Eds., Center for Applied Research and Technology, The University of Texas, San Antonio, 1978, 231.
3. **Pahren, H. R. and Jakubowski, W., Eds.,** *Wastewater Aerosols and Disease,* U.S. Environmental Protection Agency, Cincinnati, Ohio, EPA-600/9-80-028, 1980.
4. *Estimating Microorganism Densities in Aerosols from Spray Irrigation of Wastewater,* U.S. Environmental Protection Agency, Cincinnati, Ohio, EPA-600/9-82-003, 1982.
5. **Spendlove, J. C. and Fannin, K. F.,** Methods of characterization of virus aerosols, in *Methods in Environmental Virology,* Gerba, C. P. and Goyal, S. M., Eds., Marcel Dekker, New York, 1982, 261.

5a. **Fannin, K. F., Vana, S. C., and Jakubowski, W.,** Effect of an activated sludge wastewater treatment plant on ambient air densities of aerosols containing bacteria and viruses, *Appl. Environ. Microbiol.,* 49, 1191, 1985.

6. **Randall, C. W. and Ledbetter, J. O.,** Bacterial air pollution from activated sludge units, *Am. Ind. Hyg. Assoc. J.,* 27, 506, 1966.
7. **Sorber, C. A. and Guter, K. J.,** Health and hygiene aspects of spray irrigation, *Am. J. Public Health,* 65, 47, 1975.
8. **Sorber, C. A., Bausum, H. T., Schaub, S. A., and Small, M. J.,** A study of bacterial aerosols at a wastewater irrigation site, *J. Water Pollut. Control Fed.,* 48, 2367, 1976.
9. **Clark, C. S., Van Meer, G. L., Linnemann, C. C., Bjornson, A. B., Gartside, P. S., Schiff, G. M., Trimble, S. E., Alexander, D., Cleary, E. J., and Phair, J. P.,** Health effects of occupational exposure to wastewater, in *Wastewater Aerosols and Disease,* Pahren, H. R. and Jakubowski, W., Eds., U.S. Environmental Protection Agency, Cincinnati, Ohio, EPA-600/9-80-028, 1980, 239.
10. **Fannin, K. F., Cochran, K. W., Lamphiear, D. E., and Monto, A. S.,** Acute illness differences with regard to distance from the Tecumseh, Michigan, wastewater treatment plant, in *Wastewater Aerosols and Disease,* Pahren, H. R. and Jakubowski, W., Eds., U.S. Environmental Protection Agency, Cincinnati, Ohio, EPA-600/9-80-028, 1980, 117.
11. **Katzenelson, E., Buium, I., and Shuval, H. I.,** Risk of communicable disease infection associated with wastewater irrigation in agricultural settlements, *Science,* 194, 944, 1976.
12. **Northrop, R. L., Carnow, B., Wadden, R., Rosenberg, S., Neal, A., Scheaff, L., Holden, J., Meyer, S., and Scheff, P.,** Health effects of aerosols emitted from an activated sludge plant, in *Wastewater Aerosols and Disease,* Pahren, H. R. and Jakubowski, W., Eds., U.S. Environmental Protection Agency, Cincinnati, Ohio, EPA-600/9-80-028, 1980, 180.
13. **Johnson, D. E., Camann, D. E., Harding, H. J., and Sorber, C. A.,** Environmental Monitoring of a Wastewater Treatment Plant, U.S. Environmental Protection Agency, Cincinnati, Ohio, EPA-600/1-79-027, 1979.
14. **Teltsch, B. and Katzenelson, E.,** Airborne enteric bacteria and viruses from spray irrigation with wastewater, *Appl. Environ. Microbiol.,* 35, 290, 1978.
15. **Camann, D. E., Northrop, R. L., Graham, P. L., Guentzel, M. N., Harding, H. J., Kimball, K. T., Mason, R. L., Moore, B. E., Sorber, C. A., Becker, C. M., and Jakubowski, W.,** An evaluation of potential infectious health effects from sprinkler application of wastewater to land: Lubbock, Texas, 2nd Interim Rep., Southwest Research Institute, San Antonio, Texas, 1983.
16. **Camann, D. E., Northrop, R. L., Graham, P. J., Guentzel, M. N., Harding, H. J., Kimball, K. T., Moore, B. E., Sorber, C. A., Popescu, C. B., and Jakubowski, W.,** Design of the Lubbock Infection Surveillance Study of Wastewater Spray Irrigation, paper presented at the 57th Annu. Water Pollut. Control Fed. Conf., New Orleans, 1974.
17. **Sorber, C. A., Moore, B. E., Johnson, D. E., Harding, H. J., and Thomas, R. E.,** Microbiological aerosols from the application of liquid sludge to land, *J. Water Pollut. Control Fed.,* 56, 830, 1984.
18. **Bausum, H. T., Schaub, S. A., and Kenyon, K. F.,** *Viral and Bacterial Aerosols at a Wastewater Spray Irrigation Site,* Tech. Rep. 7804, U.S. Army Medical Bioengineering Research and Development Laboratory, Ft. Detrick, Maryland, 1978.
19. **Harding, H. J., Thomas, R. E., Johnson, D. E., and Sorber, C. A.,** Environmental Monitoring of Sites Conducting Land Application of Liquid Sludge, U.S. Environmental Protection Agency, Cincinnati, Ohio, EPA-600/51-81-028, 1980.
20. **Raynor, G. S. and Hayes, J. V.,** Aerosol production by irrigation equipment used for land application of wastewater, *Am. Ind. Hyg. Assoc. J.,* 37, 526, 1976.

21. **Turner, D. B.,** Workbook of Atmospheric Dispersion Estimates, Publ. AP-26, U.S. Environmental Protection Agency, Research Triangle Park, N.C., 1970.
22. **Moore, B. E., Sagik, B. P., and Sorber, C. A.,** Procedure for the recovery of airborne human enteric viruses during spray irrigation of treated wastewater, *Appl. Environ. Microbiol.*, 38, 688, 1979.
23. **Philipson, L., Albertsson, P. A., and Frick, G.,** The purification and concentration of viruses by aqueous polymer phase systems, *Virology*, 11, 553, 1960.
24. **Shuval, H. I., Fattal, B., Cymbalista, S., and Goldblum, N.,** The phase separation method for the concentration and detection of viruses in water, *Water Res.*, 3, 225, 1969.
25. **Grindrod, J. and Cliver, D. O.,** A polymer two phase system adapted to virus detection, *Arch. Gesamte Virusforsch.*, 31, 365, 1970.
26. **Shuval, H. I., Cymbalista, S., Fattal, B., and Goldblum, N.,** Concentration of enteric viruses in water by hydroextraction and two-phase separation, in *Transmission of Viruses by the Water Route*, Berg, G., Ed., Interscience, New York, 1967, 45.

Chapter 4

RECOVERING VIRUSES FROM SOILS AND AQUATIC SEDIMENTS

Gabriel Bitton

TABLE OF CONTENTS

I. INTRODUCTION

Research in environmental virology has long focused on the development and improvement of methodology for recovering viruses from water. More specifically, considerable emphasis has been placed upon the development of methods that allow the recovery of small numbers of viruses from relatively large volumes (up to 1000 ℓ) of tapwater. Two recent developments, however, have given impetus to research on methods for recovering viruses from soils and aquatic sediments. First, recent legislation (PL 92-500) has encouraged the disposal of wastewater effluents and residuals on agricultural soils. Second, it is now recognized that aquatic sediments may act as reservoirs for bacterial and viral pathogens. It has been proposed, therefore, that monitoring sediments for viruses should be an integral part of water quality assessment programs. It is recognized, moreover, that viruses readily adsorb to soils and sediments. (For review see Bitton.[1]) Therefore, in any proposed method, viruses must be efficiently eluted from soil or sediment surfaces and then concentrated from the eluate.

II. RECOVERY OF VIRUSES FROM SOILS

A. Introduction

Following the passage of PL 92-500, land disposal of municipal sewage effluents and residuals became viewed as a viable and attractive alternative to other means of waste disposal. This practice had many advantages including groundwater recharge, addition of plant nutrients, improvement of the physical properties of soils, and enrichment of soils with organic matter. However, problems associated with waste disposal on agricultural lands quickly became evident. Concern has been raised about the contamination of soil, plants, and groundwater with chemical (toxic metals, trace organics) and biological (parasites, bacteria, viruses) pollutants.[2,3] There has been particular concern with the transport and survival pattern of viruses following land disposal of wastewater effluents and residuals.[1,4,5,5a-d] Viruses are largely retained in soils by adsorption although, in some situations, viruses may be found in groundwater beneath land treatment sites.[9,9a] In view of the health risks associated with viruses during land application of human wastes, detection methods must be developed to determine the levels of these pathogens in the soil matrix. The topic of virus survival and transport through soils has been covered in recent reviews.[1,6-8]

B. Description of Available Methods

A given method for detecting viruses in soils must allow an efficient elution of viruses from soil surfaces and also provide a means for concentrating viruses in soil eluates. The concentration step is necessary because relatively low virus levels are expected in soils treated with municipal wastewater effluents and residuals. Some investigators described only elution steps when they addressed the question of virus persistence or transport patterns through soils.[10,11] Larkin and co-workers,[10] concerned with poliovirus 1 survival in plots of land sprayed with seeded sewage effluent or liquid sludge, used minimal essential medium (MEM) (containing antibiotics) at pH 8.5 to elute viruses from soils. Soil eluates were clarified by filtration through glass wool and 0.45 μm cellulose acetate membranes and portions of the concentrates were assayed for viruses on Vero cells. Of the seeded viruses, 82% were recovered. Similarly, Duboise et al.,[11] studying poliovirus distribution in soil cores, used tryptose phosphate broth to elute viruses from a sandy soil but no elution efficiency was presented by the authors. Bitton et al.[12] reported a 34% mean elution efficiency with tryptose phosphate broth from a sandy soil (Eustis Fine Sand) in Florida. Sadovski et al.[13] used two detection methods to study soil contamination by drip irrigation with sewage effluent seeded with large quantities of a poliovirus. Viruses were eluted with glycine buffer (0.02 *M*, pH 11.5) and 3% beef extract (pH 9.0), but the author presented no recovery efficiencies.

SOIL

↓

ELUTION

Suspend soil in 3–6 volumes of 3% *beef extract*, pH 9.0–9.5 (the eluent may be buffered with *glycine buffer*).

↓

Shake vigorously for 15 min (shaking may be followed by sonication).

↓

Centrifugation at 2,500xg to 8,000xg for 10–20 min.

↓

Discard soil.

↓

CONCENTRATION

***Organic flocculation*: Adjust supernatant fluid to pH 3.5. This results in flocculation.**

↓

Centrifugation at 2,500xg for 5 min.

↓

Discard supernate.

↓

Dissolve floc in 0.15M Na_2HPO_4, pH 9.0. The final pH should be 7–7.2.

↓

Assay concentrate.

FIGURE 1. General scheme for recovering viruses from soils by the beef extract method.[12,13,15]

Because concern has been raised over the inactivation of some enteric viruses (e.g., rotaviruses) at high pH levels, beef extract at pH 9 seems a desirable alternative. Moreover, viruses in beef extract may be concentrated by organic flocculation with the method of Katzenelson et al.[14] Bitton et al.[12] and Landry et al.[15] used the beef extract method, with minor differences, and reported overall (elution followed by organic flocculation) recoveries of 37.7 and 30%, respectively. In the beef extract procedure,[12,13,15] 3% beef extract solution, pH 9, is mixed with the soil sample for 5 to 15 min, the slurry is centrifuged, and the supernatant fluid is concentrated by the organic flocculation method of Katzenelson et al.[14] Further details on the technique are given in Figure 1. A beef extract method similar to those described[12,13,15] and recommended by the USEPA *Manual of Methods for Virology*[15a] is currently (1985) undergoing round-robin testing by the American Society for Testing Materials (ASTM).[17a] The effect of soil type on the performance of the beef extract method has not been thoroughly investigated.

With a similar approach (i.e., elution at pH 9.0 followed by organic flocculation), Bitton and co-workers[12] developed a recovery method based on elution with isoelectric casein at pH 9.0. The technique is similar to the beef extract method except that the organic flocculation of isoelectric casein occurs at pH 4.5 to 4.6 instead of pH 3.5, as reported for beef extract. The various steps involved in the method are shown in Figure 2. The overall recovery from four soils, including one soil with 13% clay, averaged 50%. The overall recovery of poliovirus 1 (Sabin strain), coxsackievirus B3, and echovirus 4 from a sandy soil was 59.5, 88, and 22.5%, respectively. This technique was used for monitoring polioviruses in soil cores treated with poliovirus-seeded digested sludges and exposed to natural conditions prevailing in north central Florida.[4,12] Polioviruses were detected after 35 days in numbers as low as 0.9 PFU/g of dry soil. The recovery of indigenous viruses in land disposal sites, however, has not been demonstrated yet. Purified casein or nonfat dry milk (Carnation, Atlantic and Pacific Tea Co., Montvale, N.J.) may be used in place of isoelectric casein.

	SOIL
ELUTION	Add 2–3 volumes of 0.5% *isoelectric casein* (pH 9.0) to 10–100g of soil.
	↓
	Vortex sample for 30 s and shake in inverted shaker for 15 min.
	↓
	Centrifuge at 4000 rpm for 4 min.
	↓
	Decant supernate, and centrifuge at 4000 rpm for 4 min (100-g *soil* samples).
	↓
CONCENTRATION	Reduce pH of the supernate to 4.4–4.5 flocculate *casein*.
	↓
	Centrifuge at 2000 rpm for 2 min.
	↓
	Resuspend pellet in 3–5 mL of 0.15 M Na_2HPO_4, pH 9.0. (Casein resolubilization is aided by vortexing or by magnetic stirring. *Antifoam* may be added to avoid foaming.)
	↓
DECONTAMINATION DETOXIFICATION	Centrifuge concentrate at 10,000 rpm for 10 min to remove microbial contaminants. Dialyze concentrates for 18 h at 4°C to remove toxicity to host cells.
	↓
ASSAY	Assay concentrate.

FIGURE 2. General scheme for recovering viruses from soils by the isoelectric casein method.[12]

With sludges, Berg[30] obtained better flocculation of milk protein and better virus recoveries at pH 3.5 than at pH 4.5.

So far we have reviewed detection methods where the elution step is performed at pH 9.0. In the quantitative method developed by Hurst and Gerba,[16] the elution step is carried out at pH 11.5. This technique is summarized as follows: soil is mixed with four volumes of 0.25 *M* glycine plus 0.05 *M* EDTA, pH 11.5, for 4 min. The mixture is centrifuged at 1400 × g for 5 min, and the eluate is adjusted with $AlCl_3$ to a molarity of 0.06 to neutralize the EDTA. Concentration of viruses from the soil eluate is accomplished by adjusting the pH of the eluate to 3.5, which results in the formation of a floc. The floc is collected by centrifugation and resuspended in four volumes of fetal calf serum. The supernatant fluid resulting from this centrifugation step is passed through Filterite® filters (DuoFine® series, Filterite Corp., Timonium, Md.) and the adsorbed viruses are eluated with 0.25 *M* glycine plus 0.05 *M* EDTA, pH 11.5. Prior to assay, the EDTA is neutralized by the addition of $MgCl_2$. The complete method is outlined in Figure 3.

This technique allowed the recovery of enteroviruses (poliovirus 1, coxsackievirus A9, and echovirus 7) from soils with an average efficiency of 66%. The efficiency depended, however, on soil type and generally decreased when the clay content of the soil increased.

The glycine-EDTA method was used for recovering naturally occurring viruses from soils at Flushing Meadows, Ariz., where indigenous enteroviruses were detected in soils from a rapid infiltration basin that received wastewater effluents. Viruses were detected at concentrations ranging from 1 to 101 PFU/200 g of soil.[17]

A beef extract method[15a] is undergoing round-robin testing by the American Society for Testing Materials.[17a]

	SOIL
	↓
ELUTION	**Elute viruses by mixing soil with 4 vol (100mL of 0.25M *glycine* + 0.05M *EDTA* (pH 11.5) for 4 min on magnetic stirrer; centrifuge for 5 min; add *$AlCl_3$* to molarity of 0.06, neutralize, and assay.**
	↓
CONCENTRATION	**Adjust eluate to pH 3.5 by adding 1.0M glycine, pH 2; centrifuge for 10 min.**
	↓
	Mix pellet with 4 vol of *fetal calf serum* (FCS) by rapid pipetting. Neutralize and assay by direct inoculation of floc + FCS suspension.
	↓
	Pass the supernate through 3.0- and 0.45-μm porosity filter series; elute filters with 0.25M glycine + 0.05M EDTA (pH 11.5); add *$MgCl_2$* to filter eluate to molarity of 0.12; neutralize filter eluate and assay.

FIGURE 3. Glycine-EDTA method for recovering viruses from soils.[16]

III. RECOVERY OF VIRUSES FROM AQUATIC SEDIMENTS

A. Introduction

The association of viruses with organic and inorganic particulates in the water column may lead to their accumulation in sediments, particularly in the upper layer of sediments.[18-20] On a volume basis, viruses are found in greater numbers in sediments than in overlaying water. Laboratory experiments have shown that viruses persist for longer periods of time in the sediment than in overlying water.[18,21-23] It is also known that adsorption to sediments greatly enhances virus survival, but the mechanism of this protective effect is not known. The fact that viruses accumulate and persist longer in sediments may be of public health concern. Resuspension of the upper layer of sediment may result in an increase of virus numbers in the water column. It may therefore be important to monitor aquatic sediments for their virus content. This can be accomplished only through the development of adequate detection methodology. Metcalf and Melnick[23a] recently described an apparatus for collecting estuarine sediments.

B. Description of Available Methods

In early studies on virus recovery from aquatic sediments, investigators did not address the question of efficiency of their methods.

Metcalf and collaborators[24] carried out a field survey on the presence of viruses in the Houston Ship Channel and in Galveston Bay, Tex. Their method for recovering viruses from sediment is summarized as follows: sediment is mixed with 0.05 *M* glycine, pH 8.5, and the slurry is blended and centrifuged at 2500 ×g for 10 min. The supernatant fluid (i.e., eluate) is filtered through a Cox® membrane (0.45 μm porosity filter), and adsorbed viruses are eluted from the membrane with 0.05 *M* glycine, pH 11.5. Although the first elution is carried out at pH 8.5, the second one is performed at pH 11.5. Nonetheless, enteroviruses were detected in the Houston Ship Channel at concentrations varying from 0.9 to 4.1 PFU/100 g of sediment. The enteroviruses recovered were polioviruses 1, 2, and 3, echovirus 7, and coxsackievirus B5.

SEDIMENT

↓

ELUTION — **Mix 1 volume of sediment with 3 volumes of 0.25M glycine + 0.05M EDTA (Final pH = 11.0).**

↓

Shake for 5—10 minutes.

↓

Centrifuge at 2,500xg for 4 min.

↓

CONCENTRATION — **Add** ***AlCl3*** **to supernate to molarity of 0.05 and adjust pH to 3.5.**

↓

Filter through ***Filterite filters*** **(3 μm and 0.45 μm).**

↓

Elute filters with 0.25M ***glycine buffer*** **adjusted to pH 11.5 or with** ***tryptose broth*** **+ 10%** ***fetal calf serum*** **adjusted to pH 11.5.**

↓

Neutralize eluate quickly, and assay.

FIGURE 4. Glycine-EDTA method for recovering viruses from aquatic sediments.[20,26]

DeFlora et al.[18] detected enteroviruses in sediments from coastal areas contaminated with raw sewage. As with the previous method, the recovery efficiency of the technique was not given by the authors. Their method is outlined as follows: sediment (sand or "slime" sediment) is mixed with an equal volume of sterilized seawater for a period of 30 min. After a settling period at 4°C, the eluate is mixed with a polyelectrolyte (PE 60) at pH 5.0. After 1 to 16 hr of contact, the polyelectrolyte, along with the adsorbed viruses, is collected on a filter pad (AP 20) and resuspended in a small volume of borate buffer, pH 9.0. The concentrate is centrifuged at low speed and assayed. This method was useful for recovering enteroviruses from polluted marine sediments off the coast of Italy.

Vaughn and Metcalf[25] eluted sediment samples with nutrient broth at pH 8.5, centrifuged the mixture, and assayed the eluate. This method had no reconcentration step and no recovery efficiency was given by the authors.

The first quantitative method for recovering viruses from marine and estuarine sediments was reported by Gerba et al.[26] In this technique, viruses are eluted from sediment with 0.25 *M* glycine buffer, pH 11, containing 0.05 *M* EDTA. $AlCl_3$ is added to the eluate, the pH of the eluate is adjusted to 3.5, and the eluate is filtered through a membrane. Adsorbed viruses are eluted from the membrane with glycine at pH 11.5. The method is outlined in Figure 4. The recovery efficiency of this method is approximately 50%. This technique was modified by LaBelle et al.,[20] who substituted tryptose phosphate broth with 10% fetal calf serum for glycine buffer in the final concentration step. This quantitative method was used for detecting viruses in marine and estuarine sediments in Texas[19,20] and Florida.[27] Enterovirus concentrations in the sediments ranged from 9 to 80 PFU/ℓ of sediment.

In recent work in our laboratory,[28] we tested the ability of ten eluting agents to desorb viruses from a marine sandy sediment. Virus elution from sediment surfaces ranged from 1 to 44%. The most efficient eluting agent was a 4 *M* solution of urea, pH 9.0, that contained 0.05 *M* lysine. We have also developed a method for concentrating viruses from urea-lysine eluate.[29] The complete method (elution and concentration steps) is outlined in Figure 5. This method results in low volume concentrates. The volume of concentrate from 100 g of sediment is approximately 3 mℓ.

	SEDIMENT
	↓
ELUTION	Mix sediment with 3 volumes of 4M urea in 0.05M lysine, pH 9.0.
	↓
	Vortex the mixture for 30 s, shake for 1 min, and centrifuge at 4000xg for 4 min.
	↓
	Discard sediment.
	↓
	Adjust supernate with $AlCl_3$ to molarity of 0.003. Adjust pH to 7 with Na_2CO_3.
	↓
	Mix for 5 min.
	↓
CONCENTRATION	Collect floc by centrifugation at 4000xg for 5 min.
	↓
	Resuspend floc in 3% beef extract that contains 0.1M EDTA, pH 9.0.
	↓
	Mix and centrifuge at 4000xg for 5 min.
	↓
	Neutralize supernate and dialyze against phosphate-buffered saline at pH 7.2 overnight at 4°C.
	↓
RECONCENTRATION	Adjust pH of sample to 3.5. Collect floc formed by centrifugation at 4000xg for 5 min.
	↓
	Dissolve floc in 0.15M Na_2HPO_4, pH 9.0.
	↓
	Neutralize and assay.

FIGURE 5. Urea-lysine method for detecting viruses in marine and estuarine sediments.[28,29]

The urea-lysine method was used for detecting indigenous viruses in estuarine sediments at two sampling stations (Salt Run and Matanzas River) in St. Augustine, Fla. Enteroviruses were detected at concentrations of 18 and 41 $TCID_{50}$/100 g wet sediment in Matanzas River and Salt Run, respectively.

Johnson et al.[20e] recently developed a lecithin-supplemented isoelectric casein eluent with which they recovered viruses from Mississippi estuarine sediments.

IV. SUMMARY AND CONCLUSIONS

Detection methodology for viruses in soils and sediments is now in the early stages of development and more attention should be focused in this area. Most investigators agree that elution at pH 9 is safer than elution at pH 11 because certain viruses of public health importance (e.g., rotaviruses) are sensitive to high pH levels. More efforts need to be devoted to the evaluation of existing methods under field situations.

REFERENCES

1. **Bitton, G.,** Adsorption of viruses to surfaces: technological and ecological implications, in *Adsorption of Microorganisms to Surfaces*, Bitton, G. and Marshall, K. C., Eds., John Wiley & Sons, New York, 1980, 331.
2. **Bitton, G., Damron, B. L., Edds, G. T., and Davidson, J. M., Eds.,** *Sludge: Health Risks of Land Application*, Ann Arbor, Science Publisher, Ann Arbor, Mich., 1980.
3. **Sagik, B. P. and Sorber, C. A., Eds.,** *Risk Assessment and Health Effects of Land Application of Municipal Wasteater and Sludges*, Center for Applied Research and Technology, The University of Texas, Austin, 1978.
4. **Bitton, G., Farrah, S. R., Pancorbo, O. C., and Davidson, J. M.,** Fate of viruses following land disposal of sewage sludge. I. Survival and transport patterns in core studies under natural conditions, in *Int. Symp. Viruses and Wastewater Treatment*, University of Surrey, Guilford, U.K., 1980.
5. **Farrah, S. R., Bitton, G., Hoffmann, E. M., Lanni, O., Pancorbo, O. C., Lutrick, M. C., and Bertrand, J. E.,** Survival of enteroviruses and coliform bacteria in a sludge lagoon, *Appl. Environ. Microbiol.*, 41, 459, 1981.

5a. **Bitton, G. et al.,** Virus transport and survival after land application of sewage sludge, *Appl. Environ. Microbiol.*, 47, 905, 1984.

5b. **Lance, J. C. and Gerba, C. P.,** Virus movement in soil during saturated and unsaturated flow, *Appl. Environ. Microbiol.*, 47, 335, 1984.

5c. **Lance, J. C. and Gerba, C. P.,** Effect of ionic composition of suspending solution on virus adsorption by a soil column, *Appl. Environ. Microbiol.*, 47, 484, 1984.

5d. **Dizer, H. et al.,** Penetration of different human pathogenic viruses into sand columns percolated with distilled water, groundwater, or wastewater, *Appl. Environ. Microbiol.*, 47, 409, 1984.

6. **Burge, W. D. and March, P. B.,** Infectious disease hazards of land-spreading sewage wastes, *J. Environ. Qual.*, 7, 1, 1978.
7. **Duboise, S. M., Moore, B. E., Sorber, C. A., and Sagik, B. P.,** Viruses in soil systems, *Crit. Rev. Microbiol.*, 7, 245, 1979.
8. **Gerba, C. P., Wallis, C., and Melnick, J. L.,** Fate of wastewater bacteria and viruses in soil, *J. Irrig. Drainage Div.*, 101, 157, 1979.
9. **Vaughn, J. M., Landry, E. F., Baranosky, L. J., Beckwitch, C. A., Dahl, M. C., and Delihas, N. C.,** Survey of human virus occurrence in wastewater-recharged groundwater in Long Island, *Appl. Environ. Microbiol.*, 36, 47, 1978.

9a. **Goyal, S. M. et al.,** Viruses in ground water beneath sewage irrigated cropland, *Water Res.*, 18, 299, 1984.

10. **Larkin, E. P., Tierney, J. T., and Sullivan, R.,** Persistence of poliovirus 1 in soil and vegetables irrigated with sewage wastes: potential problems, in *Virus Apsects of Applying Municipal Wastes to Land*, Baldwin, L. B., Davidson, J. M., and Gerber, J. F., Eds., University of Florida, Gainesville, 1976, 119.
11. **Duboise, S. M., Moore, B. E., and Sagik, B. P.,** Poliovirus survival and movement in a sandy soil, *Appl. Environ. Microbiol.*, 31, 536, 1976.
12. **Bitton, G., Charles, M. J., and Farrah, S. F.,** Virus detection in soils: a comparison of four recovery methods, *Can. J. Microbiol.*, 25, 874, 1979.
13. **Sadovski, A. Y., Fattal, B., Goldberg, D., Katzenelson, E., and Shuval, H. I.,** High levels of microbial contamination of vegetables irrigated with wastewater by the drip method, *Appl. Environ. Microbiol.*, 36, 824, 1978.
14. **Katzenelson, E., Fattal, B., and Hostovesky, T.,** Organic flocculation: an efficient second step concentration method for the detection of viruses in tap water, *Appl. Environ. Microbiol.*, 32, 638, 1976.
15. **Landry, E. F., Vaughn, J. M., and Penello, W. F.,** Poliovirus retention in 75-cm soils cores after sewage and rainwater application, *Appl. Environ. Microbiol.*, 40, 1032, 1980.

15a. **Berg, G., Safferman, R. S., Dahling, D. R., Berman, D., and Hurst, C.,** *USEPA Manual of Methods for Virology*, EPA 600/4-84-013, Environmental Monitoring and Support Laboratory, U.S. Environmental Protection Agency, Cincinnati, 1984.

16. **Hurst, C. J. and Gerba, C. P.,** Development of a quantitative method for the detection of enteroviruses in soil, *Appl. Environ. Microbiol.*, 37, 626, 1979.
17. **Hurst, C. J., Gerba, C. P., Lance, J. C., and Rice, R. C.,** Survival of enteroviruses in rapid infiltration basin during the land application of wastewater, *Appl. Environ. Microbiol.*, 40, 192, 1980.

17a. **Berg, G.,** personal communication, 1987.

18. **DeFlora, S., DeRenzi, G. P., and Badolati, G.,** Detection of animal viruses in coastal seawater and sediments, *Appl. Microbiol.*, 30, 472, 1975.
19. **Gerba, C. P., Goyal, S. M., Smith, E. M., and Melnick, J. L.,** Distribution of viral and bacterial pathogens in a coastal canal community, *Mar. Pollut. Bull.*, 8, 279, 1977.

20. **LaBelle, R. L., Gerba, C. P., Goyal, S. M., Melnick, J. L., Cech, I., and Bogdan, G. F.,** Relationship between environmental factors, bacterial indicators and the occurrence of enteric viruses in estuarine sediments, *Appl. Environ. Microbiol.*, 39, 588, 1980.

20a. **Goyal, S. M. et al.,** Human pathogenic viruses at sewage sludge disposal sites in the middle Atlantic region, *Appl. Environ. Microbiol.*, 48, 758, 1984.

20b. **Nestor, I. et al.,** Detection of enterovirus in sea-water and in the sand of the beaches, *Zentralbl. Bakteriol. Mikrobiol. Hyg.*, 178, 527, 1984.

20c. **Rao, V. C., Seidel, K. M., Goyal, S. M., Metcalf, T. G., and Melnick, J. L.,** Isolation of enteroviruses from water, suspended solids, and sediments from Galveston Bay, survival of poliovirus and rotavirus adsorbed to sediments, *Appl. Environ. Microbiol.*, 48, 404, 1984.

20d. **Landry, E. F., Vaughn, J. M., Vicale, T. J., and Mann, R.,** Accumulation of sediment-associated viruses in shellfish, *Appl. Environ. Microbiol.*, 45, 238, 1983.

20e. **Johnson, B. A., Ellender, R. D., and Tsai, S. C.,** Elution of enteric viruses from Mississippi estuarine sediments with lecithin-supplemented eluents, *Appl. Environ. Microbiol.*, 48, 581, 1984.

21. **LaBelle, R. and Gerba, C. P.,** Influence of estuarine sediment in virus survival under field conditions, *Appl. Environ. Microbiol.*, 39, 749, 1980.

22. **Liew, P. F. and Gerba, C. P.,** Thermostabilization of enterovirus by estuarine sediment, *Appl. Environ. Microbiol.*, 40, 305, 1980.

23. **Smith, E. M., Gerba, C. P., and Melnick, J. L.,** The role of sediments in the persistence of enteroviruses in the estuarine environment, *Appl. Environ. Microbiol.*, 35, 685, 1978.

23a. **Metcalf, T. G. and Melnick, J. L.,** Sample apparatus for collecting estuarine sediments and suspended solids to detect solids-associated virus, *Appl. Environ. Microbiol.*, 45, 323, 1983.

24. **Metcalf, T. G., Wallis, C., and Melnick, J. L.,** Virus enumeration and public health assessment in polluted surface water contributing to transmission of virus in nature, in *Virus Survival in Water and Wastewater Systems*, Malina, J. F. and Sagik, B. P., Eds.,The University of Texas, Austin, 1974, 57.

25. **Vaughn, J. M. and Metcalf, T. G.,** Coliphages as indicators of enteric viruses in shellfish-raising estuarine water, *Water Res.*, 9, 613, 1975.

26. **Gerba, C. P., Smith, E. M., and Melnick, J. L.,** Development of a quantitative method for the detection of enteroviruses in estuarine sediment, *Appl. Environ. Microbiol.*, 34, 158, 1977.

27. **Gerba, C. P., Smith, E. M., Schaiberger, G. E., and Edmond, T. D.,** Field evaluation of methods for the detection of enteric viruses in marine sediments, Am. Soc. Microbiol. Annu. Meet., Las Vegas, Nevada, 1978.

28. **Bitton, G., Farrah, S. R., Hoffmann, E. M., Feldberg, B. N., and Chou, N. J.,** Techniques for virus detection in the marine environment, Report to Office of Water Resources Research and Technology, Department of the Interior, Washington, D.C., 1981.

29. **Farrah, S. R., Scheuerman, P. R., and Bitton, G.,** Urea-lysine method for recovery of enteroviruses from sludges, *Appl. Environ. Microbiol.*, 41, 455, 1981.

30. **Berg, G.,** personal communication, 1987.

Chapter 5

METHODS FOR RECOVERING VIRUSES FROM SHELLFISH, SEAWATER, AND SEDIMENTS

Mark D. Sobsey

TABLE OF CONTENTS

I. INTRODUCTION

Fecal contamination of estuarine and coastal marine environments is of growing public health concern as coastal populations and associated development continue to increase.[1,2] Sewage treatment plant effluents and sludges, septic waste treatment and disposal systems, land runoff, and boat waste discharges continue to be major sources of such fecal contamination. People can be exposed to and become infected with enteric viruses and other enteric pathogens by ingesting fecally contaminated coastal waters and sediments in the course of primary contact recreation (swimming and skin and scuba diving) and occupational diving activities.[3-5] People may also ingest and become infected with enteric viruses by eating raw or partially cooked bivalve molluscan shellfish such as oysters and clams that have become fecally contaminated.[6-10] These shellfish can accumulate and retain enteric viruses during normal filter feeding activities because viruses and other small particles are removed from the large volumes of water they pump through their bodies.

Potentially all of the more than 120 serologically distinct human enteric viruses may be present in fecally contaminated shellfish, coastal waters, and sediments. These include polioviruses, type A and B coxsackieviruses, echoviruses, other enteroviruses (including hepatitis A virus), reoviruses, rotaviruses, adenoviruses, and the so-called small, round gastroenteritis viruses which taxonomically may be astroviruses, caliciviruses, parvoviruses, and picornaviruses.

Enteric virus contamination of coastal waters, sediments, and shellfish has been documented by recovery of viruses from environmental samples[1,7,11-16] and by the occurrence of water- and shellfish-borne enteric virus disease outbreaks, such as hepatitis A and viral gastroenteritis.[6-9,17,18] Furthermore, recent epidemiological and microbiological/virological findings have raised concerns about the reliability and validity of present coliform bacteria standards for predicting the virological quality of recreational bathing and shellfish harvesting waters. Outbreaks of hepatitis A and viral gastroenteritis have been caused by swimming in and eating shellfish from waters that were approved for these activities on the basis of coliform standards.[3,19,20] There appears to be no consistent relationship between levels of enteric viruses and coliform bacteria in marine swimming waters or in shellfish and the waters and sediments from which they are harvested.[3,11-14,16,21]

Public health concerns about the virological quality of coastal waters, sediments, and shellfish and the adequacy of coliform standards have led to the need for further studies on the occurrence, persistence, and fate of enteric viruses in estuarine and coastal marine environments and their shellfish, and the relationships, if any, between enteric viruses and indicator organisms.[10] This need has prompted considerable effort in the last decade to develop and evaluate methods for detecting enteric viruses in shellfish, seawater, and sediments.

Enteric virus levels in fecally contaminated shellfish, seawater, and sediments are relatively low compared to their levels in human feces or even raw sewage. Therefore, enteric virus detection in these environmental samples usually involves concentration and/or extraction of the viruses from representative samples of suitable size and their subsequent isolation and quantitative assay. In some cases it may be desirable to identify recovered viruses as well.

The use of methods for detecting enteric viruses in shellfish, seawater, and sediments is not yet routine because these methods are technically complex, lengthy, costly, and of unknown effectiveness for many viruses of interest. Their use is primarily for special circumstances such as the development and evaluation of new approaches to assess, control, or improve the sanitary quality of fecally contaminated estuarine or marine environments, investigation of water- and shellfish-borne disease outbreaks, and other epidemiological/microbiological research studies. Furthermore, examination of shellfish, seawater, and sediments for enteric viruses should be done only by competent and specially trained virologists who have adequate facilities and equipment for such work.

The efficiency of methods for detecting enteric viruses in shellfish, seawater, and sediments is likely to vary with the types, amounts, and conditions of viruses in the samples, the characteristics, quality, and size of the samples, and the characteristics of the virus recovery and assay procedures themselves. It is likely that only small and variable proportions of the total enteric virus populations present in field samples are detected with current procedures. Most of the methods have not been evaluated systematically for precision and accuracy through extensive inter- and intra-laboratory studies. Indeed, continued efforts to further develop and improve virus detection methods have limited efforts to standardize and critically evaluate existing ones. Furthermore, most methods for recovering viruses from shellfish, seawater, and sediments have been applied to or evaluated for only the easily cultivated, cytopathogenic enteric viruses, such as polioviruses and some other enteroviruses, which have not been strongly implicated in disease outbreaks. Most methods have not been evaluated for recovery of hepatitis A virus, human rotaviruses, Norwalk virus, and other difficult-to-cultivate viruses which have been implicated most often in water- and shellfish-borne disease outbreaks.

Before selecting methods for detecting enteric viruses in field samples of shellfish, seawater, or sediments, preliminary evaluation of the method(s) for virus recovery efficiency in seeded or spiked samples should be undertaken. This can be done by adding a known quantity of one or more test virus types to the same types and amounts of samples to be studied subsequently for recovery of field viruses, processing the samples through the recovery procedures, and assaying the final concentrates for test viruses to determine recovery efficiency.

Recently, many candidate methods for enteric virus detection in shellfish, seawater, and sediments were compiled and described in detail,[22,23] and a few were collaboratively tested.[24,25] In this chapter methods for the virological examination of shellfish, seawater, and sediments are critically reviewed and some of the most promising candidate methods are described in detail.

II. REVIEW OF METHODS FOR CONCENTRATING ENTERIC VIRUSES FROM SHELLFISH, SEAWATER, AND SEDIMENTS

A. Methods for Concentrating Enteric Viruses from Shellfish

Direct detection of enteric viruses in edible bivalve molluscan shellfish is needed for such purposes as evaluating the reliability of indicator organisms of fecal contamination, studying the effectiveness of various methods to control viral contamination of shellfish (e.g., relaying, depuration, and heat treatment), investigating shellfish-borne outbreaks of possible viral diseases, and doing epidemiological/microbiological studies to assess the risks of viral illness from shellfish consumption.

A number of approaches for recovering and concentrating enteric viruses from shellfish have been reported, but only two general approaches have proved effective and have become widely used. These are methods employing sequential steps of either: (1) extraction and subsequent concentration or (2) adsorption, elution, and concentration. These steps are intended to extract or separate viruses from the shellfish meat, produce an aqueous sample of small volume relatively free of interfering material especially toxicity to cell cultures used for virus isolation and assay, and recover most of the viruses intitially present in the shellfish. In some methods these processing steps are applied more than once.

1. Extraction and Extraction-Concentration Methods

Early studies on enteric virus contamination of shellfish employed recovery methods in which shellfish were homogenized initially in order to extract or elute viruses frm the shellfish tissues. Media for this initial extraction step included the liquor of the shellfish only,[26-29] nutrient broth[30,31] and peptone,[32] phosphate-buffered water,[33] balanced salt solution,[34] and nonfat dry milk.[35] After homogenization, samples were centrifuged to remove shellfish tissue solids, and the resulting supernates were either assayed for viruses directly[31] or further treated to separate viruses from sample components which were toxic to cell cultures used for virus recovery and assay. Treatments included extraction with ethyl ether[26,27,30,32,33] or fluorocarbon,[35,36] and acid precipitation.[29] In some cases, viruses were separated from sample components and concentrated by ultracentrifugation.[26,27] Although viruses could be recovered from shellfish with these methods, samples were often toxic to cell cultures, too large in volume for convenient and economical virus assay, or of unknown efficiency for virus recovery.

In an early systematic study of enterovirus recovery from oysters, clams, and mussels, Konowalchuk and Speirs[29] found that acid precipitation of shellfish homogenate supernates at pH 3.0 to 3.5 reduced sample toxicity to cell cultures. However, sample volumes were too large for convenient assay on monolayer cell cultures. Samples were further concentrated by ultrafiltration, but this resulted in increased toxicity to cell cultures. These methods achieved about 50 to 60% recovery of the initial viruses, with most of the virus loss attributed to the solids which were removed by initial centrifugation of the shellfish homogenates.

Kostenbader and Cliver[37] also developed extraction methods for recovering enteroviruses from oysters by adapting methods used for recovering enteroviruses from other foods. In one method, oyster tissue is homogenized in a mixture of glycine-NaOH, agamma fetal calf serum, $MgCl_2$, and fluorocarbon. The supernate containing viruses is recovered after centrifugation. In another method, minced oyster tissue is mixed with glycine-NaOH buffer and a cationic polyelectrolyte. Oyster solids and the floc that forms from polyelectrolyte treatment are removed by successive filtrations. Viruses in extracts from both methods are further concentrated by ultracentrifugation or ultrafiltration. Enterovirus recoveries from seeded samples were 85 to 98%; details of initial sample sizes, final sample volumes, and inoculum volumes for virus assay were not given. In a subsequent study employing similar procedures of polyelectrolyte treatment, filtration, and ultrafiltration, similar enterovirus recoveries were reported from 20-g oyster and clam samples concentrated to a final volume of 5 mℓ.[38]

An extraction method for enterovirus recovery from single oysters employing fluorocarbon, activated carbon, and cationic polyelectrolyte treatments to reduce toxicity has been reported.[35] Single oysters (ca. 20 g) are homogenized in a mixture of nonfat dry milk, NaCl,

$MgCl_2$, and fluorocarbon at pH 9.0. The homogenate is shaken, sonicated, and then centrifuged to remove shellfish solids. The resulting supernate is treated with granular activated carbon and cationic polyelectrolyte and then centrifuged (and possibly filtered) to remove toxic and other particulate components. The supernate (or filtrate) is assayed directly for viruses. When tested with oysters containing low levels of coxsackievirus B2, recovery efficiencies averaged 59% with a 95% confidence interval of 52 to 66%.

Recently, there have been efforts to develop quantitative extraction-concentration methods for viruses in shellfish samples of 50 to 100 g because this is considered the minimum sample size needed for reliable quantitative estimation of low level viral contamination. One such method reported to be effective for recovering poliovirus from 100-g samples of oysters is a modification of the method of Kostenbader and Cliver.[37,39,40] Viruses are extracted from oysters by homogenizing in 0.09 *M* glycine-NaOH, pH 8.8, and mixing with cationic polyelectrolyte to flocculate interfering components. After centrifuging and filtering to remove solids, the resulting sample is supplemented with beef extract to a concentration of 6%, and viruses are concentrated by acid precipitation (organic flocculation) at pH 3.5. The precipitate is recovered by centrifugation, dissolved in a small volume of phosphate buffer, pH 9.0, adjusted to neutral pH, and assayed for viruses. In comparative studies, more polioviruses were recovered from oyster homogenates with this method than with an alternative method employing adsorption, elution, and acid precipitation.

A similar extraction-precipitation method for poliovirus recovery from 50-g oyster samples was described by Richards et al.[41] Viruses are extracted by homogenizing in glycine-NaCl buffer, pH 9.5, treating with cationic polyelectrolyte, and then centrifuging to remove oyster tissue solids. The supernate is supplemented with beef extract and acid precipitated at pH 3.5. The precipitate is recovered by centrifugation, resuspended in phosphate buffer, pH 9.5, adjusted to neutral pH, and assayed for viruses after antibiotic treatment. Poliovirus recoveries from seeded oyster samples ranged from 93 to 120% in five trials.

A more complex extraction-precipitation procedure for enteric virus recovery from 100-g samples of oysters and hard and soft shell clams was developed by Metcalf et al.[42,43] Viruses are extracted from 100-g samples by homogenizing in phosphate buffered saline and then sonicating after addition of beef extract and adjustment to pH 8.0 to 9.5. After centrifuging to remove shellfish solids, viruses in the resulting supernate are concentrated by acid precipitation at pH 4.0 and low conductivity. The precipitate is recovered by centrifugation, resuspended in a small volume of phosphate buffer and then treated with cationic polyelectrolyte to precipitate toxic and other interfering components. Viruses are further concentrated by a second beef extract acid precipitation step with resuspension of the precipitate in phosphate buffer. Enteric virus recoveries from seeded samples of shellfish averaged 63%.

Another type of extraction-concentration procedure for enteric virus recovery from shellfish employs homogenization under conditions that promote virus separation from shellfish solids, filtration to remove shellfish solids, and hydroextraction against polyethylene glycol (PEG) to concentrate viruses.[44] Oyster samples of 100 g are homogenized in a mixture of Tricine buffer, $MgCl_2$, DEAE-dextran and antifoam emulsion, pH 9.0, and then stirred for 1 hr at 37°C. Shellfish solids are removed by filtration through glass wool, and the virus-containing filtrate is collected into a dialysis tube. This sample is concentrated by hydroextraction overnight against PEG, a hydroscopic material which extracts water and other dialyzable components from the sample. The concentrated sample is recovered from the dialysis tube, treated with diatomaceous earth, and centrifuged. Treatment with fluorocarbon prior to centrifugation is recommended for samples expected to be cytotoxic. The resulting supernate is treated with antibiotics, diluted in Tricine, and assayed for viruses. When tested with 100-g oyster and clam samples experimentally contaminated with representative enteroviruses, average recoveries of about 50% were obtained. This method has not been widely used because sample toxicity requires sample dilution prior to virus assay and because the hydroextraction step is cumbersome and time consuming. The developers of this method appear to have abandoned it in favor of an alternative method.[25,45]

2. Adsorption-Elution-Concentration Methods

In all of the methods described above, viruses are initially extracted from shellfish by homogenizing under conditions that promote virus separation from shellfish tissues. Although many of these methods are successful in this regard, the initial supernate samples obtained after homogenizing and centrifuging are often viscous, turbid, difficult to filter, and toxic to cell cultures. Therefore, viruses in these samples must be further separated from interfering materials and concentrated by treatments such as ethyl ether or fluorocarbon extraction, polyelectrolyte flocculation, acid precipitation, ultrafiltration, or centrifugation. This has often proved difficult because of the poor quality of the initial supernate samples.

Alternative virus recovery procedures have been developed in which shellfish are homogenized initially under conditions promoting virus adsorption to shellfish tissue solids. Upon centrifugation of the homogenate, viruses sediment with the shellfish solids, and the resulting supernatant fluid is discarded. Adsorbed viruses are then eluted from the shellfish solids by resuspension in a medium promoting virus elution. After centrifuging to remove shellfish solids, viruses in the resulting supernatant fluid are further concentrated by acid precipitation.

In the initial report of such an adsorption-elution-concentration procedure, virus adsorption to oyster meats is promoted by homogenizing in distilled water at pH 5.5 and low salt concentration (<1500 mg NaCl/ℓ).[46] After centrifuging (1900 × g) and discarding the supernate, the sediment is washed in pH 5.5 low salt buffer and centrifuged again. The supernatant is discarded, and viruses are eluted from the sediment by resuspending in high pH 0.05 *M* glycine-buffered saline. Shellfish solids are removed by centrifugation and then filtration. Viruses in the resulting filtrate are further concentrated by ultrafiltration. When this method was tested on pools of three oysters seeded with one of four enteroviruses, recovery efficiencies ranged from 24 to 97% and averaged 63%.

In subsequent studies, this method was modified to give improved recoveries of several different enteric viruses in oysters and extended to hard shell clams as well.[16,47-49] In the modified method, shellfish are homogenized in distilled water at low salt concentration (<2000 mg NaCl/ℓ) and pH 5.0 for oysters or pH 4.5 for clams in order to promote virus adsorption to shellfish solids. After centrifuging and discarding the supernate, viruses are eluted from the sedimented shellfish solids by resuspending in glycine-buffered saline at pH 7.5 instead of pH 3.5. Shellfish solids are removed by centrifugation and filtration, and viruses in the resulting filtrate are concentrated by either ultrafiltration or acid precipitation at pH 4.5. In concentrating by the latter method, the virus-containing acid precipitate is collected by centrifugation and resuspended in a small volume of phosphate buffer. Treatment of this sample with cationic polyelectrolyte precipitates toxic components which are removed by centrifugation.[50] When these methods were tested with 67- to 100-g samples of oysters and clams experimentally contaminated with low levels of poliovirus 1, reovirus 3, and simian adenovirus SV 11, virus recoveries averaged 48 and 58% from oysters and clams, respectively.

Further modifications of adsorption-elution-precipitation methods have been reported. In one modification by Ellender et al.,[12] oysters are homogenized in distilled water at pH 5.0 and conductivity <2000 mg NaCl/ℓ, centrifuged to recover virus-containing oyster solids, resuspended in glycine-NaCl, pH 7.5, and acid precipitated at pH 4.5 as described in the previous method. However, the virus-containing precipitate is then resuspended in glycine-NaCl, pH 7.5, and centrifuged to remove unwanted solids. The resulting precipitate is frozen, thawed, centrifuged to remove unwanted solids, and then sterile filtered prior to virus assay. When tested with seeded oyster samples, virus recoveries ranged from 61 to 73% and averaged 63%.

A modified adsorption-elution-precipitation method was developed for enterovirus recoveries from the Japanese cockle, *Tapes japonica*.[51] Samples of 100 g are homogenized in 700 mℓ of distilled water and 0.1% antifoam emulsion at pH 4.5 and conductivity <2000 mg/ℓ salt for virus adsorption. After centrifuging and discarding the supernate, the sediment is resuspended in glycine-saline, pH 9.5, to elute viruses and treated with cationic polyelectrolyte to flocculate toxicity. The floc is removed by centrifugation, and the resulting supernate is supplemented with beef extract and adjusted to pH 4.0 to precipitate viruses. The virus-containing precipitate is recovered by centrifugation, resuspended in a small volume of phosphate buffer, adjusted to physiological pH, and assayed for viruses. When tested with cockles contaminated with low levels of poliovirus, this method gave recoveries ranging from 42 to 104% and averaging 73%.

Recently, Sullivan et al.[45] described a modified adsorption-elution-precipitaton method for 100-g samples of oysters. Oysters (100 g) are homogenized in 150 mℓ of distilled water plus antifoam emulsion at pH 4.8. After centrifuging and removing the supernate, adsorbed viruses are eluted from the sedimented oyster solids by resuspending and mixing in a mixture of Minimum Essential Medium, nonfat dry milk, $MgCl_2$, and fluorocarbon at pH 9.1. Solids are removed by centrifugation, and the resulting supernate is diluted with distilled water and adjusted to pH 4.8 to precipitate viruses. The precipitate is recovered by centrifugation and resuspended in a small volume of phosphate buffer and cationic polyelectrolyte. After centrifuing to remove the unwanted floc that forms from cationic polyelectrolyte treatment, the supernate is assayed for viruses. When this method was tested with 100-g samples of oysters containing polioviruses, recovery efficiencies averaged 55% with a 95% confidence interval of ±2.1%. In a subsequent collaborative study of this method by several laboratories, poliovirus recoveries from seeded oysters ranged from 35 to 55%, with coefficients of variation of 35 to 50%.[25]

3. *Analysis and Recommendations*

From the literature reviewed above on methods to detect and concentrate enteric viruses from shellfish, it appears that various modifications of both extraction-concentration and adsorption-elution-concentration methods are effective for at least some shellfish and virus types. Two modifications of extraction-concentration methods (Figure 1) and two modifications of adsorption-elution-concentration methods (Figure 2) are recommended here and described in detail. These methods are recommended because they are relatively simple, require minimal laboratory equipment and supplies, and have been used successfully to detect naturally occurring enteric viruses in field samples. In addition, the effectiveness of one of the adsorption-elution-concentration methods has been verified by extensive collaborative testing.[25]

Recently, the two recommended modifications of adsorption-elution-precipitation methods (Figure 2) and one of the recommended modifications of elution-precipitation methods (Figure 1, Scheme B) were evaluated for recovery of hepatitis A virus (HAV) from experimentally contaminated oysters. Using cell culture-grown HAV, strain HM175, and input levels of about 15,000 infectious units per 50- to 100-g sample of oysters, recovery efficiencies in six trials of each method averaged 46 and 6% for the adsorption-elution-precipitation methods, Schemes A and B, respectively, and 2.1% for the elution-precipitation method, Scheme B.[52,53] In subsequent studies on the adsorption-elution-precipitation methods with wild type HAV extracted from primate feces at input levels of about 100 infectious units per 50-g sample of oysters, recovery efficiencies in 9 trials averaged 26 and 7.3% for Scheme A and B, respectively.[54] The results of these studies indicate that at least one of the three methods tested, the Scheme A adsorption-elution-precipitation method, is capable of achieving efficient recoveries of HAV from contaminated oysters.

	Scheme A:	Scheme B:
	Oysters or clams, 100 g	Oysters, 50 g
Extract:	Homogenize 1 min	Homogenize 2 min in 450 mℓ 0.05 *M* glycine-0.14 N NaCl, pH 9.5;
	Mix 5 min with 500 mℓ 0.09 *M* glycine-NaOH, pH 9.5, and 5 mℓ (oysters) or 1 mℓ (clams) 10% polyelectrolyte; adjust clams to <1000 mg NaCl/ℓ; settle 20 min	Transfer; rinse blender jar with 50 mℓ 0.05 *M* glycine-0.14 N NaCl
		Adjust homogenate to pH 9.5; add 10 mℓ of 1% polyelectrolyte; settle 15 min
	Centrifuge at 9000 ×g, 15 min	Centrifuge at 10,440 ×g, 4°C, 15 min
	Discard sediment	Discard sediment
Precipitate:	Add beef extract to supernate (6% w/v for oysters; 3% for clams)	Add 30 g beef extract powder (6% w/v)
	Adjust to pH 3.5 with 1 N HCl; mix slowly 15 min	Adjust to pH 3.5 with 1 N HCl; mix slowly 15 min
	Centrifuge at 9000 ×g, 15 min	Centrifuge at 10,440 ×g, 4°C, 15 min
	Discard supernate	Discard supernate
Dissolve:	Dissolve sediment in ca. 15 mℓ 0.15 *M* Na_2HPO_4, pH 9.5	Dissolve sediment in 20 mℓ 0.1 *M* Na_2HPO_4, pH 9.5
	Adjust to pH 7.2 to 7.4	Adjust to pH 7.2 to 7.4
	Treat with antibiotics; assay for viruses	Treat with antibiotics; assay for viruses

FIGURE 1. Elution-Precipitation methods for recovering viruses from shellfish.

In addition to the methods recommended here, other modifications of these methods also may be effective, depending on the shellfish species to be examined and virus types present. Investigators with sufficient time and resources are encouraged to evaluate candidate methods for ease and for effectiveness of virus recovery from seeded and field samples of shellfish in order to make final selection of a method and to determine virus recovery efficiency.

B. Methods for Concentrating Enteric Viruses from Sea and Estuarine Water

A number of methods to concentrate and detect enteric viruses from sea and estuarine waters have been developed over the last decade. Most of these are modifications of methods previously developed for concentrating enteric viruses from raw and finished fresh waters.[55-60] Because enteric virus levels are likely to be relatively low even in sea and estuarine waters with gross fecal contamination, the typical approach is to use two successive methods of virus concentration for relatively large samples of water. The most widely used method for initial concentration is adsorption to and subsequent elution from electronegative microporous filters. The water is acidified and usually supplemented with multivalent cation salts to promote virus adsorption, and the desired volume of conditioned water, typically 50 to several hundred liters, is filtered. Adsorbed viruses are then eluted from the filter(s) with small (less than 2-ℓ) volumes of an eluting fluid usually consisting of alkaline organic buffer or alkaline proteinaceous solution. Viruses in this eluate are then further concentrated (reconcentrated) to a volume of less than 100 mℓ by one of several alternative methods such

	Scheme A:	Scheme B:
	Oysters or clams, 100 g	Oysters, 100 g
Adsorb:	Homogenize in 700 mℓ cold distilled water	Homogenize in 150 mℓ distilled water (37°C), 1 mℓ antifoam and 2.5 mℓ 1 N HCl
	Adjust to pH 5.0 (oysters) or 4.5 (clams) with 1 N HCl; conductivity <2000 mg NaCl/ℓ	Adjust to pH 4.8 with 1 N HCl or NaOH
	Centrifuge at 2000 ×g, 15 min	Centrifuge at 2000 ×g, room temp., 15 min
	Discard supernate	
Elute:	Resuspend sediment in:	
	700 mℓ 0.05 *M* glycine-0.14 N NaCl, pH 7.5; adjust to pH 7.5	50 mℓ MEMH, 10 mℓ 10% nonfat dry milk, 2 mℓ 50% $MgCl_2$, 50 mℓ fluorocarbon and 3 mℓ 1 N NaOH; adjust to pH 9.1
	Centrifuge at 2000 ×g, 15 min	Centrifuge at 5000 ×g, 10°C, 20 min
	Recover supernate and discard sediment (and fluorocarbon, Scheme B)	
Precipitate:	Adjust supernate to pH 4.5 with 1 N HCl	Bring supernate to 250 mℓ with distilled water; adjust to pH 4.8 with 1 N HCl
	Mix 15 min	
	Centrifuge at 2000 ×g, 15 min	Centrifuge at 1500 ×g, room temp., 10 min
	Discard supernate	
	Resuspend sediment in:	
	20 mℓ 0.1 *M* Na_2HPO_4, pH 9.5; adjust to pH 7.2 to 7.5	4 mℓ 0.2 *M* Na_2HPO_4, 1 mℓ 10% polyelectrolyte; adjust to pH 7.5
	Add 10% polyelectrolyte to 0.1%	
	Centrifuge at 2000 ×g, 15 min	Centrifuge at 10,000 ×g, 10°C, 20 min
	Recover supernate and discard sediment	
	Treat supernate with antibiotics; assay for viruses	

FIGURE 2. Adsorption-Elution-Precipitation methods for recovering viruses from shellfish.

as polymer two-phase separation, precipitation of ferric or aluminum salts, hydroextraction against polyethylene glycol (PEG), organic flocculation (acid precipitation) of beef extract, casein, or skim milk, or adsorption to and elution from smaller microporous filters. The final concentrates are treated to control bacteria and mold contamination and cytotoxicity and then inoculated into cell cultures for virus recovery and assay.

1. Initial Concentration by Microporous Filter Methods

Early studies on microporous filter methods for enteric virus concentration from sea and estuarine water employed electronegative filters or other adsorbent media and water quality conditions similar to those used previously for virus concentration from tapwater and wastewaters. In one early study, a flow-through gauze sampler was used to adsorb seeded poliovirus from unconditioned seawater volumes of 38 to 341 ℓ.[61] Adsorbed viruses were eluted with about 600 mℓ of 10% fetal calf serum in alkaline buffer. With this method, poliovirus recoveries ranged from 15 to 19.4%. Metcalf et al.[62] concentrated polioviruses from 19-ℓ volumes of unconditioned, artificial seawater by adsorption to fiberglass or cellulose acetate cartridge filters. Adsorbed viruses were eluted with pH 11.5 phosphate buffered saline; virus recoveries averaged 93%.

Hill et al.[63] concentrated seeded polioviruses from 378-ℓ volumes of prefiltered estuary water adjusted to pH 4.5 and supplemented with 0.12% Mg^{2+} by adsorption to 0.45 μm porosity cartridge filters and elution with 4 250-mℓ volumes of bicarbonate buffered 5X nutrient broth, pH 9.0. Recoveries of high virus inputs averaged 70%. When a low virus input was used, viruses in eluates were further concentrated to 3 mℓ by aqueous polymer two-phase separation, and overall virus recovery was 38%. In order to concentrate viruses from turbid estuary waters, the first stage of this method was further modified by adjusting the water to pH 3.5 and adding 0.0005 *M* $AlCl_3$ and 0.01% diatomaceous earth prior to virus adsorption.[64] Diatomaceous earth was added to prevent clogging of 0.45 μm porosity disc filters composed of either cellulose nitrate or fiberglass-asbestos-epoxy. At high input levels of poliovirus in 178 to 389 ℓ of estuary water, recoveries averaged 56%, but at lower input levels, recoveries from 91 to 340 ℓ of water averaged only 1.2%.

In an effort to further improve virus recoveries from turbid estuarine waters, Sobsey et al.[65] adsorbed viruses to fiberglass cartridge filters followed by a series of 142 mm diameter fiberglass-epoxy disc filters after adjusting the water to pH 3.5 and adding 0.0005 *M* $AlCl_3$. Adsorbed viruses were eluted with 1 to 2 ℓ of pH 11.5 glycine and then further concentrated to 15 to 20 mℓ by precipitation of ferric salts. Overall recovery efficiencies of four enteroviruses from 170- to 190-ℓ volumes of seeded water averaged 41%. When similar microporous filter methods were evaluated by Loh et al.[66] for concentrating polioviruses from 150- to 190-ℓ volumes of seeded seawater, recovery efficiencies of 29 to 41% were obtained. Payment et al.[67] further modified this method by using a series of two pleated fiberglass-epoxy filter cartridges (3.0 and 0.45 μm porosity) as adsorbents because of their large surface area, high flow rates, and resistance to clogging. Optimum conditions for virus adsorption to these filters included a pH of 3.5 and 0.0015 *M* $AlCl_3$. Adsorbed viruses were eluted with 2 ℓ of pH 11.5 glycine buffer and further concentrated to 10 to 100 mℓ by precipitation of aluminum salts. With these methods, poliovirus recoveries from 100-gal volumes of seeded estuarine water averaged 53%. This method, with the substitution of a fiberglass yarn-wound cartridge filter for the 3.0 μm pleated fiberglass-epoxy filter, was further confirmed by Farrah et al.,[68] who reported recovery efficiencies of about 70% when concentrating polioviruses from 400-ℓ volumes of seeded estuary water and eluting with 3 ℓ of pH 11.5 glycine.

Because of the growing awareness that highly alkaline eluting agents such as pH 11.5 glycine inactivated at least some enteric viruses[69,70] and with the development of organic flocculation (acid precipitation) as an effective, simple, and rapid method for further concentrating enteric viruses in beef extract eluates,[71] microporous filter methods were successfully modified for concentrating viruses from sea and estuarine water using beef extract, nonfat dry milk (NFDM), and casein as eluting agents.[72] Model enteroviruses in seawater were adsorbed to a series of 3.0, 0.45, and 0.25 μm porosity pleated fiberglass-epoxy cartridge filters (Filterite®) at pH 3.5 with 0.0005 *M* $AlCl_3$. When adsorbed polioviruses were eluted with either 3% beef extract, pH 9.0, 0.5% isoelectric casein in glycine, pH

10.0, or 1% NFDM in glycine, pH 9.0, recovery efficiencies in the primary concentration step averaged 86, 100, and 71%, respectively. With NFDM, similar recovery efficiencies were obtained for polioviruses 2 and 3, coxsackievirus B3, and echoviruses 1 and 4. Viruses in these eluates were reconcentrated efficiently to even smaller volumes by organic flocculation.

Recently, Guttman-Bass and Nasser[73] found that four model enteroviruses in 5-ℓ volumes of seeded seawater could be concentrated efficiently by adsorption to fiberglass-epoxy filters (8 μm porosity; Balston®) at pH 3.5, elution with 3% beef extract, pH 9.0, and then organic flocculation of the beef extract eluates. Recovery efficencies for coxsackieviruses A9 and B1, echovirus 7, and poliovirus 1 averaged 109, 92, 75, and 93%, respectively. Viruses were adsorbed and recovered efficiently without supplementing seawater with $AlCl_3$ or other multivalent cation salts. However, sample volumes were relatively small compared to those used in other studies and may not be adequate for detecting naturally occurring enteric viruses in field samples.

Microporous filter methods have been further modified and extended to the recovery of rotaviruses from estuary water.[74] Simian rotavirus SA-11 as well as poliovirus 1 in 380-ℓ volumes of seeded estuary water were adsorbed to a series of 3.0 and 0.45 μm porosity pleated fiberglass-epoxy cartridge filters at pH 3.5 in the presence of 0.001 *M* $AlCl_3$. Adsorbed viruses were eluated with 2 ℓ of 3 or 6% beef extract, pH 10.5, and then further concentrated by organic flocculation at pH 3.5 and resuspension in alkaline glycine. Overall recovery efficiencies of SA-11 and polioviruses averaged 42 and 36%, respectively, and naturally occurring rotaviruses were recovered from sewage-contaminated estuarine water.

2. Methods for Second-Step Concentration (Reconcentration) of Viruses in Microporous Filter Eluates

Adsorbent filter eluates from primary concentration of viruses in sea and estuarine waters are generally too large in volume for practical and economical virus assay in cell cultures. Therefore, viruses in primary eluates must be further concentrated to smaller volumes by one or more of several alternative methods prior to virus assay. Although a variety of reconcentration methods are available, those which have been applied to filter eluates from sea and estuarine waters are polymer two-phase separation, precipitation of iron and aluminum salts, organic flocculation (acid precipitation) of beef extract and similar proteinaceous materials, and polyethylene glycol hydroextraction.

In an early report, Hill et al.[64] reconcentrated polioviruses in 1000 mℓ of nutrient broth-bicarbonate buffer eluate to 2 to 3 mℓ by an aqueous polymer two-phase separation technique employing sodium dextran sulfate 500. Overall virus recovery was only 1.2%, and it was not possible to determine if poor recoveries were due to deficiencies in the primary concentration procedure, the reconcentration procedure, or both. Polymer two-phase separation has not been widely used for virus reconcentration in recent years because of technical deficiencies, the need for overnight phase development, and poor virus recovery efficiencies in the hands of some investigators.

Viruses in primary filter eluates have been reconcentrated successfully by precipitation of ferric and aluminum salts.[65,67] In the ferric salt procedure, glycine eluates are adjusted to pH 3.5, supplemented with 0.001 *M* $FeCl_3$, and mixed slowly for 5 to 10 min for floc formation. The floc is recovered by centrifugation, resuspended in 20 mℓ of fetal calf serum, adjusted to pH 7.5, and centrifuged to remove residual solids. With this reconcentration method, polioviruses were recovered with an efficiency of 74%.

Virus reconcentration by precipitation of aluminum salts was reported to be a useful alternative to precipitation of ferric salts because less floc and hence smaller concentrate volumes were produced, and viruses recoveries were efficient.[67] Glycine eluates (2 ℓ) were made to 0.003 *M* $AlCl_3$ and adjusted to pH 7.0 with sodium carbonate for aluminum precipitate formation. Precipitates were collected by centrifugation, and viruses were eluted from them twice with volumes of pH 11.5 glycine equal to the floc volume. Viruses in the pooled floc eluate were re-precipitated at pH 5.5, the floc was recovered by centrifugation, and viruses were eluted from this floc with pH 11.5 glycine and sonication. Virus recovery efficiency for this procedure was not reported, but overall virus recoveries from 100-gal volumes of estuary water seeded with polioviruses, in which this reconcentration procedure was used after adsorption to and elution from microporous filters, averaged 53%.

Subsequent studies indicated that both the ferric and aluminum salt precipitation procedures sometimes resulted in the formation of excessive quantities of floc which required large amounts of eluting fluid. Furthermore, viruses were not always adsorbed efficiently in the second precipitation step of the aluminum procedure.[68] These investigators used a modified aluminum salt precipitation procedure followed by a polyethylene glycol (PEG) hydroextraction procedure to concentrate polioviruses from glycine eluates of microporous filters. For aluminum salt precipitation, eluates were made to 0.003 *M* $AlCl_3$ and adjusted to pH 7.0 with sodium carbonate. After 5 min of mixing, the resultant floc was recovered by centrifugation, and viruses were eluted from it with two or three volumes of fetal calf serum containing 1 *M* glycine. Viruses in this floc eluate were further concentrated to 20 mℓ by hydroextraction dialysis against PEG 20,000 and then dialyzed against phosphate buffered saline. Virus recovery efficiencies of the aluminum salt precipitation step and the PEG hydroextraction step averaged 71 and 80%, respectively, for a combined recovery efficiency of 57%. These methods were used successfully to recover naturally occurring enteroviruses in 400-ℓ samples of polluted estuarine water.

Although reconcentration of viruses in microporous filter eluates by polymer two-phase separation, precipitation of iron or aluminum salts, and PEG hydroextraction are effective, the development of beef extract organic flocculation as a simple, rapid, and effective reconcentration method and the successful use of beef extract as a filter eluting agent for viruses in tap water[71] soon led to the evaluation of these methods for sea and estuarine water samples. Bitton et al.[72] evaluated 3% beef extract, 0.5% casein, and 1% NFDM for organic flocculation of enteroviruses in filter eluates obtained from seawater samples. For organic flocculation of beef extract, samples were adjusted to pH 3.5, stirred for 30 min, and then centrifuged. The recovered floc was dissolved in a small volume of phosphate buffer. Casein and NFDM eluates were flocculated at pH 4.5 to 4.6 and then centrifuged. The recovered floc was dissolved in phosphate buffer and in some cases it was also dialyzed against phosphate-buffered saline. Organic flocculation of beef extract, casein, and NFDM eluates achieved poliovirus recoveries of 66, 60, and 94%, respectively. For five other enteroviruses, recoveries by NFDM organic flocculation ranged from a high of 78% for poliovirus 2 to a low of 10% for echovirus 1 and averaged 56% overall for all five viruses. These findings support those of previous studies in which there were differences in recovery efficiencies by organic flocculation of different enterovirues.[75,76]

Organic flocculation of beef extract eluates also has been successful for concentrating rotavirus SA-11 from filter eluates of seawater.[74] The beef extract concentration was 6%, flocculation was at pH 3.5, and the recovered floc was resuspended in either 0.05 *M* glycine or 3% beef extract solution. Recovery efficiency of the reconcentration step was not reported, but overall virus recovery efficiency from 380-ℓ volumes of seawater for both concentration steps averaged 42%.

Gutmann-Bass and Nasser[73] reported average recovery efficencies of 108% by organic flocculation of coxsackieviruses A9 and B1, echovirus 7, and poliovirus 1 in beef extract eluates obtained from adsorbent filters which had processed 5-ℓ samples of seawater. These results are in contrast to those of some previous studies indicating inefficient organic flocculation of some enteroviruses in beef extract eluates. It was suggested that differences observed in recovery efficiencies of some enteroviruses by organic flocculation of beef extract eluates may have been due to experimental variation when testing individual viruses

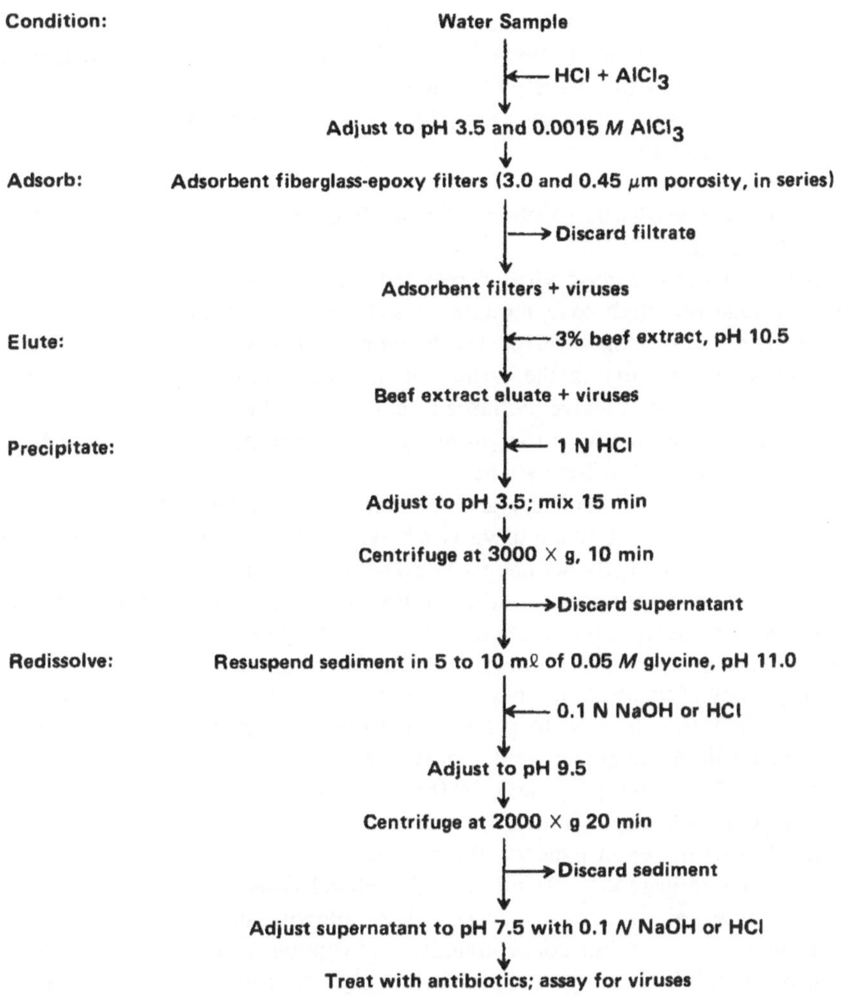

FIGURE 3. Concentration of enteric viruses from sea and estuarine water by adsorption to and elution from filters and organic flocculation.

in separate trials. In the study by Guttman-Bass and Nasser, all four viruses were mixed together, concentrated simultaneously, and then assayed individually.

3. Analysis and Recommendations

From the literature reviewed above, it is apparent that a variety of methods have been developed, evaluated, and used subsequently in the field to concentrate enteric viruses from sea and estuarine water. It is difficult to make definitive judgments on which of these methods are most effective because of differences in experimental conditions such as sample water quality and volumes processed and the types of viruses studied. Furthermore, candidate methods have not been evaluated by careful and systematic inter-laboratory collaborative testing.

Despite these difficulties, a two-step method for concentrating enteric viruses from large volumes of sea and estuarine water is recommended here (Figure 3). In the first step, viruses are concentrated from water adjusted to pH 3.5 and supplemented with 0.0015 *M* $AlCl_3$ by adsorption to 3.0 and 0.45 μm porosity fiberglass-epoxy pleated cartridge filters in series. Adsorbed viruses are eluted with 3% beef extract, pH 10.5, and concentrated further in a second step by organic flocculation at pH 3.5 and resuspension in alkaline glycine buffer.

This method is recommended because it is relatively simple, it is capable of processing large volumes, and it is a modification of well-defined and widely used methods which have been applied successfully to a variety of natural, finished, and wastewaters by a number of investigators. Furthermore, this method is effective for concentrating both enteroviruses and rotaviruses from seawater.

C. Methods for Concentrating Enteric Viruses from Marine and Estuarine Sediments

1. Review of Methods

It is now known that enteric viruses released into marine and estuarine environments accumulate to relatively high concentrations in sediments and that the association of viruses with sediments prolongs their survival. Furthermore, enteric viruses persisting in sediments may be resuspended readily in the water column, transported over long distances, and accumulated efficiently by bivalve molluscan shellfish. For these reasons, quantitative methods for the detection of enteric viruses in marine and estuarine sediments have been developed, evaluated, and used in field studies.

The general approach for recovery and concentration of enteric viruses from sediments involves initial elution or extraction of viruses by suspending and mixing (or homogenizing) in an elution fluid, removing the sediments by centrifugation, and either assaying the resulting supernate directly or further concentrating viruses in the supernate to a smaller volume by procedures such as adsorption to and elution from microporous filters, aluminum salt precipitation, PEG hydroextraction, or organic flocculation of beef extract.

In an early study Metcalf et al.[77] eluted viruses from sediments by homogenizing in 1 to 2 ℓ of glycine buffer, adjusting to pH 8.5, and centrifuging to remove sediment solids. Viruses in the resulting supernate were concentrated by adsorbing them to fiberglass-epoxy disc filters at pH 3.5 in the presence of 0.0005 *M* $AlCl_3$ and eluting with a small volume of pH 11.5 glycine. Although naturally occurring enteric viruses were detected in sediments from polluted estuarine environments, the recovery efficiency of this method was not reported. In another early report, De Flora et al.[78] eluted viruses from sediments by shaking for 30 min in an equal volume of seawater. After allowing the sediment to settle, viruses supernatant water were further concentrated by adsorption to an insoluble polyelectrolyte (PE-60) at pH 5.0,[79,80] recovery of the polyelectrolyte by filtration, and elution of viruses adsorbed to the polyelectrolyte with a small volume of borate buffer, pH 9.0. Enteric viruses were detected in marine sediments with this method, but the efficiency of virus recovery was not reported.

Subsequently, Gerba et al.[69] developed and evaluated quantitative methods for poliovirus recovery from marine sediments. Poliovirus in seeded marine sediments was eluted efficiently by 0.05 *M* ethylenediaminetetraacetate (EDTA) in glycine, pH 11.0, 10% fetal calf serum in glycine, pH 10.5, or 10% fetal calf serum in EDTA-glycine, pH 10.5. In the final form of the method, viruses were eluted from 1 part of sediment with 3 parts of EDTA-glycine, pH 11.0, by shaking for 10 min. The sediment is quickly removed by centrifugation, and viruses in the supernate are concentrated by adsorption to fiberglass-epoxy microporous filters at pH 3.5 and 0.06 *M* $AlCl_3$. Adsorbed viruses are eluted with pH 11.5 glycine or tryptose phosphate broth containing 10% fetal calf serum.[81] This method recovered poliovirus from seeded, 500-g samples of sediment with an efficiency of 50% and was used to detect naturally occurring enteroviruses in field samples. In a subsequent study, using the same method, with the exception of a different series of microporous filters for virus adsorption, the recovery efficiency of seeded poliovirus from 500-g samples was 50% and naturally occurring enteroviruses were detected in marine sediments near sewage outfalls.

Cooper et al.[82] evaluated the aforementioned method for recovering polioviruses from seeded samples of sediment and found it ineffective and cumbersome. They tested several different eluting agents for initial extraction of viruses from sediments and found beef extract-glycine-EDTA, pH 10.5, to be most effective. After centrifuging to remove the sediment, viruses in the resulting supernate were concentrated by organic flocculation of beef extract at pH 3.5. Poliovirus recovery efficiencies from 300-g sediment samples, by this method, averaged 44%, and naturally occurring enteroviruses were detected in field samples.

Bitton et al.[83] also evaluated a series of agents for eluting polioviruses from sediments because of concerns about virus inactivation by the high pH buffers used as eluting agents in some previous studies. Of the ten agents tested, highest poliovirus recoveries were obtained with 4 *M* urea plus 0.05 *M* lysine and 0.6 *M* TCA plus 0.02 *M* glycine, both at pH 9.0. In contrast, glycine-EDTA, pH 11.0, the agent used in a previously reported method,[69] gave the lowest recovery. Viruses in urea-lysine eluates were concentrated by an aluminum salt precipitation method similar to that previously used for reconcentrating viruses in microporous filter eluates from seawater and then further concentrated by organic flocculation of beef extract. When this method was tested with marine sediment seeded with poliovirus 1, overall virus recovery efficiency was 22%, and naturally occurring enteroviruses were detected in field samples.

Several agents for eluting enteroviruses from estuarine sediments were also evaluated by Tsai et al.[84] Viruses were eluted from sediments with 3 volumes of either glycine buffer, 0.5% skim milk, 0.5% casein, beef extract, or nutrient broth. In some cases eluting agents were supplemented with EDTA, citric acid, sodium oxalate, calf serum, or disodium phosphate. Although casein, beef extract, and nutrient broth were the most effective eluting agents, virus recovery efficiencies varied widely with pH, sediment composition, and virus type. A specific eluting agent that was uniformly effective under all conditions was not identified. Subsequently, these investigators evaluated lecithin from different sources for eluting enteroviruses from estuarine sediments.[85] Elution efficiency was lower from sediments with higher clay content, although other unidentified factors also influenced elution efficiency. Average enterovirus recovery efficiencies for different lecithin eluates ranged from 58 to 21%. However, eluted viruses were not concentrated to smaller volumes.

A method for concentrating enteroviruses and rotaviruses from sediments with a weak chaotropic agent, sodium nitrate, in conjunction with beef extract to enhance initial virus elution from sediments was reported by Wait and Sobsey.[86] Viruses were eluted from sediments by mixing for 15 min in 3 volumes of 2 *M* $NaNO_3$ 3% beef extract, pH 5.5. After centrifuging to remove the sediment, viruses in the resulting supernate were precipitated with 2 *M* ammonium sulfate and 0.01% Cat-floc, a cationic organic polymer. The precipitate was recovered by centrifugation, resuspended in phosphate buffer, adjusted to pH 7.2, and dialyzed against phosphate buffered saline. When tested with sediments seeded with poliovirus 1, echovirus 1, and simian rotavirus SA-11, recovery efficiencies averaged 42, 30, and 23%, respectively. Two previously reported methods, one employing elution with beef extract-glycine followed by organic flocculation[82] and the other employing elution with pH 11 glycine-EDTA followed by adsorption to and elution from filters,[81] were also tested, and they achieved much lower virus recoveries.

A method for concentrating both enteroviruses and rotaviruses from marine sediments also was reported by Goyal.[1] Viruses are eluted from sediments with 5 volumes of 3 or 6% beef extract, pH 10.5, and then concentrated by organic flocculation at pH 3.5. The floc is recovered by centrifugation, resuspended in pH 11 glycine, adjusted to pH 9.5, and centrifuged to remove undissolved material. The virus-containing supernate was adjusted to pH 7.5 for virus assay. Overall virus recovery with this method was reported to be about 50%.

Metcalf and Melnick[87] developed a method for recovering viruses from the highly mobile, uppermost layer of marine sediments, which are termed "fluffy sediments". Fluffy sediments are sucked up along with water by a modified swimming pool bottom cleaner, and the sediments are collected on 3 and 0.45 μm porosity fiberglass-epoxy pleated filter cartridges. The sediment is collected from the filters by backwash elution with 3% beef extract, and viruses are further concentrated from this eluate by organic flocculation. In model experiments with kaolin plus silt loam sediment in artificial estuary water containing poliovirus 1 and rotavirus SA-11, overall virus recovery efficiencies averaged 33 and 24%, respectively. In additional model experiments designed to minimize virus losses by inactivation and during organic flocculation, poliovirus and rotavirus recoveries were 61 and 60%, respectively. In a subsequent report in which a similar method was used to detect naturally occurring enteroviruses in fluffy sediments of Galveston Bay, more viruses were found in the fluffy sediments than in the water, the suspended solids, or the compact sediment.[88]

Several different eluting agents and concentration procedures were evaluated by Lewis et al.[89] for recovering enteroviruses from marine and freshwater sediments. Maximum virus elution from sediments was obtained with 6% beef extract at pH 9.0, averaging 15.5, 53.3, and 25.5% for recovery of seeded polioviruses from fine, sandy, and gravelly marine sediments, respectively. PEG precipitation was superior to organic flocculation of beef extract for concentrating poliovirus from sediment eluates. In a method employing 8% PEG, mixing at 4°C for 1 hr, collection of the precipitate by centrifugation, and resuspension of the precipitate in cell culture medium with 5% calf serum, poliovirus recoveries from fine and gravelly sediments averaged 76 and 92%, respectively. With these methods, overall poliovirus recoveries from fine, sandy, and gravelly marine sediments averaged 18, 56, and 6%, respectively, and naturally occurring enteroviruses were detected in marine sediments near sewage outfalls.

2. Analysis and Recommendations

As indicated by the literature reviewed above, a number of methods for concentrating enteric viruses from marine and estuarine sediments have been developed and evaluated, and some of them have been used to detect naturally occurring viruses in field samples. As with methods for concentrating viruses from shellfish and water, selection of a recommended method for concentrating viruses from sediments is made difficult by differences in experimental conditions among the studies reported in the literature. The efficiency of recovering viruses from sediments may vary with factors such as sample size, type of sediment (e.g., organic, clay, silt, sand, gravel, etc.), and virus type. In addition candidate methods have not been subjected to rigorous inter-laboratory collaborative testing.

Despite these difficulties, two methods are recommended here. Both methods are relatively simple, avoid the use of high pH levels which can cause inactivation of some viruses, and have been found to be effective for enteroviruses and rotaviruses. In these methods viruses are first eluted or extracted from sediments, the sediments are removed by centrifugation, and viruses in the resulting supernates are concentrated by a precipitation procedure (Figure 3).

In one recommended method (Scheme A), viruses are eluted from sediments with 3 volumes of 3% beef extract, pH 10.5, the sediments are removed by centrifugation, and viruses in the supernates are concentrated by organic flocculation at pH 3.5. This method is similar to a number of methods reported in the literature,[82,83,89] and the modification recommended is effective for both enteroviruses and rotaviruses.[1]

In the other recommended method (Scheme B), viruses are eluted from sediments with 3% beef extract—2 *M* $NaNO_3$, pH 5.5. The sediment is removed by centrifugation, and viruses in the supernate are concentrated by precipitation with 2 *M* ammonium sulfate and 0.01% cationic polyelectrolyte. This method is recommended because it is effective for both enteroviruses and rotaviruses and because it was more effective than a beef extract elution-organic flocculation method for concentrating viruses from organic sediments.

III. DESCRIPTION OF RECOMMENDED METHODS FOR CONCENTRATING ENTERIC VIRUSES FROM SHELLFISH, SEAWATER, AND SEDIMENTS

A. Recommended Methods for Concentrating Enteric Viruses from Shellfish

1. Equipment and Materials

- Scrubbing brushes, shucking knives, and disinfectant solution; disinfect with 70% ethanol, 10 mg sodium hypochlorite per liter or 1 *N* HCl
- Tared beaker or dish
- Blender or homogenizer; sterilize container by autoclaving
- Balance
- Graduated cylinders: 100, 500, 1000, and 2000 mℓ
- pH meter with combination electrode
- Beakers: 50, 100, 500, 1000, and 2000 mℓ
- Pipets: 1, 5, and 10 mℓ
- Centrifuge with rotors, buckets, bottles, and tubes for 25 to 500 mℓ samples, capable of operating at 10,440 × g
- Magnetic stirrer with stirring bars
- Vertical laminar-flow hood (Class II); Conduct all virus recovery and assay procedures under hood
- Conductivity meter with probe; calibrated preferably in mg NaCl/ℓ.
- Spatula: small; stainless steel or autoclavable plastic.

2. Reagents

- Hydrochloric acid, HCl, 6, 1.0, and 0.1 *N*
- Sodium hydroxide, NaOH, 6, 1.0, and 0.1 *N*
- Distilled water. Sterilize by autoclaving
- Antifoam emulsion*
- Phosphate-buffered saline (PBS), pH 7.2 containing 8.0 g NaCl, 0.2 g KCl, 0.12 g KH_2PO_4, and 0.91 g Na_2HPO_4/ℓ distilled water; adjust to pH 7.2 with NaOH or HCl and sterilize by autoclaving
- Sodium chloride, NaCl, 16% (w/v)
- 0.05 *M* glycine 0.14 *M* saline (NaCl): dissolve 3.75 g glycine and 8.18 g NaCl/ℓ distilled water, adjust to pH 7.5 or 9.5 with 6 *N* NaOH, and autoclave
- 0.09 *M* glycine, pH 9.5; dissolve 6.75 g glycine and 0.4 g NaOH/ℓ in distilled water. Adjust to pH 9.5 with 6 *N* NaOH and autoclave to sterilize
- Polyelectrolyte**: cationic soluble polymer; polydimethyl-diallyl ammonium chloride; dilute to desired concentration in distilled water and sterile filter on day of use
- Trichlorotrifluoroethane
- Eagle's Minimum Essential Medium with Hank's Salts and without phenol red or $NaHCO_3$ (MEMH); available commercially or prepare following a standard protocol[90]
- Nonfat dry milk, available commercially; to make 10% solution (w/v), dissolve 10 g to 100 mℓ in sterile distilled water
- 50% $MgCl_2$: dissolve 50 g $MgCl_2 \cdot 6H_2O$ to 100 mℓ in distilled water and autoclave to sterilize
- Disodium phosphate, Na_2HPO_4, 1, 0.15, or 0.2 *M:* dissolve 26.8 g (for 1 *M*), 40.2

* Antifoam Emulsion C, Dow Corning Corp., Midland, Mich., or equivalent.

** Cat-floc, Calgon Corp., Pittsburgh, Pa., or equivalent.

g (for 0.15 *M*), or 53.6 g (for 0.2 *M*) $Na_2PO_4 \cdot 7H_2O$/ℓ water, adjust to pH 9.5 or 9.0 with 1 *N* HCl or NaOH and autoclave to sterilize

- Beef extract powder*
- Antibiotics: stock solution (50X) of either kanamycin sulfate, 10 mg/mℓ and gentamycin, 2.5 mg/mℓ or penicillin G, 25,000 units/mℓ and streptomycin sulfate, 25 mg/mℓ; stock suspension of nystatin, 125 μg/mℓ or amphotericin B, 50 μg/mℓ; available commercially or prepare following a standard protocol (Schmidt, 1979)

3. Procedures

a. Elution-Precipitation Methods

Elution

Aseptically scrub and shuck enough oysters or clams to obtain 100 g of meat (Scheme A, Figure 1) or enough oysters to obtain 50 g of meat (Scheme B, Figure 1) and place in blender or homogenizer container.

For Scheme A, homogenize shellfish directly for about 1 min, transfer homogenate to a beaker with a magnetic stirring bar and, while mixing, add 500 mℓ cold 0.09 *M* glycine-NaOH, pH 9.5, and 5 mℓ (for oysters) or 1 mℓ (for clams) of 1% polyelectrolyte solution. Readjust to pH 9.5 with 1 *N* NaOH or HCl. Check conductivity and, if necessary, adjust to the equivalent of <1000 mg NaCl/ℓ with sterile distilled water. Mix for 5 min and let stand for 20 min.

For Scheme B, add 450 mℓ cold 0.05 *M* glycine 0.14 *N* NaCl, pH 9.5, to oysters and homogenize 2 min. Transfer homogenate to a beaker with a magnetic stirring bar, rinse blender jar with 50 mℓ of 0.05 *M* glycine 0.14 *N* NaCl, pH 9.5, and add rinse to homogenate. While mixing homogenate, add 10 mℓ of 1% polyelectrolyte (about 0.02% final concentration). Transfer mixture to centrifuge bottles and let stand 15 min.

Centrifuge samples at 9000 ×g (Scheme A) or 10,440 ×g and 4°C (Scheme B) for 15 min and collect supernate in a beaker with a magnetic stirring bar.

Precipitation

Add beef extract powder to a final concentration (w/v) of 6% for oysters (Schemes A and B) or 3% for clams (Scheme B) and mix to dissolve. Mix and add 1 *N* HCl to adjust pH to 3.5. Then mix slowly for 15 min. Transfer samples to centrifuge bottles, centrifuge at 9000 ×g (Scheme A) or 10,440 ×g and 4°C (Scheme B) for 15 min and discard supernate. Dissolve sediment in about 15 mℓ 0.15 *M* Na_2HPO_4, pH 9.5 (Scheme A) or 20 mℓ 0.1 *M* Na_2HPO_4, pH 9.5 (Scheme B). For Scheme B, dissolve sediment by stirring with a spatula and pipetting vigorously to break up sediment particles. For both schemes, adjust resuspended sediments to pH 7.2 to 7.4, treat with antibiotics (see Section IV.C.) and store appropriately until virus assay (see Section IV.B.).

b. Adsorption-Elution-Precipitation Methods

Adsorption

Aseptically scrub and shuck enough oysters or clams (Scheme A, Figure 2) or oysters (Scheme B, Figure 2) to obtain 100 g of meat and place in a blender container. For Scheme A, add 7 parts (700 mℓ) cold distilled water (w/v) and for Scheme B, add 150 mℓ distilled water (37°C) followed by 2.5 mℓ 1 *N* HCl. Blend to homogenize completely, about 2 min, in a conventional blender. Transfer homogenate to a beaker with a magnetic stirring bar. Mix homogenate and use 1 *N* HCl or NaOH to adjust to pH 5.0 (oysters, Scheme A), pH 4.5 (clams, Scheme A), or pH 4.8 (oysters, Scheme B). For Scheme A, measure conductivity and, if necessary, adjust to the equivalent of <2000 mg NaCl/ℓ by adding distilled water.

* Lablemco Powder, Oxoid USA, Inc., Columbia, Md., or equivalent.

Transfer homogenate to centrifuge bottles, centrifuge at 2000 ×g for 15 min (Scheme A) or 5000 ×g for 10 min (Scheme B), both at 4°C. Discard the supernates.

Elution

For Scheme A, resuspend sediment by homogenizing for 0.5 to 1.0 min in 7 volumes glycine-saline and adjust to pH 7.5. For Scheme B, add to the sediment the following cold (5°C) reagents: 50 mℓ MEMH, 10 mℓ 10% nonfat dry milk, 2 mℓ 50% $MgCl_2$, 50 mℓ trichlorotrifluoroethane, and 3 mℓ 1 *N* NaOH. Then mix vigorously for 5 min on a wrist action shaker or by other means, and adjust to pH 9.1 with 1 *N* HCl or NaOH. Centrifuge at 2000 ×g for 15 min (Scheme A) or 5000 ×g at 10°C for 20 min (Scheme B), and collect supernate.

Acid Precipitation

For Scheme B, bring supernate to 250 mℓ with distilled water (37°C). While mixing, use 1 *N* HCl to adjust sample to pH 4.5 (Scheme A) or pH 4.8 (Scheme B) and mix slowly for 15 min to promote floc formation. Centrifuge at 2000 ×g at 4°C (Scheme A) or 1500 ×g at room temperature (Scheme B) for 10 min and discard resulting supernate.

For Scheme A, dissolve sediment in about 20 mℓ of 0.1 *M* Na_2HPO_4, pH 9.0 to 9.5, and adjust to pH 7.2 to 7.4 with 0.1 *N* HCl or NaOH. Then add polyelectrolyte to give a final concentration of 0.1% (1 volume 10% solution per 100 volumes of sample) and let stand for 10 min. For Scheme B, resuspend sediment by vigorous mixing in 4 mℓ 0.2 *M* Na_2HPO_4 and 1 mℓ 10% polyelectrolyte. Centrifuge at 2000 ×g at 4°C for 15 min (Scheme A) or at 10,000 ×g at 10°C for 20 min (Scheme B) and collect supernate.

For both schemes, treat samples with antibiotics by adding 1 part each of kanamycin-gentamycin and/or penicillin-streptomycin and nystatin B suspension or amphotericin B solution to 50 parts of sample. Incubate at room temperature or 37°C for 1 hr and then store appropriately until virus assay (Section IV.B.).

B. Recommended Methods for Concentrating Enteric Viruses from Sea and Estuarine Water

1. Equipment and Materials

- Water pump: centrifugal or positive displacement, with electrical or gas motor, capable of delivering 0.1 to 0.6 ℓ/sec at 207 kPa; if an electric pump is used, a gas-powered electric generator or other electric source is needed; sterilize pump head by recirculating a 10 mg free chlorine per liter solution for 15 min and dechlorinate with a 50 mg $Na_2S_2O_3$/ℓ solution for 3 min
- Adsorbent filters: 25.4 cm-long, 3.0 and 0.45 μm nominal porosity, fiberglass-epoxy pleated cartridge filters*
- Filter housings: autoclavable plastic housing with an optically clear or translucent bowl and a pressure relief valve in the head**
- pH meter: with combination electrode, battery operated for field use
- Positive pressure source: up to 413 kPa, with regulator; laboratory air line, positive pressure pump, or cylinder of compressed air or nitrogen
- Hoses and connectors: nylon braid-reinforced, optically translucent plastic hose, ½ in. I.D., 7/8 in. O.D.,*** fitted with stainless steel or brass connect/disconnect male fittings.† Connect hoses to pumps, filter housings, and pressure vessels, and use to transfer water and other fluids during sample processing. Fit pumps, filter housings,

* No. DUO-FN 10-3.0 and DUO-FN 10-0.45, Duo-Fine®, Filterite, Inc., Timonium, Md., or equivalent.

** No. LMO10VP-3/4, Filterite, Inc., Timonium, Md., or equivalent.

***No. HNB-50, Newage Corp., 2300 Maryland Road, Willow Grove, Pa., or equivalent.

† No. BPHN8-8, Snap-Tite, Inc., Union City, Pa., or equivalent.

pressure vessels, and other equipment to which hoses are to be connected with brass or stainless quick connect/disconnect female fittings*

- Tubing: Flexible, thin-wall plastic or rubber, about 0.5 in O.D.; use 2 lengths of about 3 ft each fitted with brass or stainless steel quick connect/disconnect male fittings for eluting viruses from filters by recirculation; autoclave to sterilize
- Sample container: for batch processing, sterilizable plastic or metal, 100 to 200 ℓ capacity or larger, depending upon sample size
- Venturi-type proportioning injector:** fitted with stainless steel or brass quick connects/disconnects and a length of vinyl tubing as the chemical feed line; it may be necessary to use a bypass system to prevent loss of chemical feed due to back pressure from the sample water line;[91] This bypass system consists of "T" pipe fittings on the injector inlet and outlet ports connected by a length of flexible hose with an in-line shut-off/control valve; proportioning injectors available commercially will process water at flow rates of 3 to 33 ℓ/min with sample water-to-chemical feed ratios between 10:1 and 1110:1; select equipment and operating conditions providing a water-to-chemical feed ratio of 100:1
- Three-way (two-way outlet) valve:*** fitted with quick connect/disconnect fittings; use between venturi injector assembly outlet and 3 μm porosity filter inlet to direct flow of conditioned water either to or away from filters
- Peristaltic or roller (tubing) pump: operable with about 0.5 in. O.D. flexible tubing; for recirculation of elution fluid through adsorbent filters
- Centrifuge: capable of operating at 3000 × g; with rotors and buckets for 250- to 500-mℓ capacity bottles and 15 mℓ tubes
- Centrifuge bottles and tubes: 250 to 500 mℓ bottles and 15 mℓ tubes
- Magnetic stirrer: with stirring bars
- Carboys: 20 to 50 ℓ, or similar containers; autoclavable
- Beakers: 2 and 4 ℓ
- Graduated cylinders: 0.1, 1, and 2 ℓ
- Pipets: 1, 5, and 10 mℓ
- Water flow meter: with brass or stainless steel quick connect/disconnect female fittings on the inlet and outlet; connect to outlet of 0.45 μm porosity filter
- Laboratory balance

2. *Reagents*

- Hydrochloric acid, HCl, 6.0, 1.0, 0.1, and 0.06 *N*
- Sodium hydroxide, NaOH, 6, 1.0, and 0.1 *N*
- Aluminum chloride, $AlCl_3$, 0.3 or 0.15 *M*: dissolve 72.4 or 36.2 g $AlCl_3 \cdot 6H_2O$/ℓ distilled water; autoclave
- Beef extract,† 3%: dissolve 30 g beef extract powder per liter distilled water, adjust to pH 10.5 with 6 *N* NaOH and autoclave; readjust to pH 10.5 aseptically, if necessary
- Glycine, 0.05 *M*, pH 11.0: dissolve 3.75 g glycine per liter distilled water and autoclave. Aseptically adjust to pH 11.0 with 6 *N* NaOH on day of use
- Sodium thiosulfate, $Na_2S_2O_0 \cdot 5\ H_2O$
- Sodium hypochlorite, 5.25% available chlorine (household bleach)
- Antibiotics: See Section III.A.2.

* No. BPHC8-8, Snap-Tite, Inc., Union City, Pa., or equivalent.

** Models 202-P, 203-P or 204-P Dema Engineering Co., St. Louis, Mo., or equivalent.

***Three-way ball valve: brass, stainless steel or plastic; Whitey No. B-45XF8 or SS-45XF8, Crawford Fitting Co., Solon, Ohio; No. 4667K62, McMaster-Carr Supply Co., New Brunswick, N.J.: or similar valve.

† Lablemco Powder, Oxoid USA, Columbia, Md.

3. Procedures (see Figure 3)

a. Chemical Addition and Adsorption to Filters

Use either a batch or continuous flow method to chemically condition the water sample and filter it through the virus adsorbent system.

Batch Method

Connect sterile hoses to inlet and outlet of a sterilized water pump head. Place free end of inlet hose in water being sampled and free end of outlet hose in a sterile sample container. Operate pump and collect desired volume of sample (100 to 200 ℓ or more).

Place free ends of both pump hoses in sample container, making sure end of pump inlet hose is submerged. Place a pH electrode in the sample to monitor pH. Operate pump to mix sample vigorously by recirculation and slowly add 6 *N* HCl until the pH is reduced to 3.8. Continue mixing and add 1 part 0.3 *M* $AlCl_3$ per 200 parts (v/v) of water to give a final 0.0015 *M* concentration of added $AlCl_3$. If sample is not at pH 3.5, adjust by adding 6 *N* HCl or NaOH, and turn pump off.

Connect free end of pump outlet hose to inlet of 3.0 μm porosity filter, connect outlet of this filter to inlet of 0.45 μm porosity filter, connect a hose to outlet of 0.45 μm porosity filter, and position free end of this filter outlet hose to discard filtrate. Operate pump and filter entire sample at a flow rate of <1.3 ℓ/sec.

Disconnect inlet hose of 3.0 μm porosity filter from pump and connect it to a positive air (or nitrogen) pressure source. Operate pressure source to purge excess water from filters and filter housings, turn off pressure source, and disconnect from filters.

If viruses are not eluted from filters immediately, plug inlet and outlet openings of filters and store cold (on wet ice or in refrigerator).

Continuous Flow Method with Venturi-Type Injector for Chemical Conditioning of Water

Preparing HCl-$AlCl_3$ Feed Solution

Titrate 1 ℓ of sample water to pH 3.8 with 0.06 *N* HCl and note volume used. Add 10 mℓ 0.15 *M* $AlCl_3$ and, if necessary, add more 0.06 *N* HCl to adjust pH to 3.5. Note volume of acid used: total volume of HCl used determines milliliters of 6.0 *N* HCl needed per liter of 0.15 *M* $AlCl_3$ to make the HCl-$AlCl_3$ feed solution. Make at least 2.0 ℓ feed solution per 100 ℓ of sample water to be processed.

Chemical Conditioning and Filtration

Use venturi-type proportioning injector system to meter chemical feed solution during sample filtration. Use sterile hose to connect water inlet of venturi injector assembly to the sample water source (pump). Connect sterile hose with a tee fitting and a two-way outlet valve to the outlet of the venturi injector and the inlet of the 3.0 μm filter. Set two-way outlet valve to the by-pass position (no water to filters). Connect outlet of 3.0 μm porosity filter to inlet of 0.45 μm porosity filter. Connect hose and water meter to outlet of 0.45 μm porosity filter. Place chemical feed line of venturi injector into $AlCl_3$-HCl feed solution. Turn on pump, and briefly process water through venturi injector assembly so that its discharge of conditioned water is discarded and not directed to the adsorbent filters. While processing water, collect a small sample of conditioned water from the discharge of the venturi assembly outlet hose, and measure pH. If pH is not 3.5 ± 0.3, adjust screw-operated control on venturi and recheck pH. Repeat adjustment until conditioned water pH is 3.5 ± 0.3. Turn off pump to discontinue processing of sample water.

Complete water processing assembly by directing two-way outlet valve to adsorbent filters. Note and record initial reading of water meter. Add to this volume the desired volume of sample to be processed plus an additional 1% (to account for the volume of feed solution). Water pump will be turned off to stop sample processing at this meter reading.

Turn on pump to begin processing water at a flow rate of 1.3 ℓ/sec. Immediately collect a filtrate sample from filter outlet hose and check pH. If pH is not 3.5 ± 0.3, turn off pump and reposition two-way outlet valve to direct venturi assembly discharge away from filter inlet. Then operate water conditioning system, readjust venturi control, and recheck pH. Reposition two-way outlet valve to direct conditioned water to filters and turn on water pump to begin sample processing. During sample processing, periodically check pH. When desired water volume is processed, turn off pump. Disconnect venturi assembly from 3.0 μm porosity filter inlet and disconnect water meter from 0.45 μm porosity filter outlet.

Connect 3.0 μm porosity filter inlet to a positive air (or nitrogen) pressure source and pressurize to purge excess water from both filters and their housings. Turn off pressure source and disconnect.

If virus elution from filters is not done immediately, plug filter inlet and outlet openings, and store filter assemblies in the cold (on wet ice or in refrigerator).

b. Virus Elution from Filters

Assemble elution system by connecting outlet of 3.0 μm porosity filter to inlet of 0.45 μm porosity filter, and connect sterile, flexible, thin-wall hoses to 3.0 μm porosity filter inlet and 0.45 μm porosity filter outlet. Place free end of both hoses in a 4 ℓ beaker. Slip mid-section of 3.0 μm porosity filter inlet hose into the head of a peristaltic or roller pump. Add to 2 to 3 ℓ elution medium (3% beef extract, pH 10.5) to 4 ℓ beaker. Open vents on the two filter housings slightly and slowly operate pump to fill the dead space in the filter housings with elution fluid. When fluid just flows from the vents, quickly close them and increase pump speed to recirculate elution fluid from beaker through filters for 5 min. Make sure elution fluid contacts entire length of cartridge filters during recirculation. Then turn off pump and remove hose from pump head. Connect free end of 3.0 μm porosity filter inlet hose to a positive pressure source and purge excess elution fluid from filters and their housings into beaker. If organic flocculation is not done immediately, transfer sample to a sealable container and store under approved conditions until further processing.

c. Organic Flocculation

With eluate in a beaker, mix vigorously, measure pH, and slowly add 1 *N* HCl to adjust to pH 3.5. Decrease mixing speed and mix slowly for 15 min for precipitate (floc) growth. Transfer to 250- to 500-mℓ capacity centrifuge bottles and centrifuge at 3000 × g for 10 min. Decant and discard the resulting supernate. Dissolve the sedimented precipitate in 5 to 10 mℓ of 0.05 *M* glycine, pH 11.0, and bring the mixture to pH 9.5 with 0.1 *N* NaOH or HCl. Transfer to a centrifuge tube, and centrifuge at 2000 × g for 20 min. Recover the supernate and adjust to pH 7.5 with 0.1 *N* HCl or NaOH. Treat sample with antibiotics (Section IV.C.) and store for virus assay (Section IV.B.).

C. Recommended Methods for Concentrating Enteric Viruses from Sediments

1. Equipment and Materials

- Sediment core sampler: 15 to 30 cm long by 2.5 to 10 cm diameter with sterilizable head and autoclavable core tubes of plastic, glass, or corrosion-resistant metal*
- Knife or spatula (sterilizable)
- Balance
- Beakers: 0.5 or 1 ℓ
- Magnetic stirrer with stirring bars
- Graduated cylinders: 0.1, 1, and 2 ℓ
- pH meter with combination electrode
- Centrifuge with rotors and buckets for 250- to 500-mℓ capacity bottles and 15- to 50-mℓ tubes

* No. 21W0188, Wards Natural Science Establishment, Inc., Rochester, N.Y., or similar device.

- Centrifuge bottles and tubes: 250 to 500 mℓ and 15 to 50 mℓ
- Pipets: 1, 5, and 10 mℓ
- Dialysis tubing: pure, regenerated, natural cellulose tubing, molecular weight cutoff <50,000, flat width 16 to 25 mm

2. Reagents

- Hydrochloric acid, HCl: 6.0, 1.0 *N*
- Sodium hydroxide, NaOH: 6.0, 0.1 *N*
- 3% Beef extract, pH 10.5: dissolve 30 g beef extract powder* per liter distilled water; adjust to pH 10.5 with 6 *N* NaOH and autoclave to sterilize
- 3% Beef extract-2 *M* $NaNO_3$: dissolve 30 g beef extract powder* and 170 g $NaNO_3$/ℓ distilled water; autoclave to sterilize
- 4M $(NH_4)_2SO_4$: dissolve 528.6 g $(NH_4)SO_4$/ℓ distilled water and autoclave to sterilize
- Polyelectrolyte: cationic, soluble polymer; polydimethyl-diallyl ammonium chloride;** dilute to desired concentration in distilled water and filter to sterilize on day of use
- 0.05 *M* glycine-NaOH: dissolve 3.75 g glycine per liter distilled water and autoclave to sterilize; adjust aseptically to pH 11.0 with 6 *N* NaOH
- Disodium phosphate, Na_2HPO_4, 0.1 *M*, pH 9.0 to 9.5; dissolve 26.8 g $Na_2HPO_4 \cdot 7H_2O$/ℓ per liter distilled water; adjust to 9.0 to 9.5, if necessary, and autoclave
- Antibiotics: see Section III.A.2.
- Phosphate buffered saline (PBS), pH 7.5: dissolve 8.0 g NaCl, 0.2 g KCl, 0.12 g KH_2PO_4, and 0.91 g Na_2HPO_4/ℓ distilled water, adjust to pH 7.5 with 1 *N* NaOH or HCl and autoclave

3. Procedures

a. Sample Collection and Preparation

Because most viruses are likely to be in the first few centimeters of sediments, sample sediments with a relatively short core sampler. Collect several replicate cores from the same sampling site to minimize individual core variability in virus concentration and to obtain sufficient weight of sediment. Return intact cores to the laboratory and process as soon as possible, storing under approved conditions (Section IV.B.) until processed.

Aseptically remove cores from core tubes and divide into 2.5- to 7.5-cm long segments beginning at the top surface of the core. Most viruses will be found in the top segment, but deeper segments also may contain viruses and each can be processed separately. If necessary, composite corresponding segments of two or more replicate core samples from the same sampling site to obtain sufficient sediment weight. The recommended sample size is 50 to 100 g or more of sediment, wet weight.

b. Virus Elution

Place 50 g or more of sediment in a 0.5 to 1 ℓ beaker, add 5 volumes (w/v) cold 3% beef extract, pH 10.5 (Scheme A, Figure 4), or 3 volumes (w/v) 3% beef extract—2 *M* $NaNO_3$ (Scheme B, Figure 4). Mix vigorously for 5 min (Scheme A). For Scheme B, mix vigorously, adjust mixture to pH 5.5, and continue mixing for 15 min. Transfer mixture to 250- to 500-mℓ capacity centrifuge bottles and centrifuge at 500 $\times g$ for 10 min (Scheme A) or 5000 $\times g$ for 5 min (Scheme B). Collect supernate by decanting into a beaker.

* Lablemco Powder, Oxoid USA, Inc. or equivalent.

** Cat-floc-T, Calgon Corp., Pittsburgh, Pa. or equivalent.

Sediment, 50 to 100 g

Suspend in:

	Scheme A:	Scheme B:
Elute:	5 volumes 3% beef extract, pH 10.5	3 volumes 3% beef extract - 2 *M* $NaNO_3$; adjust to pH 5.5
	Shake vigorously for 5 min	Shake vigorously for 15 min
	Centrifuge at 500 ×g, 10 min	Centrifuge at 5000 ×g, 5 min
	Discard sediment	Discard sediment
Concentrate:	Adjust to pH 3.5 with 1 N HCl	Add 4 *M* $(NH_4)_2SO_4$ to 2 *M* concentration; adjust to pH 3.5 with 1 N HCl
		Add 1% polyelectrolyte to 0.01%
	Mix slowly 20 min	Mix slowly 20 min
	Centrifuge at 1500 ×g, 10 min	Centrifuge at 5000 ×g, 5 min
	Discard supernate	Discard supernate
	Resuspend sediment in:	Resuspend sediment in:
	5 to 10 mℓ 0.05 *M* glycine-NaOH, pH 11.0; adjust to pH 9.5	10 to 20 mℓ 0.1 *M* Na_2HPO_4; adjust to pH 7.2
		Dialyze overnight against PBS
	Treat with antibiotics; assay for viruses	Treat with antibiotics; assay for viruses

FIGURE 4. Concentration of enteric viruses from sediments by elution and precipitation.

c. Virus Precipitation

For Scheme A, mix sample vigorously with a magnetic mixer, measure its pH, and slowly add 1 *N* HCl to adjust to pH 3.5. Mix slowly for 20 min to promote precipitate (floc) growth.

For Scheme B, add a volume of 4 *M* $(NH_4)SO_4$ equal to the volume of sample. Mix vigorously and adjust to pH 3.5 with 1 *N* HCl. Add 1/100 the sample volume of 1% polyelectrolyte to give a final concentration of 0.01%. Mix slowly for 20 min.

Transfer sample to 250- to 500-mℓ capacity centrifuge bottles and centrifuge at 1500 ×g for 15 min (Scheme A) or 5000 ×g for 5 min (Scheme B). Decant and discard supernate.

For Scheme A, dissolve sediment in 5 to 10 mℓ 0.05 *M* glycine-NaOH, pH 11.0, and adjust to pH 9.5 with 1 *N* NaOH or HCl. Centrifuge sample at 2000 ×g for 20 min, recover the supernate, and adjust to pH 7.5.

For Scheme B, dissolve sediment in enough 0.1 *M* Na_2HPO_4, pH 9.0 to 9.5, to bring mixture to pH 7.2 to 7.4. Place sample in sterile dialysis bag, seal and dialyze against >100 volumes of phosphate-buffered saline, pH 7.5, at 4°C overnight, with one change of PBS. Treat with antibiotics (Section IV.C.) and store until virus assay (Section IV.B.).

IV. METHODS FOR RECOVERING ENTERIC VIRUSES, QUANTITATIVE ASSAY, AND IDENTIFICATION

A. Introduction

Virus assays of concentrates from shellfish, water, and sediment samples are based on the infectivity of the viruses for susceptible, live laboratory hosts. A complete description of facilities, equipment, materials, and methods for recovering enteric viruses from concentrated samples is beyond the scope of this chapter. Consult standard handbooks on general and environmental virology methods for detailed information.[22,23,60,92,93] Virus assays are beyond the capacity of most environmental microbiology laboratories. Such work should be done only by trained virologists working in specifically equipped virology laboratory facilities. Take particular care to prevent samples or inoculated hosts from becoming contaminated with viruses from extraneous sources and to prevent cross-contamination of sample concentrates or inoculated hosts. In this section, some important considerations in adapting standard virological methods to the examination of concentrate samples from shellfish, water, and sediment are reviewed and described.

B. Sample Storage

Process samples as soon as possible; hold samples for no more than 2 hr at up to 25°C or for no more than 48 hr at 2 to 10°C. Do not freeze unless samples will not be processed within 48 hr; freeze at < -70°C. Use the same storage conditions for processed samples (final concentrates) after treating with antibiotics. For long-term storage, freeze and hold at < -70°C.

C. Decontamination of Processed (Concentrate) Samples

Control bacteria and mold contamination of concentrated samples by adding 1 part each of kanamycin-gentamycin or penicillin-streptomycin solution and nystatin B suspension or amphotericin B solution to 50 parts of sample (w/v) (see Section III.A.2). Incubate at room temperature for 2 hr or at 37°C for 1 hr, and refrigerate or freeze (< -70°C) until virus assay is made. Freezing aids in destroying bacterial contaminants. To check effectiveness of treatment, place a small subsample on a general purpose medium such as plate count agar by the spread plate technique and incubate at 37°C for 24 to 48 hr.

If contamination persists, add 1/10 sample volume of chloroform, $CHCl_3$, and mix vigorously for 30 min at room temperature or homogenize for 1 to 2 min at 4 to 10 °C. To separate phases, let stand in refrigerator overnight or centrifuge at $<1000 \times g$. Remove aqueous upper layer (sample) with a pipet. Remove dissolved $CHCl_3$ from sample by bubbling with filter-sterilized air for 15 min or by placing sample in sterile, shallow container, uncovered, in a laminar flow hood or biological safety cabinet for several hours. To decontaminate shellfish samples, substitute $^1/_2$ the quantity of trichlorotrifluoroethane or treat as with $CHCl_3$.

D. Virus Recovery, Quantitative Assay, and Identification

1. Host Systems for Recovering and Assaying Viruses

The major classes of host systems for detecting enteric virus infectivity are mammalian cell cultures of primate origin and suckling mice. Unfortunately, there is no single, universal host system that will detect all or even a majority of the viruses that potentially can contaminate shellfish, water, and sediment. Some of the epidemiologically most important viruses such as HAV, Norwalk-type viruses, and human rotaviruses are difficult or impossible to recover or assay in any of the currently available, conventional host systems. However, methods for cultivation and assay of rotaviruses have been developed and applied successfully to rotavirus detection in environmental samples.[94] In addition, with recent developments in methods to cultivate and assay HAV, it is now becoming possible to detect this virus in environmental samples.[94-96]

Primary and secondary human embryonic kidney and monkey kidney cell cultures and suckling mice are preferred host systems for many enteric viruses. However, few laboratories have the facilities and financial resources to maintain and utilize those host systems. A number of convenient and sensitive alternative host systems have been suggested. Buffalo Green Monkey Kidney (BGM), a continuous line derived from African green monkey kidney cells is sensitive to a number of enteroviruses and reoviruses and is widely used as a substitute for primary monkey kidney cells for environmental samples.[97] MA104, a rhesus monkey kidney-derived continuous cell line, has also become widely used because it is sensitive to rotaviruses as well as many enteroviruses.[98-100] RD, a continuous line of human rhabdomyosarcoma cells, has been suggested as a substitute for suckling mice for detecting group A coxsackieviruses and also will detect a number of other enteric viruses.[101,102]

To obtain a wide spectrum of enteric viruses from shellfish, water, and sediment concentrates, use at least two different host systems for recovering and assaying viruses.

2. *Virus Assay Methods*

a. Quantal and Enumerative Assay Methods

There are two basic procedures for recovering and assaying viruses in cell cultures: quantal methods ($TCID_{50}$ or MPN methods) and enumerative methods (plaque assay). Quantal assays score replicate cultures inoculated with samples or sample dilutions as negative or positive for cytopathic effects (CPE). The virus titer is estimated from the resulting dose-response data. Plaque assays are more precise than quantal assays because relatively large numbers of individual infectious units are counted directly as discrete, localized areas of infection (plaques) in agar-containing cell cultures. Plaque assay methods may use cell culture monolayers or cell culture suspensions in agar, with the former most common for environmental samples. However, recent studies indicate that virus recoveries from environmental samples are greater by the agar cell suspension plaque assay than by quantal assays in liquid cultures or plaque assays in monolayer cultures.[88,102]

b. Sample Toxicity

A major problem in quantitative assay of concentrates from shellfish, water, and sediment samples is toxicity of the samples to cell cultures. Shellfish samples in particular may contain toxic dissolved and particulate shellfish tissue components. Sample concentration schemes use various treatments intended to separate viruses from these toxic components, but these treatments are not completely effective.

Sample toxicity may destory all or most of the cells in the culture so that no viruses are recovered. Because of the possibility of "false-positive", nonviral cytopathology, confirm all CPE-positive cultures in quantal assays as virus-positive or virus-negative by further passage in cell cultures. An additional toxicity problem with plaque assays is the appearance of plaque-like cytotoxic areas in the cell cultures caused by localized, non-viral toxicants.[103,104] Because of such "false-positive", plaque-like areas of toxicity, confirm all or some representative proportion of the initial plaques for viruses by further passage.

When inoculating sample concentrates into cell cultures, take precautions to control toxicity. For example, the inoculum volume per cell culture and the adsorption time for the inoculum volume on the cell monolayer may influence sample toxicity. These conditions may have to be adjusted, depending on level of toxicity in sample concentrate. Draining the inoculum volume from monolayer cultures after the adsorption period and rinsing the monolayers with buffered saline, balanced salt solution, or cell culture medium also may be helpful in controlling sample toxicity.[49]

In a study to evaluate various methods for reducing cytotoxicity in virus concentrates from environmental samples, filtration through positively charged depth filters was found to be most effective, and high speed centrifugation (8500 to 34,000 ×g, 1 hr) also was somewhat effective.[105] More recently, Richards and Weinheimer[106] reported that some brands of beef extract and skim milk caused cytotoxicity and reduced virus titers in plaque assays for poliovirus. This problem was overcome by dialysis of the samples to remove cytotoxic and other interfering components.

It is advisable to do a preliminary assay for viruses on a small volume of each concentrated sample in order to evaluate sample cytotoxicity as well as the presence or absence of other microbial contaminants. Using the assay system of choice, a few replicate host cell cultures are inoculated with different volumes of sample and incubated. Inoculated cultures are evaluated for cell survival and quality, frequency of virus recovery, and approximate virus titer. Based on the results of this preliminary assay, the remainder of the sample then can be further treated to reduce cytotoxicity or eliminate microbial contamination. Furthermore, sample inoculum volumes can be chosen to both minimize cytotoxicity and optimize the number of viruses per culture for assay purposes.

c. Mixed Cultures

Another potential problem in quantal assays is the inoculation of two or more viruses into the same culture. If one virus grows faster than the other(s), the other virus(es) may not grow sufficiently to be detected. Alternatively, if two or more viruses grow, a mixed positive culture will be scored erroneously as a single positive, thus causing an underestimate of the number of viruses in the sample. Growth of two or more viruses in the culture requires separation of the individual virus types by further passage and isolation. Such mixed cultures may go undetected unless identification of the virus isolates is attempted. Problems with mixed cultures in quantal assays can be minimized if the samples are subjected to a preliminary assay. Samples can be inoculated as sufficiently small portions or dilutions into a series of replicate cultures so that a reasonable proportion of the total cultures will remain virus-negative.

d. Inoculation and Adsorption Conditions

The methods by which cell cultures are inoculated with and exposed to shellfish sample concentrates also may influence the efficiency of virus recovery. A widely used approach for monolayer cell cultures is to inoculate portions of the sample concentrate onto the monolayers of a series of cell cultures that have been drained of their growth or maintenance medium. The inoculum is left on the monolayers for a period of one to several hours to allow viruses to adsorb to the cells. During this time, redistribute the inoculum over the monolayers either continuously with a mechanical rocker or manually every 15 to 20 min by tilting or rocking the cultures several times. In a recent study on poliovirus plaque assays, higher virus titers were obtained if inoculated cultures were rocked continuously for 2 hr than if they were held static for up to 4 hr.[106]

After adsorption, liquid (for quantal assays) or agar-containing (for plaque assay) medium is added to the cultures. Alternatively, drain the inoculum from the cultures after the adsorption period and/or rinse the monolayers with buffered saline, balanced salt solution, or cell culture medium before adding liquid or agar-containing medium. Sometimes the inoculum is not removed, and the cultures are not rinsed after adsorption.

When more than one type of cell culture is employed, most laboratories divide the entire sample into several equal portions and inoculate small volumes into a series of cultures of one type of cell (e.g., BGM or MA104). A disadvantage of this approach is that the inoculated portions are exposed to only one type of cell. If viruses in the inoculum will not grow in that cell type, they will not be detected. This limitation can be overcome by passing material from cultures that remain virus-negative after incubation into fresh cultures of another cell type. Such "blind passage" of virus-negative cultures sometimes yields additional viruses, especially in quantal assays.

In an investigation of inoculation and exposure conditions for assay of shellfish concentrates in cell cultures, Metcalf et al.[43] obtained optimum virus recoveries with inoculum volumes of <1 mℓ/45 cm^2 monolayer culture and an adsorption time of 2 hr. Virus recoveries were increased by exposing the entire shellfish concentrate to each of two or more different cell culture types by means of a sequential adsorption scheme. In this sequential exposure scheme, the inoculum volumes from initially inoculated cultures are recovered after the adsorption period, pooled, and then portions are reinoculated into another type of cell culture. The procedure can be repeated for inoculation of a third cell culture type.

3. Virus Recovery and Identification

Typically, inoculated cultures are incubated and observed periodically for the development of CPE or plaques for periods of 1 to 3 weeks or until the cells begin to degenerate as the cultures age. Material from CPE-positive cultures or plaques is confirmed for viruses by further passage into fresh cell cultures. Material from CPE-negative cultures also may be further passaged ("blind" passaged) because this sometimes yields additional virus isolates. Confirmed virus isolates can be identified using conventional serological techniques.[60,92]

REFERENCES

1. **Goyal, S. M.,** Viral pollution of the marine environment, *Crit. Rev. Environ. Control,* 14, 1, 1984.
2. **Colwell, R. R.,** *Natural Toxins and Human Pathogens in the Marine Environment,* Maryland Sea Grant College, University of Maryland, College Park, 1985.
3. **Cabelli, V. J., Dufour, A. P., McCabe, L. J., and Levin, M. A.,** Swimming-associated gastroenteritis and water quality, *Am. J. Epidemiol.,* 115, 606, 1982.
4. **Seyfried, P. L., Tobin, R. S., Brown, N. E., and Ness, P. F.,** A prospective study of swimming-related illness. I. Swimming-associated health risk, *Am. J. Public Health,* 75, 1068, 1985.
5. **Seyfried, P. L., Tobin, R. S., Brown, N. E., and Ness, P. F.,** A prospective study of swimming-related illness. II. Morbidity and the microbiological quality of water, *Am. J. Public Health,* 75, 1071, 1985.
6. **Goldfield, M.,** Epidemiological indicators for transmission of viruses in water, in *Viruses in Water,* Berg, G., Bodily, H. L., Lennette, E. H., Melnick, J. L., and Metcalf, T. G., Eds., American Public Health Association, Washington, D.C., 1976, 70.
7. **Gerba, C. P. and Goyal, S. M.,** Detection and occurrence of enteric viruses in shellfish: a review, *J. Food Prot.,* 41, 743, 1978.
8. **Larkin, E. P. and Hunt, D. A.,** Bivalve mollusks: control of microbial contaminants, *Bioscience,* 32, 193, 1982.
9. **Verber, J. L.,** Shellfish-borne disease outbreaks, Northeast Technical Services Unit, Public Health Service, U.S. Food and Drug Administration, Davisville, R.I., 1984.
10. **Richards, G. P.,** Outbreaks of shellfish-associated enteric virus illness in the United States: requisite for development of viral guidelines, *J. Food Prot.,* 48, 815, 1985.
11. **Goyal, S. M., Gerba, C. P., and Melnick, J. L.,** Human enteroviruses in oysters and their overlying waters, *Appl. Environ. Microbiol.,* 37, 572, 1979.
12. **Ellender, R. D., Mapp, J. B., Middlebrooks, B. L., Cook, D. W., and Cake, E. W.,** Natural enterovirus and fecal coliform contamination of Gulf Coast oysters, *J. Food Prot.,* 43, 105, 1980.
13. **Gerba, C. P., Goyal, S. M., Cech, I., and Bogdan, G. F.,** Bacterial indicators and environmental factors as related to contamination of oysters by enteroviruses, *J. Food Prot.,* 43, 99, 1980.
14. **Vaughn, J. M., Landry, E. F., Vicale, T. J., and Dahl, M. C.,** Isolation of naturally occurring enteroviruses from a variety of shellfish species residing in Long Island and New Jersey marine embayments, *J. Food Prot.,* 43, 95, 1980.
15. **Sobsey, M. D., Hackney, C. R., Carrick, R. J., Ray, B., and Speck, M. L.,** Occurrence of enteric bacteria and viruses in oysters, *J. Food Prot.,* 43, 111, 1980.
16. **Wait, D. A., Hackney, C. R., Carrick, R. J., Lovelace, G., and Sobsey, M. D.,** Enteric bacterial and viral pathogens and indicator bacteria in hard shell clams, *J. Food Prot.,* 46, 493, 1983.

17. **Bryan, F. L.,** Epidemiology of foodborne diseases transmitted by fish, shellfish and marine crustaceans in the United States, 1970—1978, *J. Food Prot.*, 43, 859, 1980.
18. **Appleton, H.,** Outbreaks of viral gastroenteritis associated with the consumption of shellfish, in *Viruses and Wastewater Treatment,* Goddard, M. and Butler, M., Eds., Pergamon Press, New York, 1981, 287.
19. **Portnoy, B. L., Mackowiak, P. A., Caraway, C. T., Walker, J. A., McKinley, T. W., and Klein, C. A., Jr.,** Oyster-associated hepatitis: failure of shellfish certification programs to prevent outbreaks, *JAMA*, 233, 1065, 1975.
20. **Grohmann, G. S., Murphy, A. M., Christopher, P. J., Auty, E., and Greenberg, H. B.,** Norwalk virus gastroenteritis in volunteers consuming depurated oysters, *Aust. J. Exp. Biol. Med. Sci.*, 59, 219, 1981.
21. **Cabelli, V. J., Dufour, A. P., Levin, M. A., McCabe, L. J., and Haberman, P. W.,** Relationship of microbial indicators to health effects at marine bathing beaches, *Am. J. Public Health,* 69, 690, 1979.
22. **Gerba, C. P. and Goyal, S. M.,** *Methods in Environmental Virology,* Marcel Dekker, New York, 1982.
23. **Sobsey, M. D.,** Procedures for the virological examination of seawater, shellfish and sediment, in *Laboratory Procedures for the Examination of Seawater and Shellfish,* Greenberg, A. E. and Hunt, D. A., Eds., American Public Health Association, Washington, D.C., 1984, 81.
24. **Larkin, E. P. and Metcalf, T. G.,** Cooperative study of methods for the recovery of enteric viruses from shellfish, *J. Food Prot.*, 43, 84, 1980.
25. **Tierney, J. T., Sullivan, R., Peeler, J. T., and Larkin, E. P.,** Detection of low numbers of poliovirus 1 in oysters: collaborative study, *J. Assoc. Off. Anal. Chem.*, 68, 884, 1985.
26. **Metcalf, T. G. and Stiles, W. C.,** The accumulation of enteric viruses by the oyster, *Crassostrea virginica, J. Infect. Dis.*, 115, 68, 1965.
27. **Metcalf, T. G. and Stiles, W. C.,** Enteroviruses within an estuarine environment, *Am. J. Epidemiol.*, 88, 379, 1968.
28. **Seraichekas, H. R., Brashear, D. A., Barnick, J. A., Carey, P. F., and Liu, O. C.,** Viral depuration by assaying individual shellfish, *Appl. Microbiol.*, 16, 1865, 1968.
29. **Konowalchuk, J. and Speirs, J. I.,** Enterovirus recovery from laboratory-contaminated samples of shellfish, *Can. J. Microbiol.*, 18, 1023, 1972.
30. **Hamblet, F. E., Hill, W. F., Jr., Akin, E. W., and Benton, W. H.,** Oysters and human viruses: effect of seawater turbidity on poliovirus uptake and elimination, *Am. J. Epidemiol.*, 89, 562, 1969.
31. **Di Girolamo, R. D., Liston, J., and Matches, J.,** Uptake and elimination of poliovirus by West Coast oysters, *Appl. Microbiol.*, 29, 260, 1975.
32. **Hoff, J. C. and Becker, R. C.,** The accumulation and elimination of crude and clarified poliovirus suspensions by shellfish, *Am. J. Epidemiol.*, 90, 53, 1969.
33. **Mitchell, J. R., Presnell, M. W., Akin, E. W., Cummins, J. M., and Liu, O. C.,** Accumulation and elimination of poliovirus by the Eastern oyster, *Am. J. Epidemiol.*, 84, 40, 1966.
34. **Liu, O. C., Seraichekas, H. R., and Murphy, B. L.,** Fate of poliovirus in Northern Quahaugs, *Proc. Soc. Exp. Biol. Med.*, 121, 601, 1966.
35. **Sullivan, R. and Peeler, J. T.,** Evaluation of a method for recovery of virus from oysters, *J. Food Prot.*, 45, 636, 1982.
36. **Duff, M. F.,** The uptake of enteroviruses by the New Zealand marine blue mussel, *Mytilus edulis aoteanus, Am. J. Epidemiol.*, 85, 486, 1967.
37. **Kostenbader, K. D., Jr. and Cliver, D. O.,** Polyelectrolyte flocculation as an aid to recovery of enteroviruses from oysters, *Appl. Microbiol.*, 24, 540, 1972.
38. **Kostenbader, K. D., Jr. and Cliver, D. O.,** Filtration methods for recovering enteroviruses from foods, *Appl. Microbiol.*, 26, 149, 1973.
39. **Vaughn, J. M., Landry, E. F., Vicale, T. J., and Dahl, M. C.,** Modified procedure for the recovery of naturally accumulated poliovirus from oysters, *Appl. Environ. Microbiol.*, 38, 594, 1979.
40. **Landry, E. F., Vaughn, J. M., and Vicale, T. J.,** Modified procedure for extraction of poliovirus from naturally-infected oysters using Cat-floc and beef extract, *J. Food Prot.*, 43, 91, 1980.
41. **Richards, G. P., Goldmintz, D., Green, D. L., and Babinchak, J. A.,** Rapid methods for extraction and concentration of poliovirus from oyster tissues, *J. Virol. Methods,* 5, 285, 1982.
42. **Metcalf, T. G., Eckerson, D., and Moulton, E.,** A method for recovery of viruses from oysters and hard and soft shell clams, *J. Food Prot.*, 43, 89, 1980.
43. **Metcalf, T. G., Moulton, E., and Eckerson, D.,** Improved method and test strategy for recovery of enteric viruses from shellfish, *Appl. Environ. Microbiol.*, 39, 141, 1980.
44. **Tierney, J. T., Fassolitis, A., Donsel, D. V., Rao, V. C., Sullivan, R., and Larkin, E. P.,** Glass wool-hydroextraction method for recovery of human enteroviruses from shellfish, *J. Food Prot.*, 43, 102, 1980.
45. **Sullivan, R., Peeler, J. T., Tierney, J. T., and Larkin, E. P.,** Evaluation of a method for recovering poliovirus 1 from 100-gram oyster samples, *J. Food Prot.*, 47, 108, 1984.
46. **Sobsey, M. D., Wallis, C., and Melnick, J. L.,** Development of a simple method for concentrating enteroviruses from oysters, *Appl. Microbiol.*, 29, 21, 1975.

47. **Sobsey, M. D., Carrick, R. J., Howard, B. E., and Jensen, H. R.,** Improved methods for detecting enteric viruses in oysters and clams, in *Proc. 10th Natl. Shellfish Sanitation Workshop,* Wilt, D. S., Ed., Food and Drug Administration, U.S. Department of Health, Education, and Welfare, Washington, D.C., 1977, 127.
48. **Sobsey, M. D., Carrick, R. J., and Jensen, H. R.,** Improved methods for detecting enteric viruses in oysters, *Appl. Environ. Microbiol.,* 36, 121, 1978.
49. **Howard, B. E.,** The Development and Assessment of a Method for Concentrating Enteric Viruses in the Hard Shell Clam *Mercenaria mercenaria,* Master's thesis, University of North Carolina, Chapel Hill, 1980.
50. **Wait, D. W., Kimball, H. P., and Sobsey, M. D.,** Enhancement of enteric virus recovery from shellfish by a cationic polyelectrolyte, in *Abstracts of the Annual Meeting,* American Society for Microbiology, Washington, D.C., 1981, 207.
51. **Johnson, K. M., Cooper, R. C., and Straube, D. C.,** Procedure for recovery of enteroviruses from the Japanese cockle *Tapes japonica, Appl. Environ. Microbiol.,* 41, 932, 1981.
52. **Keating, M.,** Evaluation of Methods for Recovery of Hepatitis A Virus from the Eastern Oyster, *Crassastrea virginica,* Master's thesis, University of North Carolina, Chapel Hill, 1985.
53. **Keating, M. and Sobsey, M. D.,** Evaluation of methods for recovering hepatitis A virus (HAV) from oysters, in *Abstracts of the Annual Meeting,* American Society for Microbiology, Washington, D.C., 1985, 254.
54. **Wait, D. W. and Sobsey, M. D.,** Evaluation of two methods for detecting wild type hepatitis A virus in oysters, in *Abstracts of the Annual Meeting,* American Society for Microbiology, Washington, D.C., 1986.
55. **Sobsey, M. D.,** Methods for detecting enteric viruses in water and wastewater, in *Viruses in Water,* Berg, G., Bodily, H. L., Lennette, E. H., Melnick, J. L., and Metcalf, T. G., Eds., American Public Health Association, Washington, D.C., 1976, 89.
56. **Gerba, C. P., Farrah, S. R., Goyal, S. M., Wallis, C., and Melnick, J. L.,** Concentration of enteroviruses from large volumes of tap water, treated sewage, and seawater, *Appl. Environ. Microbiol.,* 35, 540, 1978.
57. **Goyal, S. M. and Gerba, C. P.,** Concentration of viruses from water by membrane filters, in *Methods in Environmental Virology,* Gerba, C. P. and Goyal, S. M., Eds., Marcel Dekker, New York, 1982, 59.
58. **Farrah, S. R. and Bitton, G.,** Methods (other than microporous filters) for concentration of viruses from water, in *Methods in Environmental Virology,* Gerba, C. P. and Goyal, S. M., Eds., Marcel Dekker, New York, 1982, 117.
59. **Gerba, C. P.,** Methods for recovering viruses from the water environment, *Viral Pollution of the Environment,* Berg, G., Ed., CRC Press, Boca Raton, Fla., 1983, 19.
60. **Greenberg, A. E., Trussel, R. R., Clesceri, L. S., and Franson, M. A. H., Eds.,** Detection of enteric viruses, in *Standard Methods for the Examination of Water and Wastewater,* 16th ed., American Public Health Association, Washington, D.C., 1985, 946.
61. **Liu, O. C., Brashear, D. A., Seraichekas, H. R., Barnick, J. A., and Metcalf, T. G.,** Virus in water. I. A preliminary study on a flow-through gauze sampler for recovering virus from waters, *Appl. Microbiol.,* 21, 405, 1971.
62. **Metcalf, T. G., Wallis, C., and Melnick, J. L.,** Concentration of viruses from seawater, in *Advances in Water Pollution Research,* Proc. 6th Int. Conf., Jenkins, S. H., Ed., Pergamon Press, Oxford, 1973, 109.
63. **Hill, W. F., Jr., Akin, E. W., Benton, W. H., and Metcalf, T. G.,** Virus in water. II. Evaluation of membrane cartridge filters for recovering low multiplicities of poliovirus from water, *Appl. Microbiol.,* 23, 880, 1972.
64. **Hill, W. F., Jr., Akin, E. W., Benton, W. H., Mayhew, C. J., and Metcalf, T. G.,** Recovery of poliovirus from turbid estuarine water on microporous filters by the use of celite, *Appl. Microbiol.,* 27, 506, 1974.
65. **Sobsey, M. D., Gerba, C. P., Wallis, C., and Melnick, J. L.,** Concentration of enteroviruses from large volumes of turbid estuary water, *Can. J. Microbiol.,* 23, 770, 1977.
66. **Loh, P. C., Fujioka, R. S., and Lau, L. S.,** Recovery, survival and dissemination of human enteric viruses in ocean waters receiving sewage in Hawaii, *Water Air Soil Pollut.,* 12, 197, 1979.
67. **Payment, P., Gerba, C. P., Wallis, C., and Melnick, J. L.,** Methods for concentrating viruses from large volumes of estuarine water on pleated membranes, *Water Res.,* 10, 893, 1976.
68. **Farrah, S. R., Goyal, S. M., Gerba, C. P., Wallis, C., and Melnick, J. L.,** Concentration of enteroviruses from estuarine water, *Appl. Environ. Microbiol.,* 33, 1192, 1977.
69. **Gerba, C. P., Smith, E. M., and Melnick, J. L.,** Development of a quantitative method for detecting enteroviruses in estuarine sediments, *Appl. Environ. Microbiol.,* 34, 158, 1977.
70. **Sobsey, M. D., Glass, J. S., Carrick, R. J., Jacobs, R. R., and Rutala, W. A.,** Evaluation of the tentative standard method for enteric virus concentration from large volumes of tap water, *J. Am. Water Works Assoc.,* 72, 292, 1980.
71. **Katzenelson, E., Fattal, B., and Hostovesky, T.,** Organic flocculation: an efficient second-step concentration method for the detection of viruses in tap water, *Appl. Environ. Microbiol.,* 32, 638, 1976.

72. **Bitton, G., Feldberg, B. N., and Farrah, S. R.,** Concentration of enteroviruses from seawater and tapwater by organic flocculation using non-fat dry milk and casein, *Water Air Soil Pollut.*, 12, 187, 1979.
73. **Guttman-Bass, N. and Nasser, A.,** Simultaneous concentration of four enteroviruses from tap, waste, and natural waters, *Appl. Environ. Microbiol.*, 47, 1311, 1984.
74. **Goyal, S. M. and Gerba, C. P.,** Viradel method for detection of rotavirus in seawater, *J. Virol. Methods*, 7, 279, 1983.
75. **Williams, F. P. and Jakubowski, W.,** Large volume virus concentration: evaluation of the organic flocculation method for elution/reconcentration, in *Abstracts of the Annual Meeting*, American Society for Microbiology, Washington, D.C., 1978, 200.
76. **Morris, R. and Waite, W. M.,** Evaluation of procedures for recovery of viruses from water. I. Concentration systems, *Water Res.*, 14, 795.
77. **Metcalf, T. G., Wallis, C., and Melnick, J. L.,** Virus enumeration and public health assessments in polluted surface water contributing to transmission of virus in nature, in *Virus Survival in Water and Wastewater Systems*, Malina, J. F., Jr. and Sagik, B. P., Eds., Center for Research in Water Resources, The University of Texas at Austin, Austin, 1974, 58.
78. **De Flora, S., De Renzi, G. P., and Badolati, G.,** Detection of animal viruses in coastal seawater and sediments, *Appl. Microbiol.*, 30, 472, 1975.
79. **Wallis, C., Grinstein, S., Melnick, J. L., and Fields, J. E.,** Concentration of viruses from sewage and excreta on insoluble polyelectrolytes, *Appl. Microbiol.*, 18, 1007, 1969.
80. **Grinstein, S., Melnick, J. L., and Wallis, C.,** Virus isolations from sewage and from a stream receiving effluents of sewage treatment plants, *Bull. W.H.O.*, 42, 291, 1969.
81. **LaBelle, R. L., Gerba, C. P., Goyal, S. M., Melnick, J. L., Cech, I., and Bogdan, G. F.,** Relationships between environmental factors, bacterial indicators, and the occurrence of enteric viruses in estuarine sediments, *Appl. Environ. Microbiol.*, 39, 588, 1980.
82. **Cooper, R. C., Johnson, K. M., Straube, D. C., Brown, L. A., and Lysmer, D.,** Development and Evaluation of Methods for the Detection of Enteric Viruses in San Francisco Bay Shellfish, Water and Sediment, Rep. No. UCB/SERL 79-3, Sanitary Engineering Research Laboratory, University of California at Berkeley, Berkeley, 1980.
83. **Bitton, G., Chou, Y. J., and Farrah, S. R.,** Techniques for virus detection in aquatic sediments, *J. Virol. Methods*, 4, 1, 1982.
84. **Tsai, S. C., Ellender, R. D., Johnson, R. A., and Howell, F. G.,** Elution of viruses from coastal sediments, *Appl. Environ. Microbiol.*, 4, 797, 1983.
85. **Johnson, R. A., Ellender, R. D., and Tsai, S. C.,** Elution of enteric viruses from Mississippi estuarine sediments with lecithin-supplemented eluents, *Appl. Environ. Microbiol.*, 48, 581, 1984.
86. **Wait, D. A. and Sobsey, M. D.,** Method for recovery of enteric viruses from estuarine sediments with chaotropic agents, *Appl. Environ. Microbiol.*, 46, 379, 1983.
87. **Metcalf, T. G. and Melnick, J. L.,** Simple apparatus for collecting estuarine sediments and suspended solids to detect solids-associated virus, *Appl. Environ. Microbiol.*, 45, 323, 1983.
88. **Rao, V. C., Seidel, K. M., Goyal, S. M., Metcalf, T. G., and Melnick, J. L.,** Isolation of enteroviruses from water, suspended solids, and sediments from Galveston Bay: survival of poliovirus and rotavirus adsorbed to sediments, *Appl. Environ. Microbiol.*, 48, 404, 1984.
89. **Lewis, G. D., Loutit, M. W., and Austin, F. J.,** A method for detecting human enteroviruses in aquatic sediments, *J. Virol. Methods*, 10, 153, 1985.
90. **Schmidt, N. J.,** Cell culture techniques for diagnostic virology, in *Diagnostic Procedures for Viral, Rickettsial and Chlamydial Infections*, 5th ed., Lennette, E. H. and Schmidt, N. J., Eds., American Public Health Association, Washington, D.C., 1979, 65.
91. **Payment, P. and Trudel, M.,** Improved method for the use of proportioning injectors to condition large volumes of water for virological analysis, *Can. J. Microbiol.*, 27, 455, 1981.
92. **Lennette, E. H. and Schmidt, N. L.,** *Diagnostic Procedures for Viral, Rickettsial and Chlamydial Infections*, 5th ed., American Public Health Association, Washington, D.C., 1979.
93. **Berg, G., Safferman, R. S., Dahling, D. R., Berman, D., and Hurst, C. J.,** *USEPA Manual of Methods for Virology*, EPA-600/4-84-013, Environmental Monitoring and Support Laboratory, U.S. Environmental Protection Agency, Cincinnati, 1984.
94. **Gerba, C. P., Rose, J. B., and Singh, S. N.,** Waterborne gastroenteritis and viral hepatitis, *Crit. Rev. Environ. Control*, 15, 213, 1985.
95. **Sobsey, M. D., Oglesbee, S. E., Wait, D. A., and Cuenca, A. I.,** Detection of hepatitis A virus (HAV) in drinking water, *Water Sci. Tech.*, 17, 23, 1985.
96. **Sobsey, M. D., Oglesbee, S. E., and Wait, D. A.,** Evaluation of methods for concentrating hepatitis A virus from drinking water, *Appl. Environ. Microbiol.*, 50, 1457, 1985.
97. **Dahling, D. G., Berg, G., and Berman, D.,** BGM: a continuous cell line more sensitive than primary rhesus and African green kidney cells for the recovery of viruses from water, *Health Lab. Sci.*, 11, 275, 1974.

98. **Oglesbee, S. E., Wait, D. A., and Meinhold, A. F.,** MA104: a continuous cell line for the isolation of human enteric viruses from environmental samples, in *Abstracts of the Annual Meeting,* American Society for Microbiology, Washington, D.C., 1981, 219.
99. **Smith, E. M. and Gerba, C. P.,** Laboratory methods for the growth and detection of animal viruses, in *Methods in Environmental Virology,* Gerba, C. P. and Goyal, S. M., Eds., Marcel Dekker, New York, 1982, 15.
100. **Agbalika, F., Hartemann, P., and Foliguet, J. M.,** Trypsin-treated MA-104: a sensitive cell line for isolating enteric viruses from environmental samples, *Appl. Environ. Microbiol.,* 47, 378, 1984.
101. **Schmidt, N. J., Ho, H. H., Riggs, J. L., and Lennette, E. H.,** Comparative sensitivity of various cell culture systems for isolation of viruses from wastewater and fecal samples, *Appl. Environ. Microbiol.,* 36, 480, 1978.
102. **Morris, R. and Waite, M. W.,** Evaluation of procedures for recovery of viruses from water. II. Detection system, *Water Res.,* 14, 795, 1980.
103. **Leong, L. Y. C., Barrett, S. J., and Trussell, R. R.,** False-positives in testing of secondary sewage for enteric viruses, in *Abstracts of the Annual Meeting,* American Society for Microbiology, Washington, D.C., 1978, 200.
104. **Fannin, K. E., Abid, S. H., Bertucci, J. J., Reed, J. M., Vana, S. C., and Lue-Hing, C.,** Significance of reporting infectious virus or plaque forming unit concentrations from environmental samples, in *Abstracts of the Annual Meeting,* American Society for Microbiology, Washington, D.C., 1978, 200.
105. **Hejkal, T. W., Gerba, C. P., and Rao, V. C.,** Reduction of cytotoxicity in virus concentrates from environmental samples, *Appl. Environ. Microbiol.,* 43, 731, 1982.
106. **Richards, G. P. and Weinheimer, D. A.,** Influence of adsorption time, rocking, and soluble proteins on the plaque assay of monodispersed poliovirus, *Appl. Environ. Microbiol.,* 49, 744, 1985.

Chapter 6

RECONCENTRATION OF VIRUSES FROM PRIMARY ELUATES

V. Chalapati Rao

TABLE OF CONTENTS

I. INTRODUCTION

Urbanization and the resulting population centers exert an increasing demand on our water supplies. The need to reuse water raises concerns about associated public health problems. The use of sewage effluents for agricultural, aquacultural, and other purposes calls for a careful study of environmental problems associated with virus pollution. There is a need for a more thorough and systematic evaluation of the public health problems of virus transmission via recreational and drinking waters. Such an evaluation requires the availability and utilization of standard reliable methods to quantitatively detect small numbers of viruses in large volumes of waters. Current information shows that no virus recovery method now available is equally effective for all monitoring needs and environmental situations. Detection of viruses under adverse environmental conditions such as in highly turbid estuarine waters with high concentrations of organic components provides challenges quite different from those encountered with finished waters. A quantitative method should produce rapid and maximum recovery of as many different virus types as possible.

Detection of small numbers of enteric viruses in large volumes of water with adsorption-elution techniques requires between 1 and 2 ℓ of an eluting agent for efficient virus recovery. A reduction in the volume of the initial or primary eluate has to be achieved for an economical assay of the final sample in a small number of cell cultures. The second-step concentration has been referred to as "reconcentration", the efficiency of which partly determines the success of virus recovery from environmental samples. Some of the procedures initially developed for concentrating viruses from water and wastewater and later considered inadequate to deal with the problems of large sample volumes have been reexamined for their suitability in the reconcentration of viruses from primary eluates. Some of these techniques include: aqueous polymer two-phase separation,[1,2] polyethylene glycol hydroextraction,[3-5] small diameter membrane filtration,[6-8] ferric chloride[9,10] and aluminum hydroxide[9,11] precipitation, organic flocculation,[12] and magnetic iron oxide adsorption-elution.[13] A clear need has already been felt for the development of a third-step concentration procedure for techniques such as ELISA and RIA that cannot assay sample volumes greater than 50 to 100 μℓ for the detection of viruses not easily cultivatable in cell cultures such as Norwalk virus, human rotavirus, and hepatitis A virus (enterovirus 72).

II. METHODS FOR CONCENTRATING HUMAN ENTERIC VIRUSES FROM PRIMARY ELUATES

A. Aqueous Polymer Two-Phase Separation

Two-phase systems are composed of polyethylene glycol, a nonionic polymer, and sodium dextran sulfate, an anionic polyelectrolyte. At the proper ionic strength and pH, separation of the phases results because of incompatibility of the two polymers. Albertsson[14] noticed that the higher the molecular weight of polymers, the lower the concentration required for phase separation, and the bigger the difference in molecular size between two polymers, the better the separation. Temperature also influences phase separation, better separation occurring at low temperatures.

Salts are important in the distribution of viruses between the two separated phases. Salt ions distribute unequally in polymer phases. For example, positive ions are attracted to negatively charged polymers. The close spacing of oxygen atoms in the polyethylene glycol molecule makes chelation possible, and hydrogen bonding is likely to occur between anions and dextran polymers. As a result, one phase becomes electrically more negative than the other phase, and a negatively charged virus favors the more positive phase. This is the case for poliovirus separation in the sodium dextran sulfate-polyethylene glycol system. At high salt concentrations, the negatively charged poliovirus is concentrated in the more positive upper phase, but at low salt concentration poliovirus prefers the lower phase.

Table 1
RECOVERY OF POLIOVIRUS 1 FROM 100 GAL OF WATER BY MEMBRANE FILTER-POLYMER TWO-PHASE SEPARATION

Expt	Type of water	Total virus input (PFU/100 gal)	Total virus recovered (PFU)	Virus recovered (%)
1	Tap	29.6 ± 2.5	7	24
2	Tap	194.0 ± 13.0	68	35
3	Estuarine	24.0 ± 6.8	9	38
4	Tap	62.4 ± 3.2	33	53
5	Tap	24.0 ± 6.8	16	67

Adapted from Hill, W. F., Jr., Akin, E. W., Benton, W. H., and Metcalf, T. G., *Appl. Microbiol.*, 23, 880, 1972.

Two-phase separation was used by Fattal et al.[15] to concentrate small numbers of enteroviruses from 5 ℓ of filtered tapwater and filtered tapwater contaminated with sewage. Recovery efficiencies for poliovirus 1 and echovirus 7 in tapwater averaged 100 and 125%, respectively, but the recovery of natural enteroviruses from sewage-contaminated tapwater averaged only 5%. There is only one published report[2] in which this technique was used for the concentration of viruses in 1 ℓ quantities of eluates of 5x nutrient broth. The eluates were obtained from membrane cartridge filters that had been used to adsorb small numbers of poliovirus 1 (24 to 194 PFU) from 100 gal of tap or estuarine water. Results of five experiments indicated that 24 to 67% of the viruses were recovered in a final concentrate of 2 to 3 mℓ (Table 1).

B. Polyethylene Glycol Hydroextraction

When a virus-containing sample is placed in a dialysis tube made of a semipermeable membrane and is surrounded by a hygroscopic material such as polyethylene glycol, water moves freely across the membrane into the hygroscopic material, whereas viruses cannot cross the membrane and thus remain inside the dialysis tube. This process called "hydroextraction" is an ultrafiltration process and is usually carried out at 4°C for 4 to 24 hr resulting in a reduction of sample volume and concentration of the virus. The method was initially used by Cliver[3] for concentrating 100 mℓ volumes of enteroviruses suspended in various diluents to 2 mℓ by placing the samples in visking dialysis bags and hydroextracting for 2 to 3 hr against polyethylene glycol (PEG, Carbowax® 20,000). Between 10 and 30% of seeded viruses were recovered; reproducibility of results was low. Shuval et al.[4] used hydroextraction against PEG (Carbowax® 4000) to concentrate enteroviruses 100-fold from 1 ℓ samples in an 18-hr period. Virus recovery ranged from 24 to 61%. Virus loss in this procedure was attributed to adsorption of virions on the inside surface of the dialysis membrane. The concentrate was also toxic to certain cells, and either PEG molecules or other low molecular weight impurities may have entered the concentrate by counter dialysis.

Wellings et al.[16] recovered 100% of input virus by keeping the sample in a dialysis tube of 24 nm average pore diameter immersed in PEG (Carbowax® 2000) overnight at 4°C. Recovery of all of the virus was attributed to rinsing the inside of the dialysis tube with 3% beef extract in Tris hydroxymethyl aminomethane at pH 9.0. This approach was applied to recovering viruses from beef extract eluates of membrane filters through which 50 to 100 gal of well water from sites near effluent discharge sites were processed. None of the 14 hydroextracted samples yielded viruses.

Table 2
CONCENTRATION OF VIRUSES ELUTED FROM ALUMINUM HYDROXIDE FLOCS BY HYDROEXTRACTION

Trial	Final vol (mℓ)	Virus input	Virus recovery (%)
1	9	4.6×10^6	91.3
2	28	1.7×10^3	76.4
3	14	5.4×10^3	64.8
4	20	1.6×10^7	75.0

Adapted from Farrah, S. R., Goyal, S. M., Gerba, C. P., Wallis, C., and Melnick, T. L., *Appl. Environ. Microbiol.*, 33, 1192, 1977.

PEG hydroextraction was applied as a reconcentration procedure to recover viruses from glycine eluates of membrane filters through which 500 mℓ samples of sewage influent and chlorinated effluents had been filtered. In addition, beef extract eluates from sludge, soil, and mud samples obtained from sludge spray sites, drying beds, and drain fields have been tested by this procedure.[16] Viruses were detected in fully digested sludge, dried sludge cake, and mud 900 m downstream from a sewage disposal site.

Certain organic compounds and metallic ions in water samples are concentrated along with viruses on membrane filters and are found in relatively high concentrations in the filter eluates. These components form flocs at low pH and clog smaller membrane filters used for reconcentration. Farrah et al.[17] recovered viruses by hydroextracting eluates of aluminum hydroxide flocs produced in estuarine water. The eluting agent was 1 *M* glycine in fetal calf serum. Eluates were centrifuged, and supernates were seeded with poliovirus 1 and reduced in volume by hydroextraction with polyethylene glycol. The concentrated samples were dialyzed against buffered saline. Average recovery of viruses in four trials was 76.8% (Table 2). This method was also used to recover indigenous viruses from eluates of 200 ℓ samples of estuarine water. Of five samples tested, three contained viruses in numbers ranging from 3 to 33 PFU/200 ℓ.

Hydroextraction, as a second-step concentration procedure for several virus types seeded into an eluting agent (fetal calf serum) has also been evaluated by Ramia and Sattar.[18] A 100 mℓ volume of experimentally seeded 10% fetal calf serum in normal saline, pH 9.0, was placed in a dialysis sac and hydroextracted for 4 hr at 4°C with PEG 6000. The viscous material remaining in the sac was suspended in 10 mℓ of Earle's balanced salt solution.

About 91% of the input poliovirus was recovered with a 5.5-fold concentration of the samples. However, considerable variation occurred in the reduction of sample volume (11 to 33 mℓ). To overcome this problem, overnight (16 to 19 hr) hydroextraction at 4°C with PEG 6000 was attempted. This resulted in the removal of nearly all the liquid from inside the sac. The brown viscous material remaining inside the sac was suspended in 10 mℓ of Earle's balanced salt solution. The results of experiments with five laboratory-adapted strains of enteroviruses (poliovirus 1, echovirus 6, coxsackievirus A9, coxsackievirus B5, and reovirus 3) indicate recoveries of 87 to 97%.

Field samples of raw and finished waters contain a variety of chemicals and organic substances, some of which are retained by the filters and other adsorbents during sample processing. Passing eluting agents through these adsorbents to recover viruses also results in the release and accumulation of organic components in the eluates. Interference from these naturally occurring substances in the primary eluates with virus recovery during hy-

Table 3
RECONCENTRATION OF ENTEROVIRUSES FROM PRIMARY ELUATES BY HYDROEXTRACTION[a]

Virus type	Type of water filtered through talc-celite layers	Virus input in 100 mℓ eluate (PFU × 10^4)	Virus recovered in 10 mℓ concentrate (PFU × 10^4)	% Virus recovery (± SD)
Laboratory adapted stains				
Poliovirus 1	Treated water	1.4	1.27	91 ± 8
	Raw water	1.4	1.33	95 ± 4
Echovirus 6	Treated water	1.2	1.13	94 ± 10
	Raw water	1.2	1.10	93 ± 17
Field isolates				
Poliovirus 1 (vaccine strain)	Raw water	1.7	1.55	93 ± 6
Poliovirus 1 (nonvaccine strain)	Raw water	0.76	0.67	91 ± 8
Echovirus 1	Raw water	0.48	0.42	87 ± 7
Coxsackievirus B_3	Raw water	2.20	2.10	96 ± 8
Reovirus	Raw water	0.25	0.24	98 ± 7

[a] Samples of treated (500 ℓ) or raw water (40 ℓ) from the Ottawa River were passed through talc-celite layers. The layers were then eluted with 100 mℓ of 10% fetal calf serum in saline (pH 9.0). A known amount of virus under test was added to each eluate and concentrated by hydroextraction. The material remaining in the dialysis tube was resuspended in 10 mℓ of EBSS and plaque-assayed in BS-C-1 cells.

Adapted from Ramia, S. and Sattar, S. A., *Can. J. Microbiol.*, 25, 587, 1979.

droextraction was also examined by Ramia and Sattar.[18] In their experiments, 40 ℓ of raw water from the Ottawa River and 500 ℓ of treated water was passed through talc-celite layers. A total of 100 mℓ of 10% fetal calf serum in saline was then passed through each layer and collected as eluate. The eluate was seeded with virus, and the sample was hydroextracted overnight with PEG 6000. Results of these experiments with two laboratory-adapted strains of viruses (poliovirus 1 and echovirus 6) and five field isolates (poliovirus 1 [vaccine strain], poliovirus 1 [nonvaccine strain], echovirus 1, coxsackievirus B3, and a reovirus) are presented in Table 3. Virus recoveries ranged from 87 to 98%. The technique was equally efficient for concentrating laboratory-adapted and field strains of at least two major groups of enteric viruses. The promise shown by this technique has prompted its inclusion as one of the procedures for comparative round robin evaluation by the American Society for Testing Materials.[48] It appears that the type of dialysis tubing, its pore size, the eluting agent used, and the thoroughness of the elution step may influence the efficiency of virus recovery.

C. Membrane Filters

A number of filter materials and configurations have been used as virus adsorbents including cellulose nitrate membrane filters (0.45 μm porosity), cellulose acetate, cellulose nitrate and fiberglass cartridge depth filters (3.0 to 0.45 μm porosity), fiberglass-asbestos-epoxy disc filter (0.65 to 0.45 μm porosity), cellulose, modified anion-exchange resin, and inorganic filter aid (Zetaplus) filters. Usually, water is acidified and a polyvalent cation salt, such as magnesium chloride or aluminum chloride, is added and the water is passed through the filter. The viruses in the water sample adsorb to the filter surface. Adsorbed viruses are recovered from the filters with an eluting agent, usually a slightly alkaline proteinaceous solution such as serum, beef extract, or nutrient broth, or a highly alkaline buffer consisting of 0.05 *M* glycine, pH 11.5. With the above approach, several investigators used membrane filter technology to recover viruses from a variety of environmental waters and generated a wealth of information on factors that influence adsorption and elution of viruses to and from membrane filters. After a careful analysis of the published data, Mix[19] identified several important aspects of the mechanism of virus adsorption to membrane filters. They include (1) hydrophobic bonding between nonpolar aliphatic and aromatic groups on the surface of the filter and virus, (2) hydrogen bonding between polar groups on the surface of the virus and filter, and (3) salt bridging between negatively charged groups on the surface of the virus and filter and adsorbed cations obtained from the solution filtered.

Among the various eluting agents used for recovering viruses from filters, solutions with no proteinaceous material such as glycine have been used for reconcentrating viruses onto smaller diameter membrane filters.[6] Accordingly, Wallis et al.[6] suspended 1300 PFU of a poliovirus in 5 gal of tapwater, adsorbed the virus on a 293 mm membrane filter in the presence of 0.0005 *M* $AlCl_3$ at pH 4.0, washed the membrane with saline (to remove residual Al^{3+} ions and avoid subsequent gel formation with the basic eluting agent and thus avoid clogging membranes), and eluted the viruses with 300 mℓ of glycine (pH 11.5). The eluate was collected in 300 mℓ of pH 2.0 buffer to yield 600 mℓ of virus suspension at pH 6.7. The pH of the eluate was adjusted to 4.0, $AlCl_3$ was added to yield a final concentration of 0.0005 *M*, and the sample was filtered through a small diameter (25 mm) membrane of 0.45 μm porosity. The viruses were eluted from this membrane with a small volume of 0.05 *M* glycine (pH 11.5); the pH was immediately neutralized. In repeated tests, 90 to 100% of the input virus was recovered. Reconcentration of primary eluates with membrane filters is difficult when volumes of tapwater larger than 378 ℓ (400 gal) are sampled[11] or when eluates from filters used to process secondary effluents are reconcentrated.[20] In both situations, organic compounds (mostly humic acid) and metal ions also adsorb to the filters. Elution of viruses results in elution of some of these organics and metal ions also. These compunds form flocs at low pH and clog smaller diameter filters.[17] These compounds also form an insoluble precipitate when the eluate is neutralized and interfere with reconcentration on smaller filters.[21] Similar problems occur when the method is applied to small volumes of sewage or seawater. Viruses in glycine eluates (that may not permit reconcentration on membrane filters) have been successfully concentrated in inorganic precipitates of ferric and aluminum hydroxide. Membrane filters have been used by investigators for reconcentrating viruses from polluted surface water,[23] tapwater,[7] sewage,[24] turbid estuary water,[10] and potable water.[25]

Farrah and Bitton[22] used 3% lysine (pH 9.0) to elute viruses from Filterite® filters. Viruses in lysine could be adsorbed efficiently on a 25 mm diameter positively charged Zetaplus C-10 filter. Viruses on this filter were recovered with a small volume (4 mℓ) of beef extract (pH 9.0) with an efficiency of 85 to 90%.

Viruses cannot be reconcentrated from eluates that contain proteinaceous matter (beef extract, nutrient broth, tryptose phosphate broth) on small diameter membrane filters because proteins that precipitate at acid pH levels interfere. To alleviate this problem, two procedures, organic flocculation and iron-oxide adsorption and elution, have been developed.

D. Inorganic Precipitates

When a coagulant salt (e.g., aluminum chloride or ferric chloride) is added to water or wastewater, the salt dissociates and the metal ion hydrolyzes forming positively charged hydroxometal ion complexes (aluminum hydroxide and ferric hydroxide) that adsorb to the surface of negatively charged colloids, including viruses. The colloids aggregate by interparticulate attraction (Van der Waal's forces). The floc, along with the adsorbed viruses, settles down. A preformed precipitate of aluminum hydroxide produced by mixing 2 *M* Na_2CO_3 with 25 m*M* $AlCl_3$ also adsorbs virus.

1. Aluminum Hydroxide Precipitates

Wallis and Melnick[26] concentrated on preformed precipitates small amounts of poliovirus 1 and echovirus 7 from 1 ℓ samples of saline solution. They collected the precipitate on a filter surface, resuspended it in 1 mℓ of saline, and assayed the suspension for viruses. The efficiency of virus recovery averaged 82%. Subsequently, a number of investigators recovered enteroviruses from small volumes of sewage with this method. Available data from many studies were analyzed by Cookson,[27] who suggested that virus adsorption to an aluminum hydroxide precipitate could result from coordination of the virus surface by hydroxoaluminum complexes, electrostatic interactions between the negatively charged virus surfaces and positively charged aluminum hydroxide surfaces, and covalent bonding between aluminum atoms and specific sites on the virus surface. Of the three possible mechanisms, Cookson considered covalent bonding the least consistent with experimental evidence.

Aluminum hydroxide flocs formed *in situ* have been used to reconcentrate viruses from eluates of microporous filters used for processing large volumes of estuarine water.[17] These eluates were obtained by passing 400 ℓ of estuarine water through a prefilter and a 0.45 μm pleated cartridge filter and treating the filters with 1600 mℓ of glycine (pH 11). The eluates were neutralized with glycine (pH 2) and seeded with poliovirus 1. Aluminum chloride (final concentration 0.003 *M*) was added to the seeded eluates and neutralized with sodium carbonate whereupon aluminum hydroxide flocs formed. These flocs are efficient virus adsorbents. The flocs were collected by centrifugation and each was mixed with 3 volumes of 1 *M* glycine dissolved in fetal calf serum (pH 11.5). The samples were centrifuged, and the supernates were neutralized. Results of six trials are shown in Table 4. Although the recovery of viruses ranged from 81 to 100%, eluate volumes were 150 to 200 mℓ. A third step for reduction of volume was required, and this was achieved by hydroextraction.

Table 4
CONCENTRATION OF A POLIOVIRUS IN FILTER ELUATES BY ALUMINUM HYDROXIDE FLOCCULATION

Trial	Virus input (PFU)	% Virus recovered in aluminum floc
1	5.6×10^6	93
2	7.9×10^3	81
3	9.6×10^3	88
4	2.0×10^7	95
5	4.2×10^6	83
6	1.9×10^4	100

Adapted from Farrah, S. R., Goyal, S. M., Gerba, C. P., Wallis, C., and Melnick, J. L., *Appl. Environ. Microbiol.*, 33, 1192, 1977.

Table 5
CONCENTRATION OF ENTEROVIRUSES FROM TURBID ESTUARINE WATER USING CARTRIDGE FILTER-$FeCl_3$ CONCENTRATION

	Water volume (ℓ)	Turbidity (JTU)	Salinity (0/00)	Virus type	Virus input (PFU)	Virus recovered (%)
A.	170—190	4—30	4—20	Poliovirus 1	18—65 $\times 10^6$	22—76
		8—18	20—22	Poliovirus 1	18—2000	23—61
		10—18	10—21	Coxsackievirus B_3	10—6500	33—80
		10	10	Coxsackievirus A_9	25—40	28
		10—20	10	Echovirus 7	113—175	20—23
B.	378	9	7—18	Poliovirus 1	1.1—1.2 $\times 10^6$	50—57

Note: A. Data from Sobsey, M. D., Gerba, C. P., Wallis, C., and Melnick, J. L., *Can. J. Microbiol.*, 23, 770, 1977. B. Data from Payment, P., Gerba, C. P., Wallis, C., and Melnick, J. L., *Water Res.*, 10, 893, 1976.

2. Ferric Hydroxide Precipitation

Schäfer[28] developed a ferric hydroxide precipitation for concentrating viruses from 1 to 5 ℓ of surface waters. They added ferric chloride (final concentration 200 mg/ℓ) to water and coagulated by adjusting the pH to 6.0 to 6.5. They collected the precipitate on filters and recovered viruses by three successive elutions with small volumes of 3% beef extract (pH 8.0). Payment and colleagues[9] used ferric hydroxide precipitation to reconcentrate a poliovirus from glycine eluates obtained by eluting viruses from epoxy fiberglass filters that had been used for filtering 100 gal (378 ℓ) of turbid estuarine water. The virus was concentrated into a final eluate volume of 20 to 100 mℓ with virus recovery between 50 and 57% (Table 5). Subsequently, Sobsey et al.[10] determined optimum concentrations of $FeCl_3$ and optimum pH for maximum precipitate formation and highest recovery of viruses from glycine eluates. They determined that a $FeCl_3$ concentration of 0.001 *M* and a pH of 3.5 resulted in adsorption of >99% of seeded viruses to the precipitate. Of the various eluting agents examined for recovering viruses from the precipitate, fetal calf serum (pH 8) was the best, recovering 74% of the viruses.

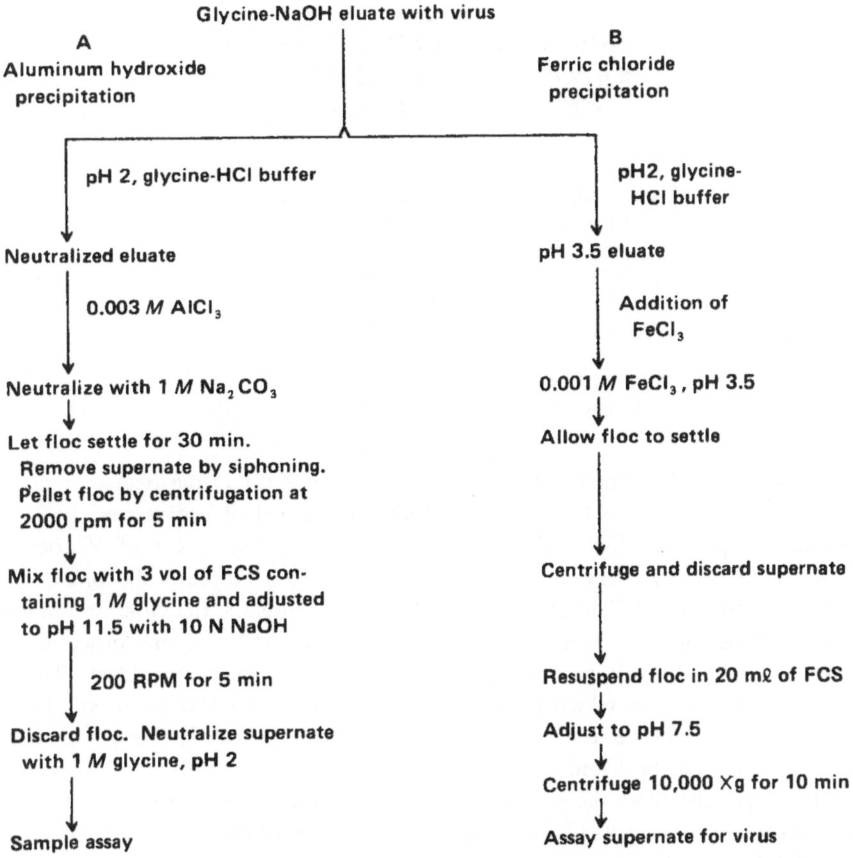

FIGURE 1. Scheme for reconcentration of viruses from eluates of inorganic precipitates. (A) Aluminum hydroxide precipitation. (Data from Farrah, S. R., Gerba, C. P., Wallis, C., and Melnick, J. L., *Appl. Environ. Microbiol.*, 31, 221, 1976.) (B) Ferric chloride precipitation. (Data from Sobsey, M. D., Gerba, C. P., Wallis, C., and Melnick, J. L., *Can. J. Microbiol.*, 23, 770, 1977.)

With these parameters, a reconcentration procedure outlined in Figure 1 was developed. In this procedure, of glycine-NaOH (pH 11.5) filtered through fiberglass filters is adjusted to pH 3.5, and $FeCl_3$ is added to a final concentration of 0.001 *M*. After 5 to 10 min of gentle mixing to maximize the formation of the floc, the sample is allowed to stand for several minutes to settle the precipitate and any viruses adsorbed on it. The supernate is decanted and the precipitate is centrifuged at 1000 $\times g$ for 10 to 15 min to sediment the precipitate. Viruses are recovered by eluting the precipitate with 20 mℓ of fetal calf serum (pH 8.0). The procedure was tested in a series of experiments with different model viruses. Recoveries of viruses under different conditions of salinity and turbidity of water ranged between 20 and 80% (Table 5). The percent recoveries do not reflect the efficiency of the $FeCl_3$ reconcentration procedure, because viruses were added to test samples of estuarine water and not to the primary eluate.

Estuarine waters that contain relatively large amounts of organic matter yield a larger volume of precipitate with ferric chloride, which in turn increases the final volume of concentrate. Aluminum hydroxide precipitates are smaller under similar conditions.

E. Organic Flocculation

The utilization of a protein solution such as 3% beef extract[29,30] for eluting viruses from adsorbent filters produces a high recovery of viruses, but does not permit the viruses to be

Table 6
ADSORPTION AND RECOVERY OF POLIOVIRUS 1 FROM 3% BEEF EXTRACT BLOCS (pH 3.5)

Virus input/150 mℓ of beef extract	% Virus recovered in	
	Sediment	Supernate
4.8×10^3	69	0
3.4×10^3	112	1.1
1.8×10^3	123	0

Adapted from Katzenelson, E., Fattal, B., and Hostovesky, T., *Appl. Environ. Microbiol.*, 32, 638, 1976.

reconcentrated on small diameter membranes because proteins in the beef extract interfere with adsorption of the viruses to the filters. Reducing the pH of beef extract to 3.5 results in flocculation of proteins. The sediment obtained by centrifuging 1 ℓ of 3% beef extract at 3000 ×g for 10 min is soluble in about 50 mℓ of 0.15 *M* sodium phosphate (pH 9.0). Katzenelson and co-workers[12] described this technique as "organic flocculation" and showed that the floc formed in beef extract could adsorb viruses and that this procedure can be utilized as a second-step concentration of primary eluates of beef extract. The authors conducted an experiment in which poliovirus 1 was seeded into 150 mℓ of sterile 3% beef extract and the pH of the beef extract was adjusted to 3.5 with 2 *N* HCl under slow stirring. They continued stirring for 30 min and centrifuged the mixture at 300 ×g for 10 min. They separated the supernate from the sediment and dissolved the sediment in 7.5 mℓ of 0.15 *M* Na_2HPO_4 (pH 9.0). The results (Table 6) showed that 69 to 123% of the virus was recovered from the solubilized sediment.

Katzenelson et al.[12] tested the practicability of the organic flocculation method for concentrating viruses from water and compared the efficiency of the method with that of the tentative standard method of Hill et al.[2] in which glycine buffer was used to elute viruses. In their tests, 1000 ℓ of Jerusalem tapwater seeded with poliovirus 1 was mixed by air bubbling and adjusted to pH 3.5. The sample was then divided into two equal volumes and each volume was filtered through a 293 mm diameter 0.45 μm Cox® fiberglass filter preceded by a fiberglass prefilter. The adsorbed viruses were eluted from one filter with 1 ℓ of 0.05 *M* glycine buffer (pH 11.5) and from the other with 300 mℓ of 3% beef extract (pH 9.0). The authors reconcentrated viruses in glycine eluates on small diameter membrane filters with the method of Jakubowski et al.[31] and reconcentrated viruses in beef extract eluates by organic flocculation. Recoveries of viruses in seven replicate experiments are shown in Table 7. Recoveries with the glycine method ranged from 30 to 47% (mean 35%); recoveries with the organic flocculation method ranged from 60 to 91% (mean 74%).

These findings have led to an increased interest in beef extract as an eluting agent for microporous filters (both positively and negatively charged), soils, sludges, sediments, and shellfish. Bitton et al.[32] proposed using isoelectric casein and nonfat dry milk to elute viruses from membrane filters. These authors concentrated the viruses in their eluates by organic flocculation at pH 4.5 to 4.6, which is the isoelectric point of casein. They used the method to concentrate viruses from tapwater, seawater, and soil. A scheme for reconcentrating viruses from beef extract and casein are outlined in Figure 2. In recent studies, organic flocculation has been used to recover viruses from marine water,[32,33] wastewater effluents,[34] river water,[35] sludges,[36,37] and sediments.[33,38-40]

Organic flocculation of beef extract eluates from estuarine sediments presented certain unique problems. Waite and Sobsey[38] reported that in seeding experiments, extraction of

Table 7
RECOVERY OF SEEDED POLIOVIRUS 1 BY ORGANIC FLOCCULATION AND BY ELUTION WITH GLYCINE BUFFER

	% Virus recovery	
Virus input/500 ℓ tap water	Glycine buffer	Organic flocculation
1.8×10^3	30	72.5
1.5×10^3	31	66.0
2.2×10^3	31	91.0
1.7×10^3	42	76.0
1.5×10	47	80.0
1.5×10	33	60.0
1.2×10	33	75.0
Mean	35	74.4

From Katzenelson, E., Fattal, B., and Hostovesky, T., *Appl. Environ. Microbiol.*, 32, 638, 1976. With permission.

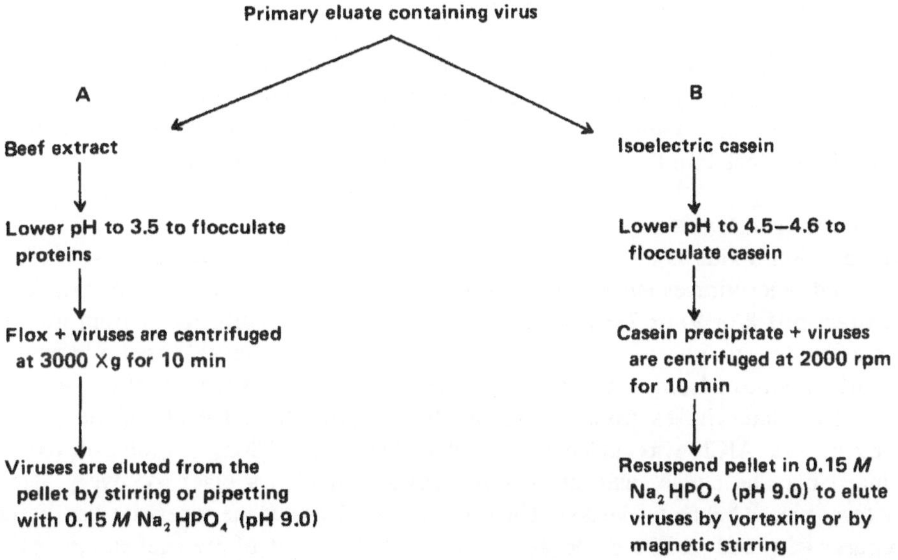

FIGURE 2. Scheme for recovering viruses from beef extract and casein by organic flocculation. (A) Beef extract. (Data from Katzenelson, E., Fattal, B., and Hostovesky, T., *Appl. Environ. Microbiol.*, 32, 638, 1976.) (B) Isoelectric casein. (Data from Bitton, G., Felberg, B. N., and Farrah, S. R., *Water, Air, Soil Pollut.*, 12, 187, 1979.)

viruses from sediments with 3% beef extract resulted in a recovery of only 9 to 18% for polioviruses and 2 to 4% for echovirus 7. The recovery of polioviruses was improved to 71% when beef extract was supplemented with a chaotropic agent such as 2 *M* $NaNO_3$. In reconcentration, 2 *M* $NaNO_3$ 3% beef extract eluate was adjusted to pH 3.5 to flocculate beef extract proteins, but only 2% of the polioviruses was recovered. The rest of the viruses remained unflocculated in the supernate. An improvement in the flocculation efficiency and virus recovery up to 24% was achieved by adding an antichaotrope, 2 *M* ammonium sulfate $[(NH_4)_2SO_4]$. Addition of Cat-floc T (a cationic polymer) to a 0.01% concentration further

3% beef extract-2 *M* $NaNO_3$ (pH 5.5)

↓

Add 4 *M* $(NH_4)_2SO_4$ to obtain 2 *M* concentration
Adjust to pH 3.5

↓

Add 1% cat-floc T to 0.01% concentration

↓

Mix slowly for 20 min

↓

Centrifuge at 5000 ×g for 5 min

↓

Discard supernate

↓

Resuspend pellet in 0.1 *M* Na_2HPO_4

↓

Adjust to pH 7.2

↓

Dialyze against PBS at 4°C for 18 to 24 hr

↓

Treat sample with antibiotics and assay

FIGURE 3. Modified organic flocculation procedure for concentrating viruses from sediment-eluates. (Data from Wait, D. A. and Sobsey, M. D., *Appl. Environ. Microbiol.*, 46, 379, 1983.)

improved flocculation and virus recovery. Final samples of virus concentrates resulting from this procedure contain high concentrations of salts producing cell toxicity during assay. Dialysis of concentrate samples against PBS for 18 to 24 hr at 4°C eliminates this problem. This modified procedure for organic flocculation is outlined in Figure 3.

F. Magnetic Iron Oxide

Magnetic iron oxide is an excellent adsorbent for polioviruses, coxsackieviruses, echoviruses, and other viruses suspended in distilled water and tapwater.[41-43] For example, 3% beef extract (pH 8 or 9) or 2% isoelectric casein elute viruses effectively from iron oxide. In studies on detecting viruses in drinking water, Rao and colleagues[13] used iron oxide both for adsorbing viruses from water and for reconcentrating viruses from beef extract that had been used to elute viruses from iron oxide. In this procedure, the pH of the eluate was adjusted to 3.0. $AlCl_3$ was added to a final molarity of 0.0005, 1 g of iron oxide was introduced into the eluate, and the mixture was stirred with a glass rod every 5 min. A magnet was placed below the beaker. The iron oxide settled, and the supernate was discarded. The viruses adsorbed on iron oxide were eluted with 4 to 5 mℓ of 3% beef extract (pH 9.0). Results from three sets of experiments with 100, 400, and 800 mℓ volumes of beef extract are shown in Table 8.

The effectiveness of iron oxide for reconcentrating viruses from primary eluates of Filterite® filters was also evaluated.[44] In these tests, 20 ℓ of treated sewage effluent and 100 ℓ of river, estuarine, and tapwater was acidified to pH 3.5, $AlCl_3$ (final concentration 0.0005 *M*) was added, and the acidified, salted mixtures were filtered through microporous filters. The filters were eluted with either 0.05 *M* glycine, 10% tryptose phosphate broth, or with 3% beef extract. The eluates were seeded with poliovirus 1 and concentrated to a final volume of 10 to 20 mℓ by an iron oxide adsorption and elution procedure (Figure 4). From 52 to 71% of the polioviruses was recovered from sewage effluent eluates, 48 to 66% from river and estuarine water eluates, and 71 to 85% from tapwater eluates. These results are summarized in Table 9.

Iron oxide adsorption-elution was also applied to reconcentration of rotavirus SA 11 from

Table 8
RECONCENTRATION OF POLIOVIRUS 1 FROM BEEF EXTRACT[a] WITH MAGNETIC IRON OXIDE[b]

Vol of beef extract (mℓ)	Quantity of iron oxide (g)	Vol. of reconcentrate (mℓ)	Virus input (PFU)	% Recovery	95% Confidence intervals
100	1	4	114	71	71 ± 6.2
			384	69	65 – 77
			88	74	
400	2	6	114	70	72 ± 16
			127	79	56 – 88
			384	66	
800	4	10	100	54	52 ± 6.2
			212	49	46 – 58
			320	52	

[a] Beef extract (Oxoid Co., London, U.K.)
[b] Iron oxide: prepared in the laboratory.

Adapted from Rao, V. C., Waghmare, S. V., and Lakhe, S. B., *Appl. Environ. Microbiol.*, 42, 421, 1981.

filter eluates of beef extract and tryptose phosphate broth (TPB).[45] In these tests, 100 mℓ volumes of the two eluates were seeded with the virus, the pH of each eluate was adjusted to 3.5, and the eluates were concentrated by an iron oxide procedure (Figure 4). The effectiveness of 3% beef extract (pH 9.5), 10% TPB (pH 8.5), and 2% casein (pH 8.5) for recovering viruses from iron oxide was evaluated. Data presented in Table 10 indicate a median recovery with casein of 88% (±14.2) compared with 66% (±2.8) for the beef extract which was second best.

Primary eluates derived from Filterite® filters through which large volumes of estuarine water are filtered contain organic and other components of water. To determine whether increasing concentrations of these components interfere in the reconcentration procedure with iron oxide, Rao and co-workers[45] seeded TPB-GLY eluates ranging in volume from 100 mℓ to 1600 mℓ with a rotavirus and determined virus recovery. Results in Table 11 suggest that recovery of viruses is favored when eluate volume is ≤800 mℓ, but the results do not indicate a significant reduction in recovery effectiveness with a 1600 mℓ volume.

The effectiveness of magnetic iron oxide for reconcentrating enterovirus 72 (hepatitis A virus) from primary eluates of Filterite® filters was also evaluated.[46] The pH of 10 ℓ of lake and tapwater was adjusted to 3.5, $AlCl_3$ (final concentration 0.5 m*M*) was added, the acidified salted waters were filtered through Filterite® filters, and the filters were eluted with 3% beef extract or 10% TPB; 100 mℓ volumes of eluate were seeded with hepatitis A virus, and the viruses in these eluates were concentrated to a final volume of 2 mℓ by the iron oxide procedure. Then, 2% casein (pH 8.5) recovered 51 to 68% of the seeded viruses from lake water and 65 to 85% from tapwater.

III. ROUND ROBIN INVESTIGATION OF METHODS

In 1972 Wallis, Melnick, and their colleagues introduced the virus adsorption-elution (Viradel) method for concentrating small numbers of viruses from water,[7] and the method has been widely used since. A modification of this procedure was introduced as a tentative standard method in the 14th edition of *Standard Methods for the Examination of Water and Wastewater.*[47] To establish the reliability of the Viradel method, the American Society for

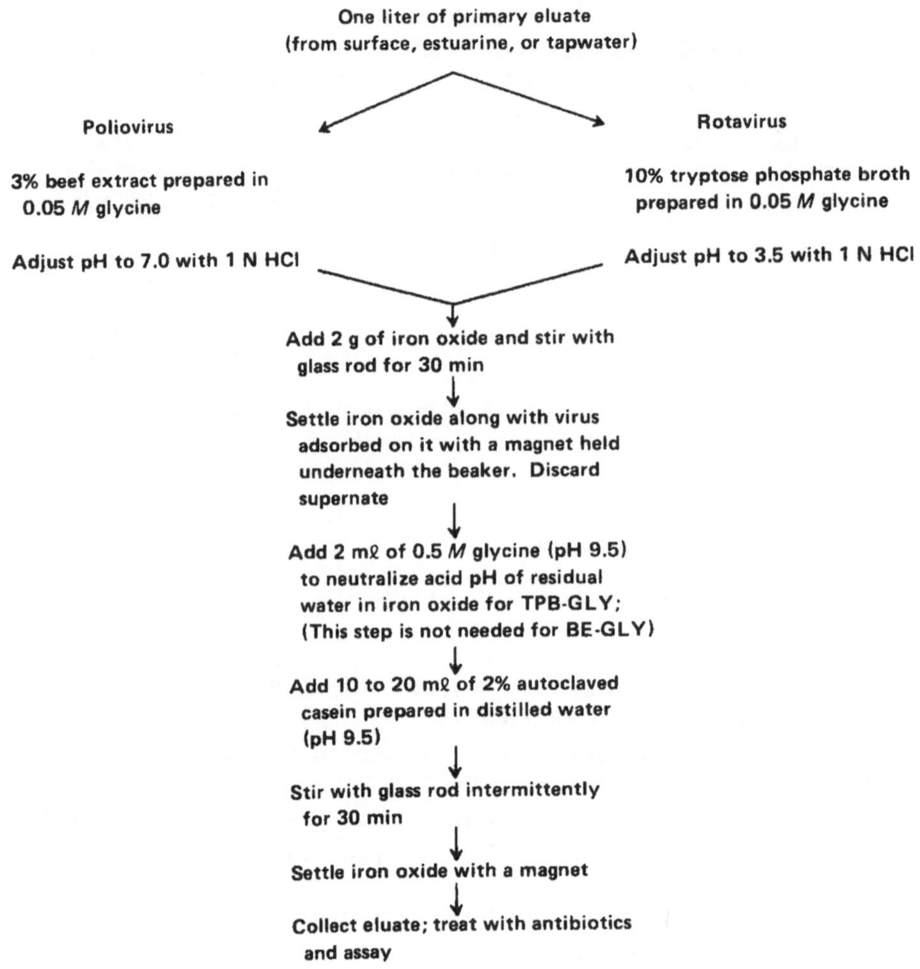

FIGURE 4. Reconcentration of poliovirus 1 and simian rotavirus SA11 from primary eluates of Filterite® filters with magnetic iron oxide procedure. (Data from Rao, V. C. and Melnick, J. L., Annu. Meet. American Society for Microbiology, Washington, D.C., 1983, QB.)

Testing Materials (ASTM) Committee on Viruses in the Aquatic Environment initiated round robin testing of the method. Six participating laboratories each processed 100 ℓ of dechlorinated tapwater seeded with small and large numbers of poliovirus 1 through cartridge-type Filterite® filters, eluted the viruses, and reconcentrated the viruses from the eluates by organic flocculation and by an aluminum hydroxide-hydroextraction procedure. Other methods for reconcentration were also tested. These included the magnetic iron oxide method and membrane disc filtration. Test results are given in Table 12.[48] Test results of the Viradel-organic flocculation method indicated that the average percentage of viruses recovered for low viral input (350 to 860 PFU) experiments was 66% with a range of 8 to 198%. For the high input experiments, the mean recovery was 26% with a range of 6 to 46%. For the Viradel-aluminum hydroxide-hydroextraction procedure, recovery of viruses ranged from 9 to 34% for low input and 3 to 18% for the high input experiments.

The iron oxide procedure was tested by a single laboratory. Results (not shown in Table 12) from three low input experiments gave 21, 45, and 64% recovery of viruses, an average of 43% of the input virus. In two experiments with large concentrations of virus input, recoveries ranged from 24 to 26%. The overall recovery for small and large input experiments was 36%.

Table 9
RECOVERY OF POLIOVIRUS 1 SEEDED INTO PRIMARY ELUATES BY MAGNETIC IRON OXIDE[a,b]

Primary eluate obtained from		Vol of sample (mℓ)	Virus input (PFU)	% Virus recovery[c]
Sewage	BE-GLY	100	10,250—18,000	71
	GLY	100	10,250—18,000	52
	TPB-GLY	100	10,250—18,000	77
Estuarine water	BE-GLY	1,000	1,800	77
	GLY	1,000		61
	TPB-GLY	1,000		48
Tapwater	BE-GLY	1,000	2,500	85
	TPB-GLY	1,000		72

Note: BE-GLY: 3% beef extract prepared in 0.05 *M* glycine; GLY: 0.05 *M* glycine; TPB-GLY: 10% tryptose phosphate broth prepared in 0.05 *M* glycine.

[a] Magnetic iron oxide used in these experiments was purchased from Fisher Scientific Co., Fair Lawn, N.J.

[b] Unpublished data from Rao et al.

[c] Viruses were recovered from iron oxide with 2% casein (pH 9.5).

Table 10
IRON OXIDE CONCENTRATION OF ROTAVIRUS SA11 FROM BE AND TPB ELUATES[a]

Virus suspending medium	Eluant	Virus input PFU	Virus recovery (%)
3% Beef extract, pH 3.5	I. 10% TPB-GL, pH 8.5	3660	34
	II. 3% BE-GL, pH 9.5	3660	64
	III. 2% casein, pH 8.5	3660	78
10% TPB-GLY	I. 10% TPB-GLY, pH 8.5	3660	86
	II. 3% BE-GL, pH 9.5	3660	68
	III. 2% casein, pH 8.5	3660	98

[a] 100 mℓ vol of 3% beef extract and 10% TBP-GLY primary eluates from Filterite® filters were seeded with 3660 PFU of SA11, adjusted to pH 3.5, and concentrated by the iron oxide techniques. Viruses adsorbed on iron oxide were eluted with 2 mℓ of the indicated eluents.

The recoveries of viruses by different methods reflect the overall efficiencies of the methods because viruses were introduced into the water samples. Although an evaluation of the efficiencies of the reconcentration procedures in these tests is not possible without data on the quantity of viruses initially present in the primary eluates, variations in the qualities of the waters tested (each laboratory used local water) and in the qualities of the eluting agents may have been responsible for the wide differences in the recoveries of viruses experienced by the different laboratories.

IV. SUMMARY AND CONCLUSIONS

Microporous filter adsorption-elution methods for concentrating human enteroviruses from large volumes of water are available. These methods seem to be applicable to the concentration of enterovirus 72 (hepatitis A virus) and rotaviruses from water and wastewater. Beef

Table 11
RECOVERY OF ROTAVIRUS SA11 FROM PRIMARY ELUATES VOLUMES VARYING FROM 100—1600 mℓ[a]

Primary eluate vol	Virus input PFU	Virus recovery (%) ±SD
100 mℓ	600—1800	59 ± 14.9
400 mℓ	600—1800	61 ± 23.8
800 mℓ	600—1800	56 ± 26
1600 mℓ	2600—5400	51.3 ± 20.4

[a] 100 gal of estuarine water, salinity 21‰, pH 8.3, were filtered through 25 cm tubular Filterite® filter assemblies of 3 μm and 0.45 μm porosity and eluated with 1600 mℓ of 10% TPB-GLY (pH 9.5). Different vol of primary eluate were seeded with 600—1800 and 2600—5400 PFU of SA11 and concentrated by the iron oxide method (0.2 g of iron oxide per 100 mℓ TPB-GLY). Viruses adsorbed on iron oxide were eluted with 4% casein (pH 8.5) (2 mℓ of casein was used for each 0.2 g iron oxide).

Table 12
AVERAGE RECOVERY OF POLIOVIRUS 1 SEEDED INTO 100 ℓ OF TAPWATER BY THE VIRADEL METHOD IN A ROUND ROBIN TEST

	Low-input experiments			High-input experiments		
Reconcentration method	% Recovery (av)	Range	Number of labs reporting	% Recovery (av)	Range	Number of labs reporting
Aluminum hydroxide-hydroextraction	9	9—11	2	3	3—5	2
	11			5		
	17	17—34	4	11	11—18	4
	22			12		
	26			14		
	34			18		
Organic flocculation	8	8—20	2	6	6—12	2
	20			12		
	49	49—63	3	26	26—46	4
	59			27		
	63			37		
				46		
	198		1			

Adapted from Melnick, J. L., et al., *Appl. Environ. Microbiol.*, 47, 148, 1984.

extract at slightly alkaline pH is preferred to glycine buffer at high pH (11.5) for eluting viruses from filters. Viruses can be reconconcentrated from primary eluates by organic flocculation and iron oxide adsorption-elution. Both are simple and inexpensive methods that produce small volumes of concentrates that are minimally toxic to cell cultures. Improvements are needed to increase the efficiency and applicability of these methods for different environmental samples. The development of a third step concentration method for further reducing sample size is needed for detecting hard-to-cultivate viruses by immunological methods.

REFERENCES

1. **Shuval, H. I., Fattal, B., Cymbalista, S., and Goldblum, N.,** The phase separation method for concentration and detection of viruses in water, *Water Res.*, 3, 225, 1969.
2. **Hill, W. E., Jr., Akin, E. W., Benton, W. H., and Metcalf, T. G.,** Viruses in water. II. Evaluation of membrane cartridge filters for recovering multiplicities of poliovirus from water, *Appl. Microbiol.*, 23, 880, 1972.
3. **Cliver, D. O.,** Detection of enteric viruses by concentration with polyethelene glycol, in *Transmission of Viruses by the Water Route*, Berg, G., Ed., John Wiley & Sons, New York, 1967, 109.
4. **Shuval, H. I., Cymbalista, S., Fattal, B., and Goldblum, N.,** Concentration of enteric viruses in water by hydro-extraction and two-phase separation, in *Transmission of Viruses by the Water Route*, Berg, G., Ed., John Wiley & Sons, New York, 1967, 45.
5. **Wellings, F. M., Lewis, A. L., Mountain, C. W., and Pierce, L. V.,** Demonstration of virus in ground water and effluent discharged onto soil, *Appl. Environ. Microbiol.*, 29, 751, 1975.
6. **Wallis, C., Henderson, M., and Melnick, J. L.,** Enterovirus concentration on cellulose membranes, *Appl. Microbiol.*, 23, 476, 1972.
7. **Wallis, C., Homma, A., and Melnick, J. L.,** Apparatus for concentrating viruses from large volumes, *J. Am. Water Works Assoc.*, 64, 189, 1972.
8. **Sobsey, M. D., Wallis, C., Henderson, M., and Melnick, J. L.,** Concentration of enteroviruses from large volumes of water, *Appl. Microbiol.*, 26, 529, 1973.
9. **Payment, P., Gerba, C. P., Wallis, C., and Melnick, J. L.,** Methods for concentrating viruses from large volumes of estuarine water on pleated membranes, *Water Res.*, 10, 893, 1976.
10. **Sobsey, M. D., Gerba, C. P., Wallis, C., and Melnick, J. L.,** Concentration of enteroviruses from large volumes of turbid estuary water, *Can. J. Microbiol.*, 23, 770, 1977.
11. **Farrah, S. R., Gerba, C. P., Wallis, C., and Melnick, J. L.,** Concentration of viruses from large volumes of tap water using pleated membrane filters, *Appl. Environ. Microbiol.*, 31, 221, 1976.
12. **Katzenelson, E., Fattal, B., and Hostovesky, T.,** Organic flocculation: an efficient second-step concentration method for the detection of viruses in tap water, *Appl. Environ. Microbiol.*, 32, 638, 1976.
13. **Rao, V. C., Waghmare, S. V., and Lakhe, S. B.,** Detection of viruses in drinking water by concentration on magnetic iron oxide, *Appl. Environ. Microbiol.*, 42, 421, 1981.
14. **Albertsson, P.,** Two phase separation of viruses, in *Methods in Virology*, Maramorosch, K. and Koprowski, H., Eds., Academic Press, New York, 1967, 303.
15. **Fattal, B., Katzenelson, E., and Shuval, H. I.,** Comparison of methods for isolation of viruses in water, in *Virus Survival in Water and Waste Water Systems*, Malina, J. F. and Sagik, B. P., Eds., The University of Texas, Austin, 1974, 19.
16. **Wellings, F. M., Lewis, A. L., and Mountain, C. W.,** Demonstration of solids-associated virus in waste water and sludge, *Appl. Environ. Microbiol.*, 31, 354, 1976.
17. **Farrah, S. R., Goyal, S. M., Gerba, C. P., Wallis, C., and Melnick, J. L.,** Concentration of enteroviruses from estuarine water, *Appl. Environ. Microbiol.*, 33, 1192, 1977.
18. **Ramia, S. and Sattar, S. A.,** Second-step concentration of viruses in drinking and surface waters using polyethylene glycol hydroextraction, *Can. J. Microbiol.*, 25, 587, 1979.
19. **Mix, T. W.,** The physical chemistry of membrane-virus interaction, *Dev. Ind. Microbiol.*, 15, 136, 1974.
20. **Farrah, S. R., Wallis, C., Schaffer, P. T. B., and Melnick, J. L.,** Reconcentration of poliovirus from sewage, *Appl. Environ. Microbiol.*, 32, 653, 1976.
21. **Gerba, C. P., Farrah, S. R., Goyal, S. M., Wallis, C., and Melnick, J. L.,** Concentration of enteroviruses from large volumes of tap water, treated sewage and seawater, *Appl. Environ. Microbiol.*, 35, 540, 1978.
22. **Farrah, S. R. and Bitton, G.,** Low molecular weight substitutes for beef extract as eluents for poliovirus adsorbed to membrane filters, *Can. J. Microbiol.*, 25, 1045, 1979.
23. **Metcalf, T. G., Wallis, C., and Melnick, J. L.,** Environmental factors influencing isolation of enteroviruses from polluted surface waters, *Appl. Microbiol.*, 27, 920, 1974.
24. **Homma, A., Sobsey, M. D., Wallis, C., and Melnick, J. L.,** Virus concentration from sewage, *Water Res.*, 7, 945, 1973.
25. **Jakubowski, W., Hoff, J. C., Anthony, N. C., and Hill, W. F., Jr.,** Epoxy-fiberglass adsorbent for concentrating viruses from large volumes of potable water, *Appl. Microbiol.*, 28, 501, 1974.
26. **Wallis, C. and Melnick, J. L.,** Concentration of viruses on aluminum and calcium salts, *Am. J. Epidemiol.*, 85, 459, 1967.
27. **Cookson, J. T.,** The chemistry of virus concentration by chemical methods, *Dev. Ind. Microbiol.*, 15, 160, 1974.
28. **Schäfer, E.,** Quantitative recovery of poliovirus 2 in surface waters, *Gas/Water Assoc. Sewage*, 112, 109, 1971.

29. **Berg, G., Dahling, D. R., and Berman, D.,** Recovery of small quantities of viruses from clean waters on cellulose nitrate membrane filters, *Appl. Microbiol.*, 22, 608, 1971.
30. **Rao, V. C., Chandorkar, U., Rao, N. U., Kumaran, P., and Lakhe, S. B.,** A simple method for concentrating and detecting viruses in waste water, *Water Res.*, 6, 1565, 1972.
31. **Jakubowski, W., Hill, W. F., and Clarke, N. A.,** Comparative study of four microporous filters for concentrating viruses from drinking water, *Appl. Microbiol.*, 30, 58, 1975.
32. **Bitton, G., Felberg, B. N., and Farrah, S. R.,** Concentration of enteroviruses from seawater and tap water by organic flocculation using non-fat dry milk and casein, *Water, Air, Soil Pollut.*, 12, 187, 1979.
33. **Rao, V. C., Seidel, K. M., Goyal, S. M., Metcalf, T. G., and Melnick, J. L.,** Isolation of enteroviruses from water, suspended solids, and sediments from Galveston Bay: survival of poliovirus and rotavirus adsorbed to sediments, *Appl. Environ. Microbiol.*, 48, 404, 1984.
34. **Landry, E. F., Vaughn, J. M., Thomas, M. J., and Vicale, T. J.,** Efficiency of beef extract for the recovery of poliovirus from waste water effluents, *Appl. Environ. Microbiol.*, 36, 544, 1978.
35. **Payment, P., Trudel, M., and Pavillanis, V.,** Efficacy of adsorption-elution technique on poliovirus type 1 using fiberglass filters: application to virological analysis of 100 mℓ to 1000 ℓ of water, *Can. J. Microbiol.*, 24, 1413, 1979.
36. **Glass, J. S., Van Sluis, R. J., and Yanko, W. A.,** Practical method for detecting poliovirus in anaerobic digestor sludge, *Appl. Environ. Microbiol.*, 35, 983, 1978.
37. **Goddard, M. R., Bates, J., and Butler, M.,** Recovery of indigenous enteroviruses from raw and digested sewage sludges, *Appl. Environ. Microbiol.*, 42, 1023, 1981.
38. **Waite, D. S. and Sobsey, M. D.,** Method for recovery of enteric viruses from estuarine sediments with chaotropic agents, *Appl. Environ. Microbiol.*, 46, 379, 1983.
39. **Lewis, D., Loutit, M. W., and Austin, F. J.,** A method for detecting enteroviruses in aquatic sediments, *J. Virol. Methods*, 10, 153, 1985.
40. **Goyal, S. M., Adams, W. N., O'Malley, M. L., and Lear, D. W.,** Human pathogenic viruses at sewage disposal sites in the Middle Atlantic Region, *Appl. Environ. Microbiol.*, 48, 758, 1984.
41. **Rao, V. C., Sullivan, R., Read, R. B., and Clarke, N. A.,** A simple method for concentrating and detecting viruses, *J. Am. Water Works Assoc.*, 60, 1288, 1968.
42. **Bitton, G., Poncorbo, O., and Gifford, G. E.,** Factors affecting the adsorption of poliovirus to magnetite in water and waste water, *Water Res.*, 10, 973, 1976.
43. **Atherton, J. G. and Bell, S. S.,** Adsorption of viruses on magnetic particles, *Water Res.*, 17, 943, 1983.
44. **Rao, V. C. and Melnick, J. L.,** Recovery of Poliovirus and Rotavirus from Sewage by Adsorption on Magnetic Iron Oxide, Annu. Meet. American Society for Microbiology, Washington, D.C., 1983, Q8.
45. **Rao, V. C., Metcalf, T. G., and Melnick, J. L.,** Recovery of rotaviruses from estuarine water on Galveston Bay, *Appl. Environ. Microbiol.*, 50, 1985, in press.
46. **Rao, V. C., Rao, T. V., Metcalf, T. G., Dahling, D. R., and Melnick, J. L.,** Reconcentration of hepatitis A virus and rotavirus from primary eluates of microporous filters, Water Quality Technology Conf., Am. Water Works Assoc., Dec. 1985.
47. Detection of viruses in water and waste water, *Standard Methods for the Examination of Water and Waste Water, Section 913*, 14th ed., American Public Health Association, Washington, D.C.
48. **Melnick, J. L., Safferman, R., Rao, V. C., Goyal, S., Berg, G., Dahling, D. R., Wright, B. A., Akin, E., Stetler, R., and Sorber, R.,** Round robin investigation of methods for the recovery of poliovirus from drinking water, *Appl. Environ. Microbiol.*, 47, 144, 1984.

Chapter 7

MECHANISM OF ADSORPTION AND ELUTION OF VIRUSES TO AND FROM SURFACES

Thomas W. Mix

TABLE OF CONTENTS

I. INTRODUCTION

Since the Second World War, much progress has been made in microporous membrane technology and in the application of that technology to biological analysis and purification. Significant progress has been made also in ultrafiltration applied to concentration, purification, and fractionation of colloids and of mixtures that contain dissolved macromolecular components. In this chapter, the structures and properties of membranes, flow management, and membrane surface adsorption characteristics, as these related to viruses, will be discussed.

II. MEMBRANE STRUCTURE

Three types of membranes used for reverse osmosis (RO), ultrafiltration, and microporous filtration and their characteristics are described in Table 1. Differentiation among the three on the basis of size-scale of particles separated is largely arbitrary. Reverse osmosis implies membrane separations of solutes, the molecular dimensions of which are within one order of magnitude of those for the solvent in which osmotic effects can be significant. Reverse osmosis membranes are dense, anisotropic, hydrophilic gel structures. Such membranes are thin skinned and 1 to 20 μm in thickness, and the skin is the actual filter. Solvent (water) and solutes are transported through this skin by molecular diffusion under reaction to a concentration or activity gradient. The flux of water through a reverse osmosis membrane increases as pressure increases. Solute flux through the membrane is almost unaffected by pressure. To compensate for the osmotic effect of high concentrations of solute, and to obtain effective water fluxes, high pressure drops across reverse osmosis membranes are necessary.

Ultrafilter membranes are anisotropic structures. A typical hollow fiber is shown in Figure 1*. These membranes are filters in a more traditional sense than reverse osmosis membranes. The skin, which is only 0.2 to 2.0 μm thick, is a rigid, highly voided structure that contains random interconnected pores ranging from about 10 Å to about 100 Å in diameter. Water flows through these pores viscously because of an hydraulic pressure gradient in contrast to the diffusive flow through the reverse osmosis gel. The structure of ultrafilters is highly porous. The structure of RO membranes is more dense. The membrane thickness of ultrafilters is about 75 to 200 μm. Dissolved solutes pass through the pores of ultrafilters along with solvent if the solute particles are smaller than the pores, and the particles are rejected by the skin if their dimensions are greater than the pores. Because the skin is thin and the pores in the support structure are large, particles are not trapped within the membrane. They either pass through the skin or they are rejected by it. Osmotic effects are small because of the much higher solute molecular weights. With both reverse osmosis and ultrafiltration, high flux rates are maintained with high shear flow to minimize concentration polarization. Thus, solutes rejected by the membrane surface are transported rapidly away from the membrane so that these rejected particles cannot clog the membrane. A high shear flow may be created by uniform laminar flow through a multiplicity of thin channels as, for example, through hollow fibers.

High shear flows for concentration polarization minimization are not as useful with the membrane (Millipore) type of microporous filter because these are depth filters (Figures 2 and 3). These membranes are highly voided (70 to 80% voids), reticulated structures uniform throughout. (Compare Figure 2, a scanning electron micrograph of the front surface, with Figure 3, a scanning electron micrograph of the rear surface of the same membrane.) Because of the large pore sizes and high void content in these membranes, water fluxes through them may be large. Figure 4 shows another type of microporous membrane filter which has a

* All figures appear following page 133.

Table 1
TYPES AND CHARACTERISTICS OF MEMBRANE FILTERS

	Microporous filtration	Reverse osmosis	Ultrafiltration
Solute size/pore size	250 Å (0.025 μm)/14 μm	5—10 Å	10 Å—100 Å
Membrane type	Hydrophilic and hydrophobic	Hydrophilic gel	Hydrophobic and hydrophilic
Flow	Viscous	Diffusive	Viscous
Structure	Homogeneous Porous membrane	Anisotropic Dense membrane	Anisotropic Porous membrane
Skin thickness (μm)	125—150 (Millipore) 5—10 (Nucleopore)	1—20	0.2—2.0
Membrane thickness (μm)	As above	75—200	75—200
Water flux (mℓ/cm min)	0.2—1000 @ 13.5 psi	0.03—0.06 @ 800—1200 psi	0.02—5.0 @ 55 psi
Configuration	Flat sheet Tube	Flat sheet Hollow fiber tube	Flat sheet Hollow fiber tube

uniform pore size obtained with ionizing radiation and chemical etching. These membranes are only 5 to 10 μm thick and have a high water flux despite their much smaller void content.

Membranes are used for concentrating and purifying viruses in two principal ways. First, in sequential adsorption and elution steps controlled by adjustment of the chemical environment and, second, in concentration by ultrafiltration through a membrane that rejects viruses without adsorbing them. Reverse osmosis membranes reject not only viruses but much smaller molecules as well, including all salts.[15,26,28] Microporous membranes are most useful in adsorption schemes for virus concentration.

III. ADSORPTION OF VIRUSES ONTO MEMBRANES

In determining whether viruses adsorb onto membranes, a number of factors must be considered. First is the nature of the membrane and virus surfaces. Viruses, whether they have a simple protein coat or an envelope containing lipids and carbohydrates, as well as protein, have regions which are nonpolar, polar, and ionic. An understanding of how these regions interact with similar regions on membrane surfaces aids in understanding adsorption.

A. Hydrophobic Bonding

Side groups of several of the amino acids in proteins are nonpolar, lack affinity for water, and are likely to serve as a source of attraction for other nonpolar groups in neighboring molecules. Kauzmann[14] called this tendency of nonpolar groups to aggregate hydrophobic bonding and pointed out that this cohesion between nonpolar side chains is an important factor in protein structure. The energy required to separate these nonpolar side chains is considerably greater than one would calculate from van der Waals attractions alone since it involves increasing the interfacial contact area between these side chains and water and, therefore, the disruption of hydrogen bonds in water molecules. The water molecules surrounding nonpolar solutes such as simple hydrocarbons in aqueous solution have been shown to be more ordered than in the bulk solution.[9] Aggregation of nonpolar solutes into micelles reduces the interfacial area of contact between solutes and water and therefore reduces the volume of ordered water molecules surrounding the solutes and increases the system entropy. The tendency for hydrophobic surfaces in an aqueous medium to cluster and bond together may therefore be said to be entropy driven because the higher entropy state has a lower free energy and is therefore thermodynamically favored.[17] Although most of the hydrophobic groups in a virion in aqueous medium are probably buried within the molecule, there are hydrophobic patches exposed to the water that are potential binding sites for hydrophobic adsorption of the virion to nonpolar regions in membrane surfaces.[6,14] Thus, even in the absence of a lipid-containing envelope, strong hydrophobic bonding can take place between virions and a hydrophobic membrane surface. Within such a hydrophobic bond, London dispersion forces cause like groups to cluster (i.e., aromatic groups and aliphatic groups will cluster). Several factors promote such hydrophobic bonding to membrane surfaces, including:

1. The presence of hydrophobic ligands as dangling chains attached to the membrane
2. The use of a high concentration of ions, high in the Hofmeister series, to quench the effect of electrostatic charges on the virus and membrane and eliminate Coulombic interaction
3. An increase in temperature that thermodynamically favors the higher entropy state
4. Adjustment of the pH to close to the isoelectric points for the virion and membrane
5. Similar isoelectric points for the virion and membrane

By altering the pH of an aqueous medium so that it is close to the isoelectric points of the virion and membrane, where these are similar, the electric charges surrounding the virion and membrane are neutralized. Because the charges are not spread uniformly over the virion and membrane surfaces but are concentrated where there are charged groups, there will still be, on a microscale, variations in charge over these surfaces, but these are balanced at the isoelectric point resulting in no net charge some distance from the surface. This condition of no net charge is highly conducive to both virion aggregation and adsorption. Several proteins have more than one isoelectric pH and, in many cases, the net charge of a protein will be small over a broad range of pH because of the balance of groups present.

Shaltiel et al.[23] indicated that the addition of long dangling alkyl chains to a surface to increase hydrophobic bonding can result in sufficiently strong bonds that elution is difficult to achieve. In order that desorption may efficiently be achieved, they recommended that hydrophobic adsorption should be carried out using as adsorbents gels or membrane materials with short alkyl chains attached and that a high concentration of salts high in the Hofmeister series should be employed during adsorption to compensate for the comparatively low intrinsic capacity of the adsorbent. Where virions are being recovered from large quantities of water, this procedure may be impractical.

Hydrophobic bonding of virions to surfaces appears to have been achieved in several instances. Satellite necrosis virus (STNV) was adsorbed on amphiphilic gels from a 4 *M* NaCl solution and desorbed by removing salt from the buffer consistent with hydrophobic bonding.[19] Viruses adsorb to activated carbon apparently by hydrophobic bonding also.[5] Early studies by Wallis and Melnick[31] of virus adsorption to cellulose nitrate membranes appear to have involved hydrophobic bonding in those instances where sodium chloride or other univalent cations were effective in generating adsorption. Such singly charged cations can adsorb to negatively charged virion and membrane surfaces, neutralizing the repellant effect that results when both surfaces are negatively charged. Cliver[2] has shown that fluorinated vinyl and acrylic polyvinyl chloride-type membrane filters (0.45 μm Millipore-type) adsorb enteroviruses. These also appear to be cases where hydrophobic bonding is taking place.

B. Polar and Electrostatic Bonding

Proteins, like many of the amino acids and polypeptides from which they are formed, may be highly polar, even more so than water. In a recent study polar and electrostatic forces were found to dominate in the synthesis of protein segments that elicited antibodies for protection against viruses, whereas peptides that were hydrophobic were ineffective.[22] This suggests that a virion in its normal conformation in aqueous solution predominantly exposes hydrophilic segments and that adsorption through polar, hydrogen-bonding, and electrostatic forces is simpler to achieve than hydrophobic bonding. Hydrogen bonding between peptide linkages in the viral protein is a key factor determining structure. Adsorption through hydrogen bonding between groups in the viral proteins and groups on the membrane surface can occur if the hydrogen bonds are sufficiently polar to displace the water-hydrogen bond. Electrolytes may play a significant role in viral adsorption.

Both cations and anions adsorb to protein and, on doing so, change the net electric charge of the virion and its isoelectric point. Anions are likely to adsorb below the isoelectric pH of the virion where it is positively charged and cations are likely to adsorb above the isoelectric pH of the virion where it is negatively charged. At neutral pH where both virion and membrane are likely to be negatively charged, multiple charged cations may adsorb to both surfaces generating a salt bridge between the virion and membrane surface. Virion adsorption due to the formation of such salt bridges is a principal mechanism for adsorbing viruses from water. Nucleophilic hydroxyl groups on cellulose ester membranes that are not completely esterified may provide appropriate cation adsorption sites[18] as can fiberglass and many other membrane materials.[35] From the work of Cliver,[2] Berg et al.,[1] Hill et al.,[12] Wallis and Melnick,[31] and Wallis et al.,[32-35] cation adsorption follows the Hofmeister series in which adsorption increases with decreases in the hydrated radii of the ions. Transition elements (manganese, iron, copper, zinc, and molybdenum) adsorb much more strongly to proteins than alkaline earth elements, but the bonding of the transitional elements can be covalent and irreversible.[7] At higher salt concentrations, cation adsorption can be sufficient to lead to reversal of charge on the virion and repulsion of viral particles from each other and from similarly charged surfaces. Still higher salt concentrations quench electrostatic charge effects as noted above under hydrophobic bonding. Sufficiently high salt concentrations will cause salting out of virions.

The charged nature of a virion at a pH removed from its isoelectric point suggests that virions should be adsorbable by an oppositely charged surface. Thus a virion should be adsorbable by a cationic exchange surface at a pH below its isoelectric point and by an anionic exchange surface at a pH above its isoelectric point. Sobsey et al.,[24,25] Farrah et al.,[7] and Hill et al.[13] found virions to be adsorbed by membrane surfaces at a pH below the isoelectric point for the membrane material but above the isoelectric point for the virions, resulting in an electrostatic attraction between the membrane surface and the virion. Such adsorption has been obtained with a wide variety of membrane materials including epoxy-fiberglass, epoxy-fiberglass-asbestos, acrylonitrile-polyvinyl chloride copolymer, and cellulose esters.[35]

Polymers and membranes may be classified as acidic or basic depending on their isoelectric points. If the isoelectric point is below seven, they will be negatively charged at a neutral pH and may be considered basic. If the isoelectric point is above seven, they will be positively charged at a neutral pH and may be considered acidic.

Fowkes et al.[8] noted that an indication of the nature of the charge on a specific polymer can be derived from the hard and soft acid and base concepts of Pearson[20] and the related work of Drago et al.[3,4]

Where interfering organic and proteins molecules are not present, virions can be adsorbed directly on anion exchange resins above the isoelectric pH of the virus.[5,24]

Virions also adsorb on other basic materials at neutral pHs including ferric hydroxide and aluminum hydroxide flocs or gels,[35] aluminum alginate membranes soluble in sodium citrate,[10] and calcium phosphate.[21] All these adsorptions are based on electrostatic attraction.

C. Competitive Adsorption

Organic molecules and proteins, if present in an aqueous medium with virions, are also apt to be negatively charged at neutral pH and will have a lower molecular weight and a higher diffusivity than the virions in most instances. Such molecules will, as a result, be preferentially adsorbed and will interfere with virion adsorption by occupying the available adsorption sites. These interfering adsorbents can be removed by pretreatment of the aqueous medium with an anion exchange resin which preferentially adsorbs these materials. Operating conditions for this pretreatment must be selected with care because once the interfering materials have been removed, virions will adsorb on the anion exchange resin if there is excess capacity or if the contact times are too long. Wallis and co-workers,[35] however, have effected this separation routinely, suggesting that the interfering materials have significantly higher diffusivities and/or charges than the virions.

If desired, virion adsorption may be inhibited by precoating the membrane surface with competitive adsorbents such as serum or whey proteins or polyvinyl pyrrolidone.[11,12,36] These high molecular weight adsorbents form many bonds with the membrane surface, as do virions, making it very difficult for a virion to displace them even if the virion is more strongly adsorbed.

D. Effect of Operating Conditions on Adsorption

Increase in temperature, in general, increases adsorption suggesting, as Zittle[37] has indicated, a degree of unfolding of the protein on the surface of the adsorbent because this type of reaction is endothermic. Scutt[22] has shown that virion adsorption decreases with increase in flow rate through a filter. This is to be expected for laminar flow, which probably predominates within a microporous filter. Virion adsorption decreases with decrease in the membrane area through which a solution is filtered because there are a finite number of adsorption sites per unit membrane area. In general, factors that favor virion solvation and dispersion oppose adsorption, and factors that favor virion aggregation favor adsorption. van Oss et al.[29] noted that the chemical environment may influence whether virions have appendages which protrude and aid adsorption.

E. Elution

Factors which favor elution of viruses adsorbed to membranes include:

1. Raising (or reducing) the pH to a point well above or below the isoelectric points for both the adsorbing surface and the virus so as to induce repulsive electrostatic forces between the virion and the membrane
2. Use of a buffering agent to maintain the pH and to act as a source of ions for displacement of the virions
3. Reducing the concentration of salts high in the Hofmeister series to enable a repulsive electrostatic effect to be generated by pH change
4. Addition of chaotropic ions (e.g., CCl_3COO^-, SCN^-) in a concentration sufficient to counteract association, folding, and coiling of virions in water by disrupting hydrogen bonds and thereby promote solvation;[16] the concentration must not be high enough to quench electrostatic effects which aid elution
5. Addition of a chelating agent (e.g., EDTA and its salts or citric acid) to chelate metal ions that have formed or might form salt bridges between the virions and the membrane surface
6. Addition of a polarity-reducing solvent (e.g., isopropanol or ethylene glycol) to adjust the liquid surface tension value of the aqueous medium so it is intermediate between those for the membrane and the virion.[30]

Wallis et al.,[35] in summarizing the literature on concentration of viruses from water by membrane chromatography, indicated that elution with glycine (typically 0.05 *M*) at high pH (11.5) was frequently reported an effective means of desorbing viruses. The high pH induces strong negative charges in both the membrane surface and the virion and thereby electrostatic repulsion between the two. The glycine acts as a buffer and provides a source of anions to displace the virions. A wide range of materials including amino acids, proteins, and surfactants are effective buffers and elutants, but the pH for elution must not be so high (or low) that it will destabilize the virions. A pH close to this destabilization pH minimizes the volume of elutant required to effect desorption. The displacing anion should have a high diffusivity to give a sharp elution front. Although exposure to high pH levels is now known to be destructive to many enteric viruses, the high pH of the glycine buffered solution and the low molecular weight and high diffusivity of glycine and its high charge density make it an effective elutant.

F. Virus Recovery Practices

Membrane filter adsorption/elution has had the greatest use thus far in detecting viruses in water because it is adaptable to recovering viruses from large quantities of drinking water, turbid waters, or waters containing large quantities of organics, and can be used in the field. For concentrating viruses from tapwater, a number of systems that are variants of a virus concentration technique developed by Wallis et al.[33] enable efficient concentration of viruses. These in general use electrostatic forces through addition of multivalent cations (e.g., Mg^{2+}, Al^{3+}) to adsorb viruses via salt bridges to membranes or fiberglass filters or via flocculation on multivalent cation hydroxides followed by membrane filtration. Elution may be effected with glycine elution at high pH or by other means. For recovering viruses from seawater, the flocculation technique is particularly effective.

IV. SUMMARY AND CONCLUSIONS

Virus adsorption on membranes is most easily accomplished via electrostatic and polar forces, but can also result from hydrophobic bonding. Competitive adsorbents can often be removed in a pretreatment. Elution requires a significant change in the chemical environment. Equipment and procedures for recovering viruses from a broad range of waters have been developed.

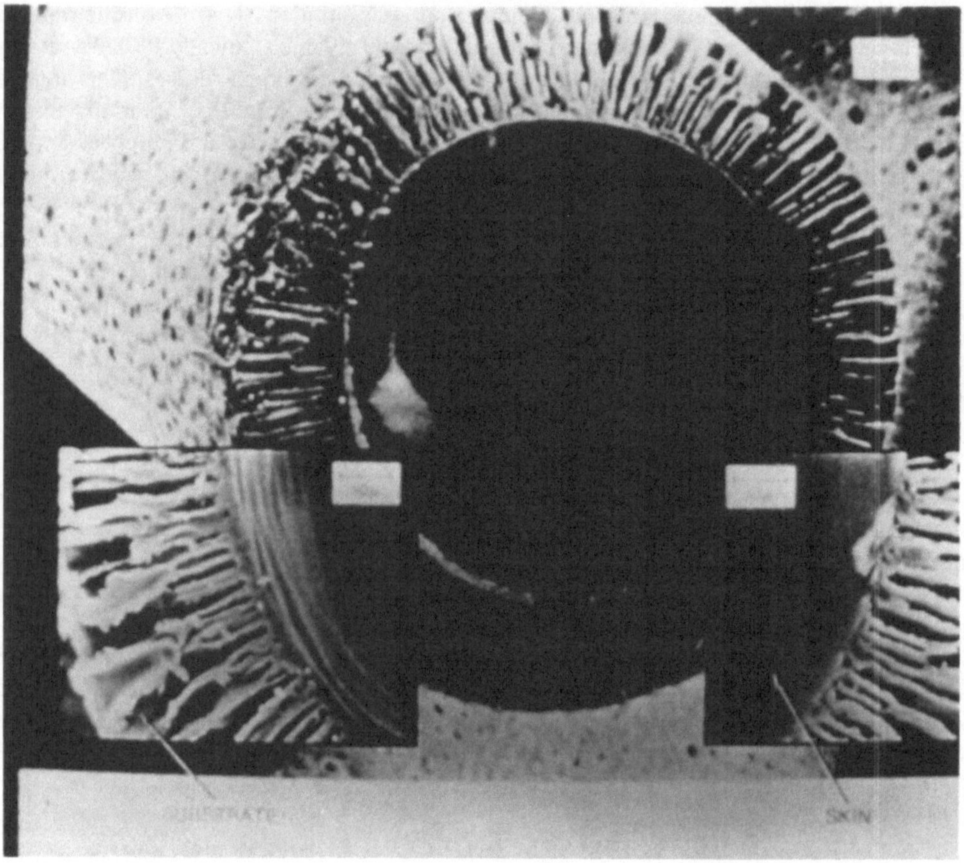

FIGURE 1. Scanning electron micrograph of ultrafiltration-type hollow fiber filter (Amicon-type PM-10).

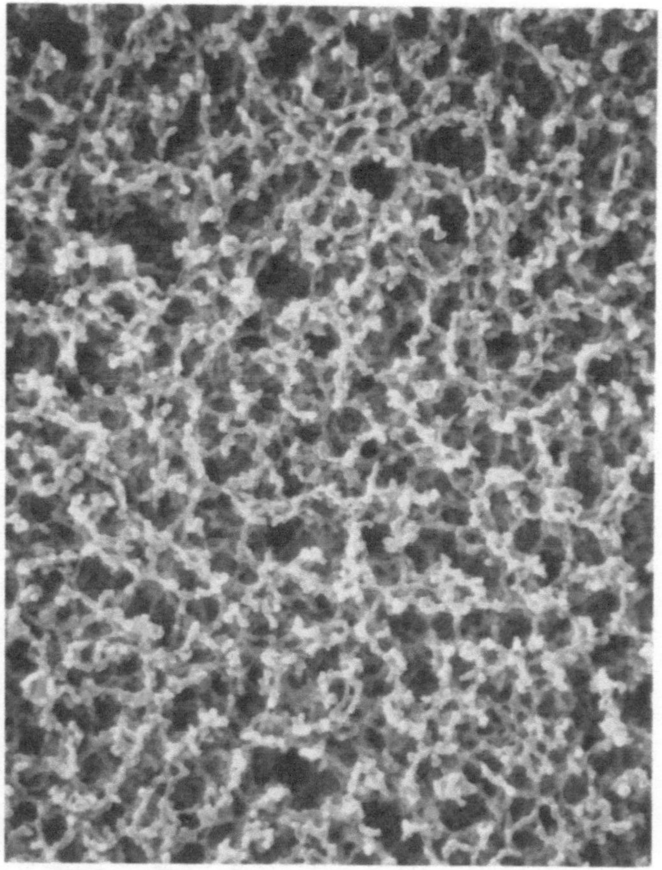

FIGURE 2. Scanning electron micrograph of a membrane-type filter (0.22 μm Millipore filter; magnification approximately × 5000.)

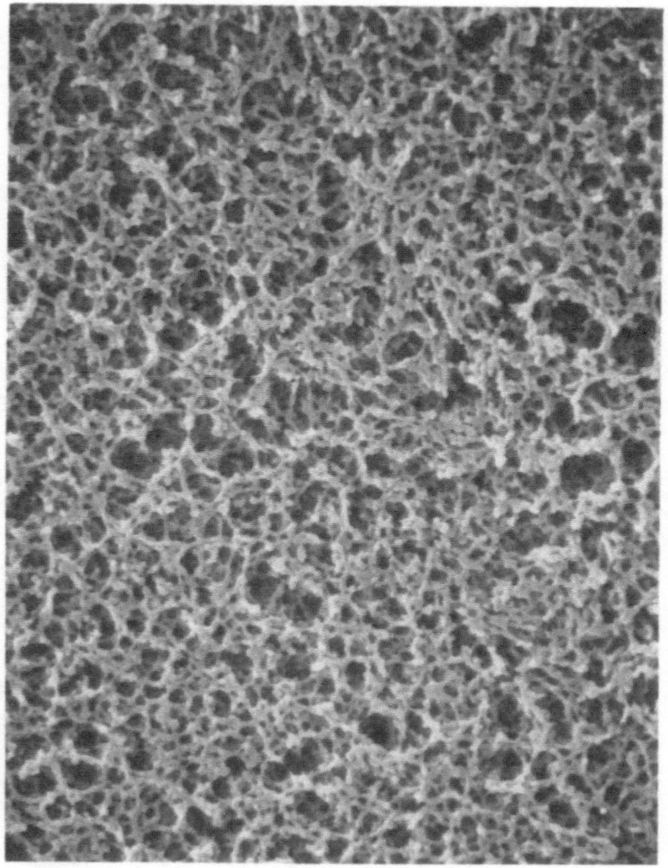

FIGURE 3. Rear view of same membrane shown in Figure 2. (Magnification unchanged.)

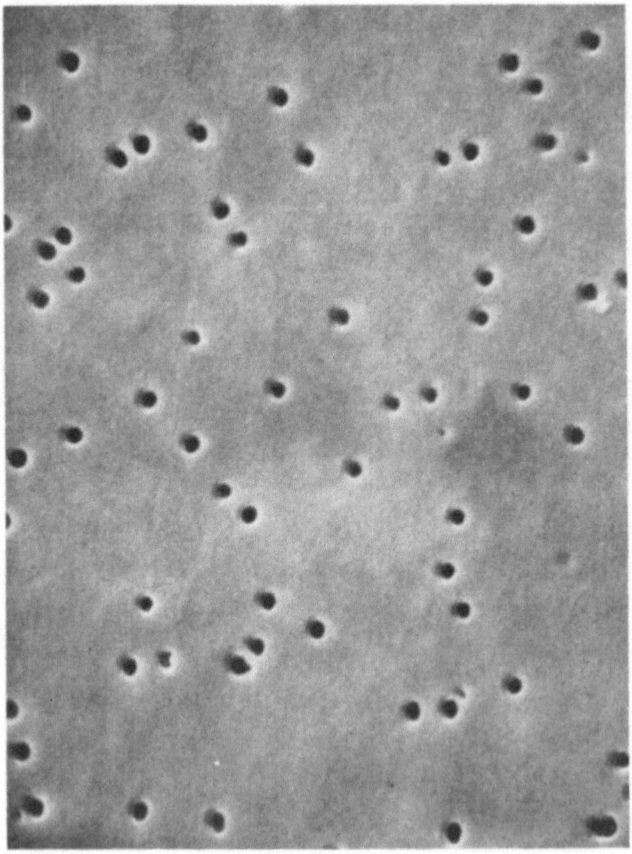

FIGURE 4. Scanning electron micrograph of a uniformly porous membrane filter made with ionizing radiation and chemical etching (Nucleopore-type, 0.4 μm pore size.)

REFERENCES

1. **Berg, G., Dahling, D. R., and Berman, D.,** Recovery of small quantities of viruses from clean waters on cellulose nitrate membrane filters, *Appl. Microbiol.*, 23, 608, 1971.
2. **Cliver, D. O.,** Virus interaction with membrane filters, *Biotechnol. Bioeng.*, 10, 877, 1968.
3. **Drago, R. S., Parr, L. B., and Chamberlain, C. S.,** Solvent effects and their relationship to the E and C equation, *J.A. C.S.*, 93, 3203, 1977.
4. **Drago, R. S., Vogel, G. C., and Neeham, T. E.,** A four-parameter equation for predicting enthalpies of adduct formation, *J.A. C.S.*, 93, 6014, 1971.
5. **Drewry, W. A.,** Virus removal, water and wastewater, National Technical Information Service, PB-210857, 1972.
6. **Edsall, J. T. and Wyman, J.,** *Biophysical Chemistry,* Academic Press, New York, 1958.
7. **Farrah, S. R., Gerba, C. P., Wallis, C., and Melnick, J. L.,** *Appl. Environ. Microbiol.*, 31, 221, 1976.
8. **Fowkes, R. M., Tischler, D. O., Wolfe, J. A., and Halliwell, M. J.,** Acidbase complexes of polymers with solvents, *Org. Coat. Appl. Polym. Sci. Proc.*, 46, 1, 1981.
9. **Franks, F., Ed.,** *Water: A Comprehensive Treatise,* Vol. 6, Plenum Press, New York, 1979.

10. **Gartner, H.**, Retention and recovery of polioviruses on a soluble ultrafilter, in, *Transmission of Viruses by the Water Route*, Berg, G., Ed., Interscience, New York.
11. **Hahn, R. J., Hatlen, J. B., and Kenny, G. E.**, Comparative poliovirus permeability of silver, polycarbonate, and cellulose membrane filters, *Appl. Microbiol.*, 19, 317, 1970.
12. **Hill, W. F., Jr., Akin, E. W., Benton, W. H., and Metcalf, T. G.**, Virus in water. II. Evaluation of membrane cartridge filters for recovering multiplicities of poliovirus from water, *Appl. Microbiol.*, 23, 880, 1972.
13. **Hill, W. F., Jr., Jakubowski, W., Akin, E. W., and Clarke, N. A.**, *Appl. Environ. Microbiol.*, 31, 254, 1976.
14. **Kauzmann, W.**, Some factors in the interpretation of protein denaturation, *Adv. Protein Chem.*, 14, 1, 1959.
15. **Klein, E., Smith, J. K., Morton, F., and Sweet, B. H.**, Concentration of virus from water by osmotic ultrafiltration. II. Mass transport aspects, *Water Res.*, 5, 1067, 1971.
16. **Kristiansen, T.**, Matrix-bound antigens and antibodies, in *Affinity Chromatography*, Proc. Int. Symp., Hoffman-Ostenhof, O. et al. Eds., Pergamon Press, Oxford, 1978.
17. **Lyklema, J.**, Molecular interactions in affinity chromatography, in *Affinity Chromatography and Related Techniques*, Gribnau, T. C. J., Visser, J., and Nivard, R. J. F., Eds., Elsevier, New York, 1982.
18. **Ott, E., Spurlin, H. M., and Grafflin, M. V.**, *Cellulose and Cellulose Derivatives, Part I*, Interscience, New York, 1963.
19. **Pahlman, S.**, Adsorption of proteins at high salt concentrations on hydrophobically interacting matrices, in *Affinity Chromatography*, Hoffman-Ostenhof, O. et al., Pergamon Press, Oxford, 1978.
20. **Pearson, R. G., Ed.**, *Hard and Soft Acids and Bases*, Dowden, Hutchinson & Ross, Stroudsburg, Pa., 1973.
21. **Philipson, L.**, Chromatography and membrane separation, in *Methods in Virology, Part II*, Maramorosch, and Koprowski, Eds., Academic Press, New York, 1967, 179.
22. **Scutt, J. E.**, Virus retention by membrane filters, *Water Res.*, 5, 183, 1971.
23. **Shaltiel, S., Halperin, G., Er-el, Z., Tauber-Finkelstein, M., and Amsterdam, A.**, Homologous series of hydrocarbon-coated agarose in hydrophobic chromatography, in *Affinity Chromatography and Related Techniques*, Gribnau, T. C. J., Visser, J., and Nivard, R. J. F., Eds., Elsevier, New York, 1982.
24. **Sobsey, M. D. and Jones, B. L.**, Concentration of poliovirus from top water using positively charged microporous filters, *Appl. Environ. Microbiol.*, 37, 588, 1979.
25. **Sobsey, M. D., Wallis, C., Henderson, M., and Melnick, J. L.**, *Appl. Microbiol.*, 26, 529, 1973.
26. **Sorber, C. A., Malina, J. F., Jr., and Sagik, B. P.**, Virus rejection by the reverse osmosis-ultrafiltration process, *Water Res.*, 6, 1377, 1972.
27. **Sutcliffe, J. G., Scinnick, T. M., Green, N., and Lerner, R. A.**, Antibodies that react with predetermined sites on proteins, *Science*, 219, 660, 1983.
28. **Sweet, B. H., McHale, J. S., Hardy, K. J., Morton, F., Smith, J. K., and Klein, E.**, Concentration of virus from water by osmotic ultrafiltration. I. Biological aspects, *Water Res.*, 5, 823, 1971.
29. **van Oss, C. J., Gillman, C. F., and Good, R. J.**, The influence of the shape of phagocytes on their adhesiveness, *Immunol. Commun.*, 1(6), 627, 1972.
30. **van Oss, C. J., Absolom, D. R., and Neumann, A. W.**, Role of attractive and repulsive van der Waals forces in affinity and hydrophobic chromatography, in *Affinity Chromatography and Related Techniques*, Gribnau, T. C. J., Visser, J., and Nivard, R. J. F., Eds., Elsevier, Amsterdam, 1982.
31. **Wallis, C. and Melnick, J. L.**, Concentration of enteroviruses on membrane filters, *Virology*, 1, 472, 1967.
32. **Wallis, C., Henderson, M., and Melnick, J. K.**, Enterovirus concentration on cellulose membranes, *Appl. Microbiol.*, 23, 476, 1972.
33. **Wallis, C., Homma, A., and Melnick, J. L.**, Apparatus for concentrating viruses from large volumes, *J. Am. Water Works Assoc.*, 64, 189, 1972.
34. **Wallis, C. et al.**, A portable virus concentrator for testing water in the field, *Water Res.*, 6, 1249, 1972.
35. **Wallis, C., Melnick, J. L., and Gerba, C. P.**, Concentration of viruses from water by membrane chromatography, *Annu. Rev. Microbiol.*, 33, 413, 1979.
36. **Yoshino, K. and Morishima, T.**, Filtration of viruses through Millipore membranes coated with polyvinylpyrrolidone, *Arch. Gesamte Virusforsch.*, 35, 399, 1971.
37. **Zittle, C. A.**, Adsorption studies of enzymes and other proteins, *Adv. Enzymol.*, 44, 319, 1953.

Chapter 8

MECHANISMS OF ADSORPTION AND ELUTION OF VIRUSES TO SOLIDS IN THE NATURAL ENVIRONMENT

G. Wolfgang Fuhs

TABLE OF CONTENTS

I. INTRODUCTION

Adsorption of animal viruses to natural surfaces is a common phenomenon. In fact, most viruses in the environment exist in the adsorbed state. Invasion of a host by viruses is initiated by its adsorption to the host's mucous membranes. After release from the host's cells and excretion into the environment, most virus particles (virions) are still in aggregates or associated with cell debris. When the cell debris decays, the fate of each virion is determined by environmental factors, among them are those which determine whether the virion will exist in monodisperse form or as part of an aggregate, whether it will exist free or adsorbed to other solids, and whether it will retain its integrity so as to remain infective. Recovery procedures for viruses must therefore be concerned with freeing virions from solid matter that interferes with successful recovery and growth in cell cultures. If viruses are to be quantified, the dispersion of virus aggregates is also desirable. If the relative strength of some of the adsorptive forces and the physical lability of virus particles are considered, it becomes clear that the recovery and accurate quantification of viruses from environmental samples can be difficult.

Adsorbing surfaces of practical importance include organic materials, minerals, and soils. Organic materials include cell debris, residues of vegetation, organic coatings, and surface films, sewage bacteria, protozoa, and viruses. Soils are mixtures of minerals and organic matter. Soils occur not only *in situ* but also as suspended matter in runoff, leachate, and wastewater and as sludge, filter material, and coagulant floc in water and wastewater treatment.

Adsorption to natural materials has been a major topic, sometimes implicitly, of several reviews of the literature on the fate and persistence of viruses (e.g., Drewry and Eliassen,[1] Bitton,[2] Gerba et al.,[3] DuBoise,[4] and Fuhs and Taylor[5]). Theoretical aspects of the adsorption process are dealt with in the literature but rarely with specific application to viruses. One such significant contribution is the work by Murray and Parks.[6] This chapter summarizes the theory in this field and more recent experimental findings.

II. THEORETICAL AND GENERAL ASPECTS

The major forces and properties that determine adsorption of viruses to natural surfaces and other colloids are well known. The practical application of this knowledge, however, meets with difficulties because many natural surfaces are chemically complex and, as Renn[7] has noted, nearly always carry adsorbed organic materials. The constituents of the virus surface are also known in principle, but environmental factors affect the conformation of the coat protein molecules in ways that are not well understood; viruses also carry other adsorbed chemical species. The major forces at work are London-van der Waals and coulombic (electrostatic double-layer) forces (DLVO theory, after Derjaguin and Landau[8] and Verwey and Overbeek[9]). If we also consider gravitational forces, the total potential energy of interaction over a distance x between the surfaces is the sum of the three:

$$\phi_t(x) = \phi_{vdw}(x) + \phi_{dl}(x) + \phi_g(x) \tag{1}$$

The gravitational component is

$$\phi_g(x) = 4/3[\pi a^3(\rho_p - \rho_f)\, g\, x] \tag{2}$$

where ρ_p and ρ_f are the densities of the particle and fluid, respectively, g is the gravitational acceleration, and a the particle radius.

A. London-van der Waals Forces

Because of the unequal radii of the virus particles and most environmental particulates, we consider here the interaction between a semi-infinite plate and a sphere of radius a whose center is separated from the surface of the plate by a distance x + a. The contribution from unretarded van der Waals forces to the total potential energy of interaction is[10]

$$\phi_{vdw}(x) = A_{123}\left(\frac{1}{6}\ln\frac{x + 2a}{x} - \frac{1}{3}\frac{a}{x}\frac{x + a}{x + 2a}\right) \quad (3)$$

A_{123} is the combined Hamaker coefficient, which is a characteristic of the two bodies (subscripts 1 and 2) and the intervening medium (subscript 3). Within certain limitations, particularly with very polar surfaces (see Prieve and Ruckenstein[10] for details and references), the combined Hamaker coefficient can be estimated from the individual coefficient A_{11}, A_{22}, and A_{33} as

$$A_{123} = 1.6\,(A_{12} + A_{33} - A_{13} - A_{23}) \quad (4)$$

where the constant, 1.6, is characteristic for interactions in water and, generally, $A_{ij} = A_{ii}^{1/2} \times A_{jj}^{1/2}$.[10] The Hamaker coefficients are related to the densities of the bodies, n_1 and n_2, in molecules per unit volume, by

$$A_{12} = \pi^2 n_1 n_2 B_{12} \quad (5)$$

B_{12}, the London constant, can be obtained from optical dispersion data for the two phases 1 and 2 (i.e., the variation of index of refraction with frequency) or from measurements of dielectric constants (Lifshitz method).[11] As it is related to the polarizability of the atoms in each phase, it is related to the size of the atoms and therefore again to their mass. Individual Hamaker coefficients are small for water, for highly hydrated colloids, and for hydrocarbons; they are large for metal oxides. For water, A_{33} was calculated as 5.0×10^{-13} erg, and for leukocytes, $A_{11} = 6.5 \times 10^{-13}$ erg.[10] The existing uncertainties are evident from the lower value of 6×10^{-14} erg given by Weiss et al.[12] for the hypothetical case of interaction of a protein sphere with a protein slab. Equation 3 shows that van der Waals forces, which are usually, but not necessarily, attractive forces, diminish rapidly with increasing distance of the bodies.

B. Coulombic Forces — The Electrochemical Double Layer

Coulombic forces result from charge development on the surfaces involved. In minerals, replacement of ions by others of similar radius but different valency (isomorphic replacement) accounts for pH-independent charge development. Such replacement, e.g., of Si^{4+} by Al^{3+} or of Al^{3+} by Mg^{2+}, is common in aluminosilicate clay minerals and conveys a structural negative charge to the surface of these minerals. Other structural charges occur, e.g., at the edges of clay crystals, from exposed Al^{3+} ions, and other irregularities. Materials that derive their charges from these structural features are known as constant-charge materials.

Oxides and hydroxides dissociate according to

$$R\text{–}OH \rightleftharpoons R\text{–}O^- + H^+$$

$$R\text{–}OH_2^+ \rightleftharpoons R\text{–}OH + H^+$$

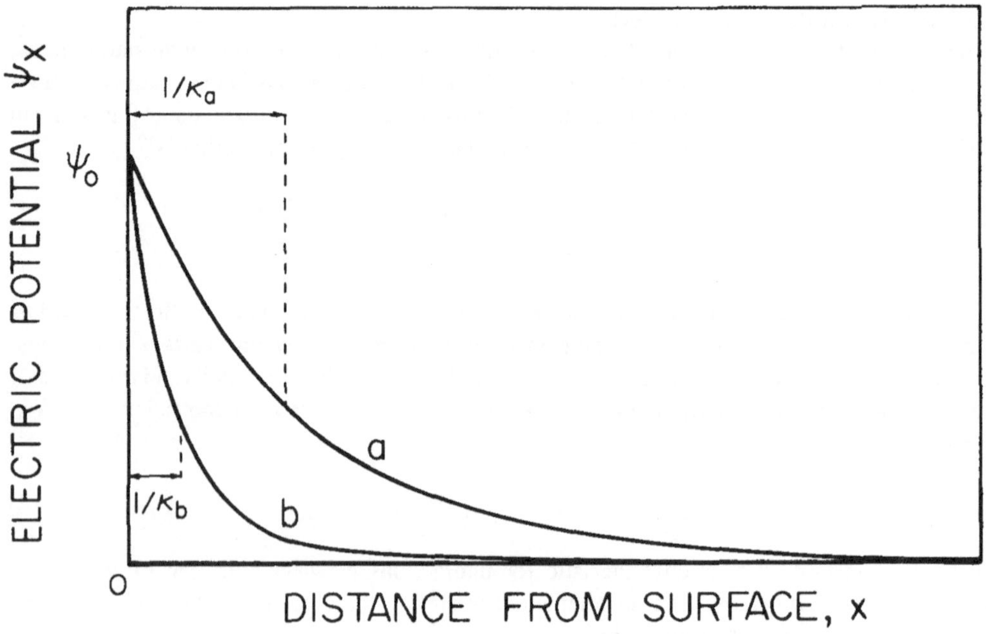

FIGURE 1. Variation of the potential with distance in the diffuse layer at the charged surface/aqueous interface. (a) At low electrolyte concentration; (b) at high electrolyte concentration. $1/\kappa$ is the double-layer thickness. (From Fuhs, G. W. and Taylor, D. H., *Water Reuse,* Middlebrooks, E. J., Ed., Ann Arbor Science Publishers, Ann Arbor, Mich., 1982, 603. With permission.)

where R represents the solid. The surface components of organic colloids, including viruses, dissociate primarily according to

$$R\text{–}COOH \leftrightarrows R\text{–}COO^- + H^+$$

$$R\text{–}NH_3^+ \leftrightarrows R\text{–}NH_2 + H^+$$

Moreover, phenolic hydroxyl groups dissociate to form weak acids.

These charges are pH dependent. Positive charges develop at low, negative charges at high pH. The pH at which there is no net charge is termed the point of zero charge or PZC. The surface charge is proportional to the difference between the surface densities of cationic and anionic sites. Materials of this type are known as constant-potential materials.

In electrolyte solutions, the surface charges of both kinds of materials cause a redistribution of electrolyte ions in the layer immediately adjacent to the solid. Within this layer, the electrochemical double-layer, oppositely charged ions (counterions) are attracted to the surface, while ions of the same charge (coions) are repelled. The potential in this region decays exponentially with the distance x from the solid (Figure 1):

$$\psi_x = \psi_\delta \exp(-\kappa x) \tag{6}$$

where ψ_δ is the potential at the distance of closest approach of the ions. The rate of decay is controlled by the Debye parameter κ which has the dimension of a reciprocal distance:

$$\kappa = [(8\pi z^2 e^2 c)/(\epsilon k\ T)]^{1/2} \tag{7}$$

Here z is the electron valence; e, the electron charge; c, the ionic concentration in the surrounding solution; k, the Boltzmann constant; and T, the absolute temperature. It can be seen from Equation 7 in combination with Equation 6 that the potential decays more rapidly with distance from the solid when the electrolyte concentration is increased and also when monovalent ions are replaced by divalent and trivalent ions. This relationship is shown in Figure 1.

As certain ions may become more firmly associated with the surface, e.g., by hydrogen or covalent bonding, an additional layer, the Stern layer, may form close to the surface and may modify, even invert, the surface charge. This in turn will affect to some extent the charge distribution in the electrochemical double-layer. As a result, the pH of effective electroneutrality, the isoelectric point (IEP), measured at the shear surface between the solid and the surrounding liquid, may be different from the PZC.

Upon approach of two particles of like charge, the double layers interpenetrate. In the case of constant-potential materials the surface potential is determined by the concentration of potential-determining ions according to Nernst's law. As a result of double-layer interaction the surface charge should decrease.[13] By contrast, during the approach of constant-charge materials the surface charge remains constant, and as a result the energy of interaction increases. Kinetic factors are at work, altering the energy of interaction upon approach as various ions are exchanged by the interactions of the double-layers, particularly in the case of constant-potential materials. The relationship for the double-layer forces between a sphere and a plate is

$$\phi_{dl}(x) = \pm (\epsilon a/4)[(\psi_1 + \psi_2)^2 \ln(1 \pm e^{-\kappa x}) + (\psi_1 - \psi_2)^2 \ln(1 \mp e^{-\kappa x})] \quad (8)$$

where ϵ is the dielectric constant of the medium and ψ_1 and ψ_2 the surface potentials of the isolated bodies. The upper sign in the equation refers to approach at constant surface potential, the lower to approach at constant surface charge.[10]

The sum of the potential energies from both sources, $\phi_t(x)$ forms the energy profile shown in Figure 2. In the figure, the van der Waals forces are considered attractive, consistent with the experience that Hamaker coefficients for the surface and the virus (n_{11}, n_{22}) are both greater than that for water. In close proximity of the bodies, the sum of the components strongly favors adsorption. This region is called the primary minimum of the potential surface energies. At a distance of a few nanometers, there may exist an energy maximum that acts as a barrier to adsorption. It may be followed at a distance of approximately 10 nm by a weaker secondary minimum (Figure 2). Increasing the Hamaker coefficient decreases the maximum and increases the depth of the secondary energy minimum.[14] Reducing the surface potential, e.g., by compressing the electrochemical double-layer as it occurs at increased ionic strengths, also decreases the maximum but does not greatly affect the secondary minimum. Double-layer compression allows the surfaces to approach so closely that repulsion is overcome by van der Waals attraction, which increases sharply at small separations. The energy maximum also decreases very markedly and the secondary minimum weakens as the diameter of the sphere decreases.[14]

Particles the size of bacteria often cannot penetrate the energy maximum and are held in the secondary minimum in a loosely adsorbed state.[16] Bacteria may establish first contact with the surface by appendages of small diameter (pili, fimbriae) that can penetrate the energy barrier, and ultimately become attached by the secretion of exopolymer. For viruses, the energy of Brownian motion is the same order of magnitude as the energy maximum. As a result, the viruses, and particularly the small viruses, are much more likely bound in the primary minimum, where attraction is strong.

C. The Effect of Coatings

Most natural surfaces and interfaces carry adsorbed microlayers of biologic origin.[7] They

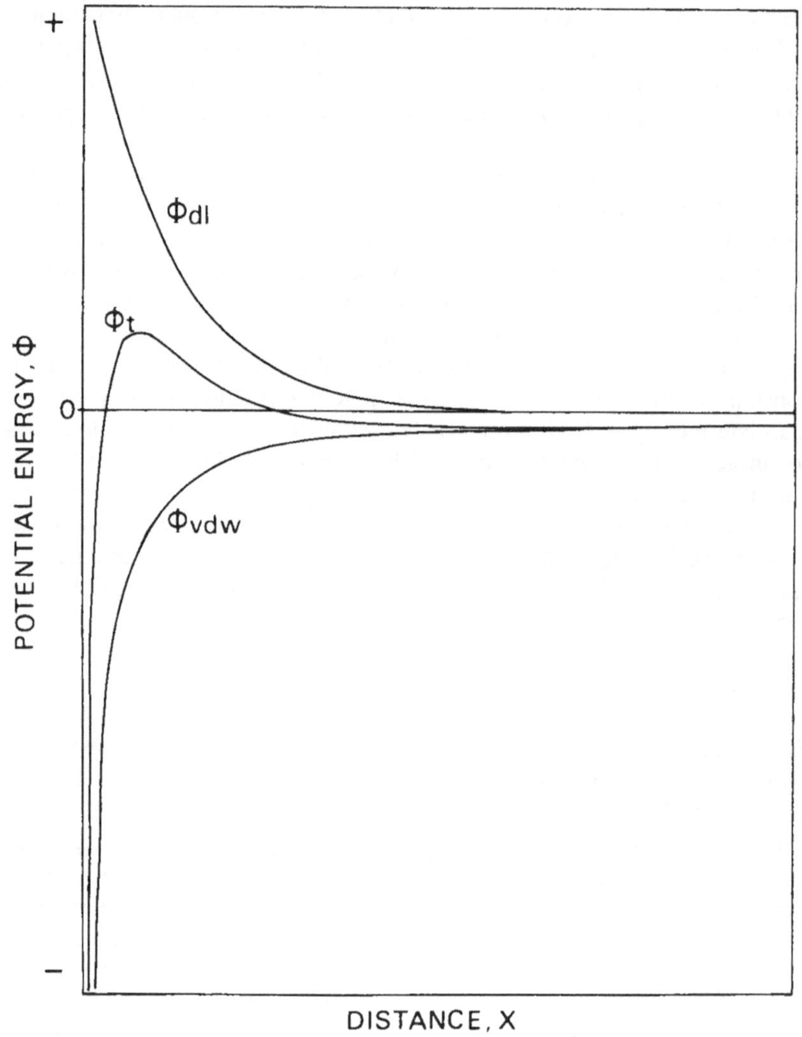

FIGURE 2. Potential energy of interaction on approach of a sphere to a plate of like charge. ϕ_{dl} and ϕ_{vdw} , individual components attributable to double-layer and van der Waals interactions, respectively; ϕ_t, total interaction.

consist of polar molecules (polar lipids, proteins, fatty acids) whose orientation depends on the surface-free energies of the materials in contact with each other. Interpenetration of dangling hydrocarbon chains leads to steric repulsion, which is the expression of a decrease in free energy resulting from a decrease in entropy.[13] Coating of a dense particle with a lighter layer reduces the effective Hamaker coefficient of the particle, and attraction is reduced. The reverse occurs when a light particle is coated by a denser layer. A coating can never reverse attraction into repulsion.[17]

D. Adhesion

In close contact, i.e., upon deposition in the primary minimum, the water layer between bodies may be displaced. In this case, adhesion and the potential for desorption are determined by the surface and interfacial energies of the two bodies and of each body and the aqueous medium, subject to uncertainties caused by short-range molecular interactions in aqueous media that are highly polar.[18-20] Interfacial energies can be determined from contact angle measurements by Young's equation:

$$\gamma_{SA} = \gamma_{LA} \cos\vartheta + \gamma_{LS} \qquad (9)$$

where γ in ergs cm^{-2} represents an interfacial energy, and the subscripts refer to the solid/air, liquid/air, and solid/liquid interfaces, respectively. In measurements, one may either (1) obtain the critical surface tension γ_c according to Zisman[21] by extrapolating surface angle measurements with a series of homologous liquids to cos ϑ = 1, or (2) measure the contact angle with a single suitable liquid such as α- monobromonaphthalin and calculate γ_S from the equation:

$$\cos\vartheta = 2\varphi(\gamma_s/\gamma_L)^{1/2} - 1 \qquad (10)$$

where γ_L is known and φ for practical purposes is unity.[22]

The strength of adhesion is expected to increase with decreasing γ_c, i.e., increasing hydrophobicity. Hydrophilic surfaces are stabilized by oriented water molecules on their surface, and solvent-mediated repulsion results.[20] There have been contact angle measurements on films of bacteria and on layers of mammalian cells, but none on layers of viruses. Surface energy data on a number of minerals are available from the published literature. The mean surface energies of certain natural rocks and soils can possibly be determined on highly compressed pellets by using the methods of Zisman and Becher, but the significance of such measurements will be limited if the adsorption of viruses by such solids is due primarily to highly adsorbing components.

E. Adsorption Isotherms

Adsorption isotherms are useful in expressing results of virus adsorption experiments. Of the various expressions the Freundlich isotherm has proven the most appropriate in many cases.[23-28] It has the form:

$$X = a(C_e)^b \qquad (11)$$

where X is the amount of material absorbed per unit adsorbent; C_e is the concentration of the particles in suspension at equilibrium; and a and b are constants for any given system.

The Freundlich isotherm is an empirical expression. It is thought to reflect heterogeneity of the surface or the particle population, or both, with regard to the affinity of each for the other.[29] Heterogeneity of surfaces in soils is self-evident. Heterogeneity in virus populations adsorbed to soils was indicated in experiments by several workers.[27,30,31] The plot of the logarithm of the amount adsorbed to the amount remaining in suspension is linear, often over several orders of magnitude. Taylor et al.[28] (personal communication) noted, however, that the isotherms obtained by varying the total quantity of viruses added and by varying the amount of substrate, while both of the Freundlich form, are not necessarily collinear.

Freundlich plots of virus adsorption to surfaces often show slopes of approximately 1.0 indicating that the distribution of viruses between the surface and the liquid is independent of the concentration of viruses in the suspension (Henry's law). However, values for the slope as low as 0.2 and higher than 1.0 (up to 1.35) have been observed.[32-35] Values greater than 1.0 indicate that the affinities of viruses to a substrate increase with their concentration. As surface coverage is usually low (see below), aggregation of viruses and destabilization of the virus suspension, rather than multilayer adsorption, may be responsible for this effect.

When surfaces approach saturation, the Langmuir isotherm represents the adsorption process:

$$C_e/X = C_e/C_{max} + 1/(\alpha\, X_{max}) \qquad (12)$$

where X_{max} is the apparent surface maximum concentration, α is a constant, and X and C_e are as defined before.[33,36-38] In two soils of approximately equal sand, silt, and clay contents, approximately 10^8 to 10^9 plaque-forming units of phage were adsorbed per gram of soil, suggesting a capacity consistent with monolayer adsorption.[39] Coliphage adsorbed on coal may cover up to 0.1% of the available surface.[37] In our studies of poliovirus adsorption to Ottawa sand, maximum surface coverage was only 1%;[33] poliovirus on montmorillonite covered up to 2.6% of the available surface.[35]

F. Rate of Adsorption

The rate of adsorption of virus particles to a solid surface depends on the concentration of particles in suspension and the velocity of the particles under the effects of gravity, Brownian motion, and shear forces, as modified by the potential gradient $\phi(x)$. When double-layer repulsion was minimized by the addition of 0.1 *M* NaCl, Valentine and Allison[40] found that the rate of virus adsorption to an aluminum surface was independent of turbulence of the liquid. The number of particles adsorbed increased linearly with the square root of the elapsed time. The process was therefore consistent with Brownian theory, i.e., it was diffusion-controlled. For adsorption of small spherical particles of radius a, where C is the concentration of particles at time t relative to C_0, the concentration at time 0:

$$C/C_0 = \exp\{-4\pi n\ a\ D[t + R(t/\pi D)^{1/2}]\} \qquad (13)$$

where n is the number of adsorbing particles, and D is the diffusion coefficient, which is given by

$$D = k\ T(6\pi\ \eta\ a)^{-1} \qquad (14)$$

Empirically, in the initial phase of adsorption, the ratio of cells adsorbed was 1.13 $(D\ t/d)^{1/2}$, where d is the thickness of the liquid film.

The rate of adsorption as a function of van der Waals, double-layer, and gravitational forces was considered by Ruckenstein and Prieve.[41] The particle flux J, defined as dn/dt, is determined by Brownian motion and the total interaction force $\phi(x)$. The number of particles remaining in suspension decays exponentially with time:
where the time constant, τ, is exponentially sensitive to the height of the energy barrier, i.e., the difference in energy between the maximum and the secondary minimum, $\phi_{max} - \phi_{min}$ (Figure 2). The calculations were carried out for mammalian cells. For these, Brownian motion alone is sufficient only for adsorption into the secondary minimum. Shear and gravitational forces in combination with inertia are required to overcome the energy maximum, or barrier. As indicated earlier for viruses, particularly for the smaller forms, the energy of Brownian motion is of the order of the energy required to penetrate the energy barrier.

With a suspension of poliovirus (28 nm particle diameter) and Ottawa sand in artificial freshwater (4.1 meq ℓ^{-1}), three logs of virus were removed in 1 hr, whereas with the larger reovirus (70 nm) removal was only 40% in 1 hr.[42] Poliovirus adsorption to kaolinite or magnetite was virtually complete in 20 min.[26,43] Adsorption of phage ϕx-174 to a silt loam at 0.02 ionic strength decreased hyperbolically with time. Equilibrium was reached more slowly after exchangeable calcium was removed from the substrate.[34]

G. Desorption

Reversibility of virus adsorption is generally incomplete. LaBelle and Gerba[44] recovered little more than 1% of adsorbed viruses (1.5% in the case of rotavirus) from estuarine sediments at any salinity. In freshwater and soil environments, however, deionized water removes up to 10% of adsorbed viruses from soil columns.[25,28,45] In our experiments with soils and minerals suspended in artificial freshwater, we recovered between 2.5 and 6.8% of the adsorbed viruses (Sturman et al., in preparation). Desorption is prevented by the addition of Ca^{2+}.[46] Organic dissolved matter and high pH increase desorption, a fact which is exploited in the recovery of viruses from filters and solid matter. In our studies, the greatest recovery, in the order of 10%, occurred with solutions of hexametaphosphate.

This behavior suggests double-layer repulsion as the mechanism for desorption. At high pH, negative charges are developed at the surfaces. Hexametaphosphate has anionic properties and conveys negative charges to the surfaces by adsorption. From the results, no conclusions can be drawn about the deposition of some of the more easily desorbed viruses, particularly of the larger reoviruses, in the shallow secondary minimum. Viruses adsorbed into the primary minimum have a much higher probability of becoming permanently attached to the surface. In seawater, the energy barrier is likely to be nonexistent, so that adsorption into the primary minimum is likely. This adsorption would be largely irreversible as is borne out in the experiments. Duboise et al.[45] found that recovery decreases sharply after the material carrying the adsorbed viruses is dried. As the water film which separates the viruses from the surface evaporates, the particle is drawn to the surface, where other bonds with the substrate may form. Dissociation of the particle from the surface depends on the strength of these bonds and on the ability of the solvent to penetrate and wet the virus-solid interface. This aspect, which is of considerable practical interest, has not been investigated systematically. Murray and Parks[6] found reversibility of poliovirus adsorption to metal surface except when the surface caused disintegration of the virus particles. The calculated separation was 1 nm, suggesting adsorption into the primary minimum.

H. Adsorption of Virus Particles to Other Virus Particles — Aggregation

Viruses replicate intracellularly and are liberated by rupture or lysis of the host cell within the host organism. When liberated, virus particles occur in the form of aggregates, which in turn are associated with cell debris. When excreted into the intestinal tract, these clumps and aggregates may or may not dissociate upon contact with other solid matter and under the influence of shear forces. In sewage and in receiving waters, the virus encounters a dilute, low-electrolyte environment, which to some extent can be expected to weaken interparticle bonds. Most viruses, however, appear to remain or become associated again with sewage particulate matter. In these environments, the virus particles are exposed to organic solids, including other virus particles; aggregation is the adsorption of virus particles to virus particles. In the aquatic environment, aggregation is not likely to be important because the probability of virus-virus collisions is reduced by dilution. Viral aggregates and firmly particle-associated virus particles appear to persist in this form, while free virus particles adsorb to any suitable surface. Even with the low surface coverage observed in experiments, the small numbers and mass of virus particles in the environment are too small for competition among virus particles for free surfaces, but there is competition for surfaces between viruses and other adsorbates. There is also competition of available surfaces for virus particles. In our batch adsorption experiments[33] with 2 $m\ell$ volumes of virus suspension in 50-$m\ell$ centrifuge tubes of relatively inert material, interference from the adsorption of viruses to the container was significant unless a substantial amount (50 to 200 mg) of adsorbent was present.

Aggregation of monodisperse virus particles is a common occurrence in concentrated suspensions in dilute aqueous media. Such aggregation is observed upon dilution into distilled water at pH levels below neutral.[47-50] The addition of divalent and trivalent cations promotes aggregation at low pH when the virus particle can be expected to be positively charged. Apparent inconsistencies could be resolved if aggregation measurements were combined with measurements of ζ potential. In seawater coagulation of colloidal matter occurs by double-layer compression and is responsible for virus titer loss.[51]

III. MEASURED PROPERTIES OF SUBSTRATES AND VIRUSES AND OTHER FACTORS

A. Measured Properties of Substrates

1. Surface Area

Measurements of surface area are important in expressing virus adsorption in quantitative terms. The gas adsorption method (Brunauer-Emmett-Teller or BET Method) for some materials gives values that are probably too high for virus adsorption studies. Gas penetrates into pores too small for viruses and even appears to penetrate the seven to ten AE lattice of the ordered clay minerals. More representative measurements can be obtained by mercury porosimetry.[52,53] Material is dried in a vacuum and placed in contact with mercury under increasing pressure. Each pressure reading corresponds to a minimum pore size penetrated by the mercury, and the decrease in total volume equals pore volume. A pore size distribution is obtained, and, assuming that the pores are cylindrical, the distribution of pore surface as a function of pore diameter is calculated. Table 2 lists surface areas for crushed and sieved samples of substrates, described in Table 1, measured by the two methods and by microscopic projected-area and size distribution analysis with an image analyzer.

2. Surface Charge

The measured quantity is the particle charge at the hydrodynamic shear surface, determined as ζ potential, e.g., by particle microelectrophoresis. Measurements of microelectrophoretic mobility of several minerals as a function of pH were performed by Taylor and co-workers[56,57] (Figure 3). They confirm both the pH-independent structural charge in the highly ordered clays and the pH-dependent charge development in the oxide minerals and soils containing various amounts of organic material. The IEP of aluminum oxide is near nine. Soils are generally electronegative over the pH range of interest (Figure 3).

The adsorption of a polyelectrolyte, polydiallyldimethylammonium chloride (PDADMA), to a substrate may be used to estimate surface charge (Table 2). Because the polyelectrolyte carries a positive charge, the measurements are inversely related to the affinity of the substrate to negatively charged colloids, including viruses.[33,58]

3. Cation Exchange Capacity

Cation exchange capacity (CEC) is a measurement of the counterions adsorbed to negatively charged mineral surfaces. Determinations of CEC are routine in soil chemistry. The positively charged counterions tend to facilitate the adsorption of viruses to materials such as montmorillonite, which carry strong structural negative charges.[43] If exchangeable cations are removed from a substrate, the affinity to viruses is reduced.[34,59-61] According to others, CEC parallels virus adsorption occasionally but not consistently.[62,63] This is so because CEC measures only a supplemental property, or a modification of a more fundamental property, of the substrates. CEC significance must be weighed in those terms.

4. Equilibrium pH

Equilibration with distilled water or with aqueous media simulating wastewater effluent or rainwater may provide information on the excess of basic or acidic groups in a natural substrate.[55] Generally the acid substrate is the better adsorbent.[64] More important, however, is the pH of the suspension in relation to the IEP of the substrate. Adsorption of electronegatively charged colloids is favored at pH levels lower than the IEP, i.e., at positive ζ potentials.

5. Organic Content

Organic content of substrates can be estimated in several ways. Weight loss upon ignition

Table 1
LISTING OF REFERENCE SUBSTRATES[a] NO., NAME, ORIGIN, AND DESCRIPTION

Minerals

M01	Dolomite, Penfield, N.Y. — $CaMg(CO_3)_2$, a sedimentary, crystalline rock, 10% was acid insoluble, weakly alkaline in water
M02	Montmorillonite, Cameron, Ariz. — API[b] mixed-layer aggregate no. 31, highly ordered 2:1 clay, relatively pure, low IEP, attracts divalent cations, alkaline in solution
M03	Montmorillonite, Clay Spur, Wyo. — API no. 26, smectite of marine origin, more basic than A, higher organic content
M04	Quartz, Herkimer Co., N.Y. — "Herkimer Diamond", clear, crystalline rock, with occasional inclusions of hematite, Fe_2O_3
M05	Calcite, Rosiclare, Ill. — Highly crystalline $CaCO_3$, but contained some silicates and other matter
M06	Kaolinite, Bath, S.C. (Lamar Pit) — API no. 5, 1:1 structured aluminum silicate from a sedimentary deposit, 95% pure, weakly acidic
M07	Kaolinite, Macon, Ga. — API no. 4, 93% pure, essentially free of organic matter, similar to M06, but weakly alkaline in suspension
M08	Halloysite, Eureka, Utah (Dragon Iron Mine) — API no. 13, hydrated kaolinite, significantly more disordered than kaolinite, 93% pure
M09	Dickite, St. George, Utah (Bull Valley) — API no. 16, disordered form of kaolinite, 96% dickite and nacrite (a similar mineral)
M10	Attapulgite, Attapulgus, Ga. — API no. 64, palygorskite, fibrous 2:1 layer clay mineral, contained only traces of organic and calcareous matter, high surface area
M11	Glauconite, Birmingham, N.J. — A phyllosilicate similar to illite (see S03 below), very pure
M12	Magnetite sand, Marysville, Calif. — ca. 65% magnetite (Fe_3O_4) and 25% hematite (αFe_2O_3) and ilmenite ($FeTiO_3$)

Soils

S01	Bituminous shale, LeRoy, N.Y. — Sedimentary rock of clay-sized particles, containing hydrocarbon with 0.4% organic carbon as hydrocarbon and 43% CO_3 as $CaCO_3$
S02	Fossiliferous limey shale, Summit, N.Y. — Sedimentary rock of marine origin, containing at least 6% $CaCO_3$
S03	Illite-bearing shale, Rochester, N.Y. — Sedimentary rock 83% illite, a 2:1 phyllosilicate and ca. 3.7% $CaCO_3$ releases Ca and Mg but remains neutral in suspension
S04	Argillaceous shale, Rochester, N.Y. — Sedimentary rock of consolidated clays, weakly basic, contains 1.6% CO_3
S05	Arenaceous shale, Green Co., N.Y. — Sedimentary rock of consolidated sands
S06	Calcareous shale, Lima, N.Y. — Sedimentary rock containing at least 5% Ca and 13% CO_3
S07	Lake Ontario dune sand, Monroe Co., N.Y. — A soil, although mostly quartz, it contained discrete calcareous particles (ca. 10% CO_3) and organic matter
S08	Lake Ontario beach sand, Rochester, N.Y. — A water-transported soil, predominantly coarse silica sand, some calcareous (4.5% CO_3) sand
S09	Lake Superior dune sand, Autrain, Mich. — A wind-blown soil, mostly quartz, few calcareous and organic components, some iron oxide
S10	Colonie sandy loam, Monroe Co., N.Y. (Irondequoit Bay, 15—30 cm) — Entisol, alfic udipsamment, mixed, mesic
S11	Genesee silt loam, Monroe Co., N.Y. (Industry, N.Y. 15—30 cm) — Inceptisol, fluventic eutrochrepts, fine-loamy, mixed, mesic
S12	Genesee silt, Monroe Co., N.Y. — Same site as S11, 30—60 cm
S13	Brown sandstone, Medina, N.Y. — Slightly calcareous sedimentary rock of consolidated quartz sand
S14	Quartz conglomerate, Olean, N.Y. — Calcareous conglomerate of sand and gravel-size quartz, high iron content in this sample possibly from contamination during grinding
S15	Kame conglomerate, Monroe Co., N.Y. — Poorly consolidated sand and gravel, from Palmyra gravelly outwash, contains at least 30% calcareous matter, one of the most basic materials used
S16	Adobe, Deming, N.M. — Unclassified heavy-textured clayey soil with about 2.5% CO_3 and about 1% organic content

Table 1 (continued)
LISTING OF REFERENCE SUBSTRATES[a]
NO., NAME, ORIGIN, AND DESCRIPTION

S17	Muck, Wayne Co., N.Y. (15—30 cm) — Histosol, terric medisoprists, euic, mesic, composed of 50% well-decomposed, finely divided organic matter and silt
S18	Loess, Kansas City, Kansas — Unclassified, unconsolidated wind-blown soil containing some clay

[a] All substrates except M04 (quartz) were obtained from Ward's Natural Scientific Establishment, Rochester, N.Y.; M04 was a gift from R. M. Bryce, New York State Geological Survey, Albany.
[b] API, American Petroleum Institute.

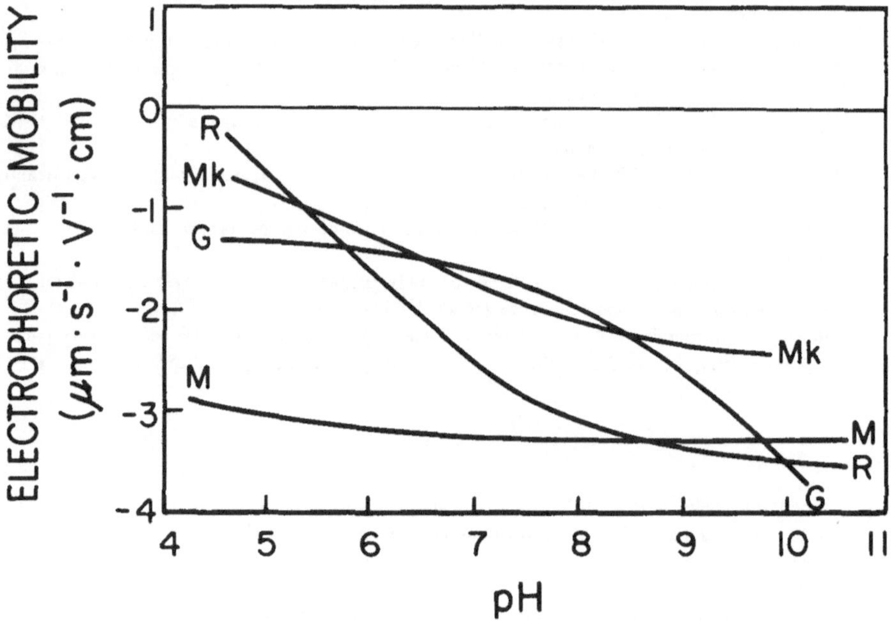

FIGURE 3. Variation with pH of electrophoretic mobility of soils in synthetic freshwater. Abbreviations: (M) Arizona montmorillonite, (R), Riverhead, Long Island, N.Y. sandy loam, (G) Genesee silt, and (Mk) Wayne County, N.Y. muck. (From Taylor, D. H., Moore, R. S., and Sturman, L. S., *Appl. Environ. Microbiol.*, 42, 976, 1981. With permission.)

and organic carbon content are established methods. Measurements of protein content (actually measurements of certain functional groups of amino acids), or microbial exopolymers, or of humic materials may serve as surrogate measurements — or perhaps even as more specific measurements — of the materials that may make up or cover the surface of the substrates. Some of these measurements were used in the experiments discussed below. Low-temperature ashing in an oxygen plasma may be used to remove organic surface materials while maintaining the chemical and physical integrity of the substrate.[64a] As noted above, organic material exerts few van der Waals forces, while carboxylic acids and phenolic and amino groups contribute to surface charge development. By chelating heavy metals, the organic materials contribute to the complex surface chemistry of natural materials. As heavy metal content and organic content have opposite effects on virus adsorption, this illustrates the difficulty of predicting adsorptive behavior from a chemical analysis of substrates. The strong correlation between organic content and chelated (oxalate-extractable) metals is shown in Table 4.

Table 2
PHYSICAL AND CHEMICAL PROPERTIES OF THE REFERENCE SUBSTRATES

No.	Name of substrate	Sp. gr.	BET	Surface Por(R)	Por(P)	Im.	pH	Mobility at pH	Ign. loss	LTA loss	Carbon Total	Org.	Nitrogen total	Protein	PDADMA Ads.
M12	Magnetite sand (Calif.)	3.77	5.70	0.17	0.20	0.09	8.7	−1.48	n.a.	0.18	0.13	0.04	n.a.	60	4.89
S13	Brown sandstone (N.Y.)	2.45	5.60	0.31	0.40	0.26	8.3	−1.51	0.84	0.70	0.19	0.12	0.02	n.a.	2.50
S02	Fossil. limey shale (N.Y.)	3.02	4.50	0.36	0.36	n.d.	8.8	−1.47	0.79	0.77	0.93	0.14	0.02	45	2.73
S14	Quartz conglomerate (N.Y.)	2.91	3.20	0.11	0.11	0.14	8.9	−1.81	n.a.	0.36	0.14	0.12	n.a.	n.a.	10.19
M04	Quartz (N.Y.)	2.15	0.54	0.13	0.23	0.16	8.3	−1.46	n.a.	0.03	0.03	0.01	0.01	75	1.13
S15	Kame conglomerate (N.Y.)	2.47	3.10	0.36	0.36	0.17	9.6	−2.26	0.56	0.27	5.39	0.66	0.03	n.a.	0.61
S01	Bituminous shale (N.Y.)	1.46	3.80	0.71	0.86	0.13	8.4	−1.50	2.06	0.39	8.96	0.42	0.01	50	1.28
S08	Lake Ontario beach sand (N.Y.)	2.60	0.73	0.10	0.13	0.31	8.7	−1.51	0.93	0.56	1.21	0.28	0.02	1100	10.35
M09	Dickite (Utah)	2.68	2.70	1.69	1.79	0.09	8.6	−1.45	13.97	3.05	0.22	0.09	0.04	675	0.91
M08	Halloysite (Utah)	1.80	42.00	2.60	3.16	0.15	6.6	−1.64	8.28	0.14	0.29	0.15	0.04	265	1.06
M01	Dolomite (N.Y.)	2.71	0.85	0.23	0.28	0.09	8.7	−1.20	1.06	0.01	13.1	1.10	0.02	45	0.78
S05	Arenaceous shale (N.Y.)	2.56	3.50	0.44	0.58	0.32	8.7	−1.66	0.63	0.34	0.62	0.11	0.01	35	2.19
S06	Calcareous shale (N.Y.)	2.93	9.40	1.50	1.85	0.33	8.6	−1.71	2.98	0.59	3.14	0.24	0.05	125	1.37
M05	Calcite (Ill.)	2.52	4.00	0.58	0.67	0.14	8.5	−1.55	1.1	0.15	4.38	0.52	0.01	50	2.67
S03	Illitic shale (N.Y.)	2.25	55.00	1.89	2.63	0.17	7.3	−1.03	5.8	2.85	2.06	1.36	0.11	845	3.97
S18	Loess (Kan.)	2.29	25.00	0.42	0.57	0.06	7.8	−1.31	2.53	1.10	0.18	0.13	0.02	n.a.	4.36
S09	Dune sand (Mich.)	2.54	1.20	0.26	0.26	0.23	8.4	−1.34	n.a.	0.37	0.15	0.08	0.01	60	2.64
M07	Kaolin (Ga.)	1.92	12.00	3.02	3.94	0.62	8.1	−1.23	12.18	1.12	0.05	0.01	0.03	150	0.70
M06	Kaolinite (S.C.)	2.60	16.00	4.35	5.43	0.16	5.7	−1.32	2.14	0.95	0.19	0.01	0.01	475	0.41
S12	Genesee silt (N.Y.)	3.04	11.00	0.68	0.81	0.12	8.0	−1.31	12.06	0.83	0.02	0.15	0.01	125	2.31
S10	Colonie sandy loam (N.Y.)	2.98	5.30	0.40	0.56	0.21	8.3	−1.31	2.42	1.14	0.72	0.36	0.06	1165	3.58
S04	Argillaceous shale (N.Y.)	2.33	29.00	2.89	3.79	0.12	8.4	−2.79	4.35	0.77	0.41	0.09	0.08	825	1.30
M10	Attapulgite (Ga.)	1.47	118.00	4.04	7.28	0.19	8.2	−1.66	8.93	5.49	0.15	0.06	0.05	265	2.11
S07	Dune sand (N.Y.)	2.50	4.10	0.43	0.56	0.17	8.6	−1.69	0.79	0.36	2.32	0.20	0.004	60	1.18
S17	Muck (N.Y.)	1.59	1.20	0.19	0.19	0.11	7.6	−1.16	51	34	27	20.00	1.72	17,000	88.42
M03	Montmorillonite (Wyo.)	2.45	32.00	0.94	2.37	0.16	9.3	−3.00	4.01	1.66	1.01	0.16	0.08	n.a.	28.35
S16	Adobe (N.M.)	2.34	34.00	0.52	0.82	0.11	8.3	−1.11	1.47	0.59	0.26	0.57	0.01	440	4.15
M11	Glauconite (N.J.)	2.59	54.00	1.74	2.52	0.11	7.7	−3.36	4.32	1.84	0.14	0.10	n.a.	n.a.	3.43
S11	Genesee silt loam (N.Y.)	2.20	4.20	0.68	0.87	0.26	8.1	−1.32	5.36	3.36	2.11	1.65	0.20	4750	4.40

Table 2 (continued)
PHYSICAL AND CHEMICAL PROPERTIES OF THE REFERENCE SUBSTRATES

				Surface							Carbon				
No.	Name of substrate	Sp. gr.	BET	Por(R)	Por(P)	Im.	pH	Mobility at pH	Ign. loss	LTA loss	Total	Org.	Nitrogen total	Protein	PDADMA Ads.
M02	Montmorillonite (Ariz.)	2.20	41.00	1.23	1.77	0.18	9.1	−3.40	5.24	1.87	0.19	0.06	0.02	75	18.53

Note: The substrates are ordered according to overall virus adsorption capacity expressed as the average of the logarithms of the distribution coefficients as shown on Table 3; n.a. = not analyzed; Sp. gr. = specific gravity in gram per cubic centimeter; BET = surface area by nitrogen gas adsorption method, in square meter per gram; Por = surface area calculated from Hg porosimetry data for pores up to 70 nm (R, for reovirus) and 30 nm (P, for picornaviruses); Im. = surface area from particle size distribution measured with a light microscope and projected area in IMANCO image analyzer; pH, in equilibrium with synthetic freshwater according to Peech (1965); Mobility by particle microelectrophoresis in dilute synthetic freshwater suspension adjusted to the equilibrium pH; the units are (micrometer per second)/(volts per centimeter); Ign. = weight loss upon ignition at 550°C, 30 min; LTA = weight loss upon low temperature ashing in an oxygen plasma, all in %; Carbon: total C (i.e., including carbonate carbon) and organic C by elemental analysis, in %; total N by elemental analysis, in %; Protein, apparent protein by Lowry method, in microgram per gram; PDADMA ads. = adsorption of polydiallyldimethylammonium chloride in milligram per square meter measured and expressed on a basis of surface area determined by Hg porosimetry to include 30 nm diameter pores.

6. Metal Oxide Content

Metal oxide content of minerals and soils affects virus adsorption by affecting both van der Waals attraction and charge development. Relevant are metal oxides adsorbed to or exposed at the particle surface. Acid digestion solubilizes these without significantly degrading the matrix of a silicate mineral. On the other hand, carbonate and oxide minerals are completely digested by acids. Careful interpretation of the measurements is therefore necessary.

B. Measured Properties of Viruses

Compared to other natural surfaces, animal viruses appear to be a relatively homogeneous group of hydrophilic colloids, as their surface consists always of protein. Viruses that are human pathogens and transmitted by the water route are an even less diverse group. In large part, they consist of spherical or icosahedral particles, i.e., picornaviruses with a hydrated diameter near 20 to 30 nm[65] and reoviruses with a hydrated diameter of approximately 70 to 80 nm. Theoretically the larger viruses should adsorb more strongly.[6] Even so, a number of properties of interest have yet to be determined. The Hamaker coefficients and surface energies for viruses have only been assumed to resemble those of organic and biologic materials of similar density (see above). A greater difficulty, however, lies in the complexity of the charge mosaic on the exterior surface of each virus particle, which varies with the composition and conformation of the coat proteins. These mosaics are subject to genetic variation from strain to strain, even within a single "species", and are modified further by ambient pH and by adsorption of antibodies in the intestinal tract of the organisms excreting the virus particles.

A considerable difficulty already lies in the accurate determination of surface charge of most viruses. Electrophoresis on a macroscopic scale is conducted in media different from freshwater or soil solution and may give results that cannot be used to predict virus behavior in these media. Only reovirus is sufficiently large to be detectable by direct particle microelectrophoresis. Reovirus 3 (Dearing strain) has shown a single IEP at pH 3.8 $\pm$ 0.1. This value was obtained in 1 m*M* $NaHCO_3$[66] and agrees with earlier results obtained by isoelectric focusing.[50]

Of great importance, and still incompletely understood, are apparent changes in the conformation of certain or all enteroviruses that cause dramatic shifts in the IEP. This shift was first discovered by Mandel[67] in a strain of poliovirus and later observed by others in other poliovirus strains and in strains of echovirus 1.[50,68-70] According to these studies, poliovirus has two IEPs. One of these, depending on the strain, lies between pH 6.5 and 8.5, the other between 3.8 and 4.8. The Farouk strain of echovirus 1 has an IEP at pH 5.6 and 5.1. The conformation with the higher-pH IEP is established in high-pH media, the other at low pH. The conversion is at least partly reversible. In many experiments reported in the literature, the virus suspension was probably heterogeneous with respect to the IEP, and this can explain to some extent the observed variability of the results (Figure 4).[57] Determination of the IEP of hepatitis A (enterovirus 72) virus would be important in controlling this important waterborne agent.

C. Measured Properties of Suspending Medium

Relevant properties here are ionic strength, more specifically the content in cations and their valency, and pH.

1. pH

The effect of pH on surface charge development is well established.[56,57] For two colloids, adsorption is inhibited when the electrokinetic potentials of adsorbate and adsorbent (ψ_1 and ψ_2) are both high and of the same sign. Adsorption is facilitated upon reduction of the charge

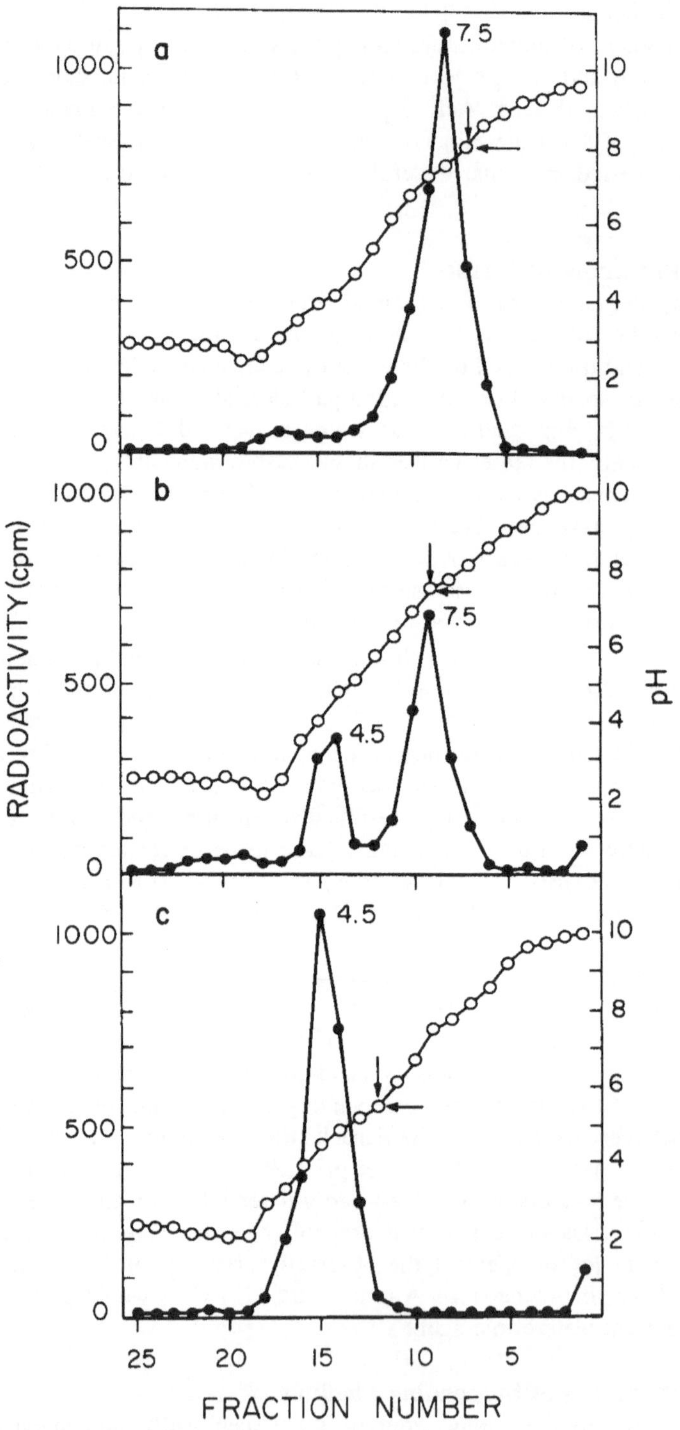

FIGURE 4. Effect of initial position on the distribution of radioactivity after isolectric focusing of poliovirus in a sucrose gradient. The ampholyte mixture generated a pH gradient from 2 to 10. The arrows indicate the positions in the column where the virus was added. The initial pH in each case was 7; the final pH values were (a) 8, (b) 7.5, (c) 5.5. Solid circles: cpm. Open circles: pH. (From Taylor, D. H., Moore, R. S., and Sturman, L. S., *Appl. Environ. Microbiol.*, 42, 976, 1981. With permission.)

of either material because the double-layer repulsion is reduced (ψ_1 or $\psi_2 \rightarrow 0$). Adsorption is rapid when the charge of one is zero (i.e., at the IEP) or when the charges are of the opposite sign. It follows that mutual adsorption of materials of similar IEP will be strong in the pH region of the IEP. For materials with differing IEP, conditions will be favorable at pH near either IEP and most favorable in the pH region between the two. This situation is illustrated in numerous examples in the literature.[2,3,5,64] Qualitative correlation of adsorption data with measurements of surface charge development (electrophoretic mobility, ζ potential, IEP) may be found in the experiments by Taylor et al.[28] on adsorption of reovirus (IEP at pH 3.8) to two clays. Adsorption to allophane (IEP at pH 6.5) was strong between pH 7 and 4; adsorption to montmorillonite (IEP below pH 3) under the conditions of the experiments was significant only at pH 4 or lower. Enterovirus adsorption data are often difficult to interpret because of uncertainty about the distribution of IEP in the viral suspension. With several minerals and soils, including montmorillonite, in synthetic freshwater with 0.01 *M* Ca^{2+}, adsorption became significant in a pH range of 0.5 unit when pH was reduced from values between 10 and 7.5, dependent on substrate.[57] Adsorption to montmorillonite at pH 9 indicates that the IEP of the poliovirus in these experiments was high (see above).

2. Electrolyte Content

Increased electrolyte concentrations reduce double-layer repulsion. Specifically they promote adsorption by reducing the height of the energy barrier, then facilitating deposition of adsorbed particles in the primary energy minimum (see above). Electrolyte effects are most pronounced where double-layer effects are strong, i.e., where strong and like charges prevent adsorption. Therefore, the electrolyte effect is strongest with low-IEP viruses or low-IEP substrates or combinations of both at high pH. Negative surface charge, measured as electrophoretic mobility, decreases by a factor of 1:10:1000 as the valency of the ion increases from 1 to 2 to 3 (Schulze-Hardy rule).[66,71] Additions of calcium increase virus adsorption. The optimum concentration is 0.01 *M*,[62] although in our hands concentrations of 1 m*M*, such as are found in many freshwaters, exerted about half the effect observed at 0.01 *M*.[57] The high ionic content of seawater (35.5% salt, 600 meq ℓ^{-1}, anions or cations) promotes adsorption and coagulation of suspended solids. The association of viruses with estuarine and marine particulate matter and the effect of this association on virus survival have been reviewed elsewhere.[72-75]

3. Organic Solutes

As a rule, natural waters contain one to several grams of dissolved organic carbon per liter. Most of it is in the form of humic acids which are weak, negatively charged polyelectrolytes subject only to slow biodegradation. The remainder is a mixture of the excreta and intermediate products of decomposition of living matter including poly- and oligopeptides, amino acids, sugars, and heterocyclic bases, with individual concentrations in the micromolar range or lower. Of these, the humic acids and amino acids adsorb to natural surfaces and compete or interfere with the adsorption of viruses or, at high concentrations, cause desorption of adsorbed virus, particularly in the absence of significant concentrations of divalent cations.[26,27,76-78] Freshwater humic polymers have an IEP near three.[79] They ionize increasingly above this pH due to their carboxylic acid groups. Phenolic groups dissociate only in the pH range 8.5 to 12; this dissociation is greatly affected by the presence of the ionized carboxylic groups.[80]

IV. ADSORPTION OF VIRUSES TO REFERENCE MATERIALS AND SOILS*

An extensive batch adsorption study has been done with five enteric viruses and numerous mineral and soil substrates. This study is reported in detail below.

The viruses used in this study were prepared and radiolabeled. Poliovirus 2 (vaccine strain P712-ch-2ab), reovirus 3 (Dearing strain), echovirus 7, and coxsackievirus B5 were produced and assayed in monolayer cultures of HeLa, L2, Vero, and FL amnion cells, respectively. Echovirus 1 was produced in Vero cells and assayed in FL amnion cells. Production and purification of radiolabeled poliovirus and reovirus were carried out as described elsewhere (Moore et al.[33,58] and Fuhs et al.[64a] Cell cultures infected with echoviruses 1 and 7 or coxsackievirus at multiplicities of 5 to 25, 10 to 20, or 1, respectively, were labeled wtih [^{35}S]methionine, 4 to 7.5 μCi/mℓ (>400 μCi/m*M*) at 2 to 6 hr postinfection in Dulbecco's modified Eagle's medium with 5% fetal bovine serum, either lacking methionine (echovirus 7) or containing the usual amount (echovirus 1) or 1/20 the usual amount of methionine (coxsackievirus).

Cell-associated virus particles were harvested at 24 hr. Echovirus was purified by Genesolv-D extraction and isopycnic centrifugation in CsCl, similarly to reovirus.[58] Echovirus 7 and coxsackievirus B5 were purified at 4 to 10°C as follows: the cells were pelleted by centrifugation in a GSA rotor at 4000 × g for 20 min and then the pellet was resuspended in Dulbecco's phosphate-buffered saline (PBS) either with Ca^{2+} and Mg^{2+} (coxsackievirus) or without additions (PBS_{def}, echovirus 7), treated with 0.5% Nonidet P40 (NP40, Shell) and disrupted in a Dounce homogenizer. The lysate was clarified in a SS34 rotor at 1500 × g or 7700 × g for 10 min and the cell-associated virus was pelleted at 130,000 × g for 2.5 hr in a Beckman L3-50 or L5-65B ultracentrifuge. This pellet was resuspended in PBS_{def} containing 0.5% NP40 with a Duall-style Teflon tissue grinder. The coxsackievirus was purified by isopycnic sedimentation three times in preformed CsCl gradients and dialyzed against PBS_{def} after each sedimentation. Resuspended echovirus 7 was incubated overnight and rehomogenized with 0.5% NP40. The resuspended pellets were clarified by centrifugation at 7000 × g for 30 min and the virus particles in the supernate were purified by isopycnic sedimentation on preformed CsCl gradients similar to the coxsackievirus. Measurements of poliovirus and reovirus infectivity by plaque titration were done as described elsewhere.[33,58] For the other viruses, the overlay medium for plaque assay contained Eagle's minimal essential medium supplemented with 5% fetal bovine serum and 0.8% Noble agar (echoviruses 1 and 7) or 0.75% agarose (coxsackievirus).[85]

The IEP of the reovirus strain was about pH 3.8 as determined by particle electrophoresis;[66] those of the poliovirus, coxsackievirus and echovirus 1 and 7, maintained at higher pH, were 7.5, 8.1, 5.4, and 8.6, respectively, as determined by isoelectric focusing. The lower IEP of the poliovirus was 4.5 (Figure 4).

The mineral and soil substrates were ground either with a tungsten-carbide mill or with mortar and pestle. The fraction passing a 2-mm sieve was used. Suspending medium was an artificial freshwater prepared from KCl, $MgSO_4$, $CaCl_2$, and $NaHCO_3$ and contained (in meq ℓ^{-1}): Na^+, 1.25; K^+, 0.1; Mg^{2+}, 0.75; Ca^{2+}, 2.0; Cl^-, 2.1; SO_4^{2-}, 0.75; and HCO_3^-, 1.25. The ionic strength of this medium was 0.003. The pH was adjusted to 7, but the weak buffering of the medium was usually overcome by the dissociation of acid or basic groups in the substrate.

* Work described in this section was performed with grant support from the U.S. Environmental Protection Agency by G. W. Fuhs, L. S. Sturman, D. H. Taylor, R. S. Moore, M. M. Reddy, and Min Chen. D. H. Taylor described the substrates used in the study; M. M. Reddy and Dale Boshart did the chemical analyses at the Chemical Sciences Laboratory of the Center for Laboratories and Research; E. Canelli did the specific gravity determinations and part of the microelectrophoretic measurements on the substrates; and K. V. Krishnamurty suggested the extraction method for chelated metals. Mercury porosimetry determinations were carried out under contract with the USEPA by Carborundum Corporation Laboratories, Niagara Falls, N.Y.

The substrates and their origin are listed in Table 1. They were selected to represent important mineral constituents of U.S. soils and major soil types. Tables 2 and 3 show their physical and chemical properties. In Table 4 significant correlations are shown between a number of chemical properties, particularly among the concentrations of several trace metals and measurements of organic matter. Moreover, there are strong correlations between certain trace metal fractions and organic matter, presumably as a result of chelation. These relationships are important for correct interpretation of the results of virus adsorption experiments.

Batch adsorption experiments were carried out as described earlier.[33] Essentially a 0.1-mℓ portion of viral suspension containing 10^8 to 10^9 PFU and 10^3 to 10^4 CPM of viruses was added to a slurry of 500 mg of substrate in 1.9 mℓ of the solution in a 50-mℓ Nalgene (no. 3119-0050) Oak Ridge-type polypropylene centrifuge bottle. The containers were shaken at 300 rpm on a gyratory shaker for 1 hr at 4°C followed by separation of substrate and suspending medium at 12,000 × g. The pH of the suspending medium was measured, and samples were removed in duplicate or triplicate for measurements of infectivity and radioactivity. The amount of virus adsorbed was calculated as the difference between the amount recovered from controls and the amount found in the supernate of samples containing substrate. Controls were substrate-free. With echovirus 1, adsorption to the container walls was significant, so that viral loss from suspension in the substrate-free controls was greater than viral loss with the poorest adsorbent (muck). Here, the controls were the muck samples, assuming 10% adsorption by the muck. The variability between duplicates was typically 10% of the mean for infectivity and 15% for radioactivity.

Viral adsorption was expressed either as log removal under the conditions of the experiment,[81] or as the distribution coefficient a (Equation 11, assuming that the exponent b equals 1; see below), or as the free energy of adsorption ($-\Delta G$) in kJ mol^{-1} calculated according to Murray[82] and assuming a particle-to-PFU ratio of 1000 and a displacement from the surface of the substrate of 50 and 139 water molecules by picornavirus and reovirus particles, respectively. Distribution coefficients and free energies were calculated on the basis of available surface rather than weight of the substrate. In the other two methods, the available surface was used. Surface measurements by mercury porosimetry were considered the most appropriate and were used throughout.

We have already reported the results of batch adsorption experiments expressed on a basis of radioactivity or infectivity and substrate weight.[5,33,42,58,81] In the following discussion, only infectivity data are considered. Removals of viruses from suspension ranged from 40 to 99.998% (from 0.22 to 4.70 logs, median for all substrates and viruses near 2.0 logs).[81] The log removal data for 30 substrates were normally distributed. Reovirus showed the strongest adsorption (2.94 ± 0.79 logs), echovirus 7 the weakest (1.73 ± 0.71 logs). Behavior of the five viruses studied was generally well correlated except for poliovirus and reovirus.[81] With few exceptions, the ranking of the adsorbing substrates resembled that given below for adsorption data calculated on a basis of surface area (Table 5). Several substrates showed little variation between viruses; others showed substantial variation. Differences of up to two logs between the strongest and the weakest adsorbed viruses were found; with adobe, halloysite, and attapulgite, the difference was almost three logs.

On a weight/volume basis, highly significant ($p<0.01$) negative correlations existed for log removals of most viruses and, for the average log removal of all five viruses, with organic carbon content and with PDADMA adsorption. Up to 45 to 55% of the variance in log removals could be explained by either one of these variables.[81]

The surface available for adsorption is a key variable. Correlation between surface area of the substrates as measured by the BET technique was very poor except, for unknown reasons, for reovirus. In particular, the minerals with very high surface areas failed to show expected strong adsorption. This was explained by the excessive penetration of the gas relative to the part of the surface available to the virus. In surface area determinations by

Table 3
ADDITIONAL CHEMICAL PROPERTIES OF THE REFERENCE SUBSTRATES

		Nitric acid-extractable metals (mg/g)					Nitric acid extraction (μg/g)						Oxal. extr.	
No.	Name of substrate	Fe	Al	Ca	Fe+Al+Ca	Mg	Cr	Cu	Pb	Mn	Ni	Zn	Cu	Zn
M12	Magnetite sand (Calif.)	155	4.85	3.45	163.30	3.05	445	34	45	980	92.0	160	11.3	6.8
S13	Brown sandstone (N.Y.)	22.00	7.05	1.10	30.15	2.50	30	230	35	1000	12.5	53	52.5	2.5
S02	Fossilif. limey shale (N.Y.)	30.50	13.00	6.50	50.00	4.45	10	14	35	890	30.0	82	6.2	n.a.
S14	Quartz conglomerate (N.Y.)	22.00	1.65	905.00	928.65	0.30	55	58	35	100	9.5	31	10.0	4.0
M04	Quartz (N.Y.)	14.00	0.69	0.16	14.85	0.08	40	135	20	100	12.5	77	5.8	1.2
S15	Kame conglomerate (N.Y.)	40.00	2.70	87.50	130.20	8.50	135	97	40	565	42.5	70	n.a.	9.3
S01	Bituminous shale (N.Y.)	1.00	5.25	180.00	186.25	4.70	20	n.a.	50	490	28.5	39	4.8	1.2
S08	Lake Ontario beach sand (N.Y.)	14.50	2.35	9.10	25.95	4.25	45	64	40	200	11.0	110	8.7	27.0
M09	Halloysite (Utah)	56.00	6.00	0.46	62.46	0.23	5	5	25	5	47.5	150	2.5	1.2
M08	Dickite (Utah)	1.00	120.00	0.20	121.20	0.23	205	150	30	440	41.5	29	n.a.	1.2
M01	Dolomite (N.Y.)	12.00	0.75	250.00	262.75	100	40	60	25	600	20.0	60	10.0	2.5
S05	Arenaceous shale (N.Y.)	45.00	1.30	3.50	49.80	5.65	165	150	35	705	57.0	165	9.3	3.0
S06	Calcareous shale (N.Y.)	26.00	15.50	50.00	91.50	4.70	40	56	50	420	38.5	61	2.5	2.8
M05	Calcite (Ill.)	28.50	11.00	81.00	120.50	6.45	40	630	43	2000	21.5	120	6.8	4.8
S03	Illite-bearing shale (N.Y.)	30.00	16.00	6.15	52.15	5.80	50	90	70	280	54.0	92	8.5	4.8
S18	Loess (Kan.)	18.00	2.00	1.55	21.55	4.10	20	27	35	470	25.5	n.a.	4.0	1.2
S09	Dune sand (Mich.)	28.00	2.05	1.20	31.25	0.67	100	88	40	200	16.5	35	10.3	1.2
M07	Kaolinite (S.C.)	0.51	3.55	0.19	4.24	0.10	10	16	25	3	2.5	18	2.5	1.2
M06	Genesee silt (N.Y.)	0.67	7.50	0.16	8.32	3.85	20	21	35	360	20.0	80	3.7	n.a.
S12	Kaolin (Ga.)	25.00	12.50	0.53	38.03	0.14	20	5	25	6.5	7.0	15	2.5	1.2
S10	Colonie sandy loam (N.Y.)	17.00	10.00	2.70	29.70	3.05	30	19	35	260	11.0	60	10.3	1.2
S04	Argillaceous shale (N.Y.)	20.00	26.00	2.40	48.40	7.65	30	130	20	200	33.5	63	3.3	2.5
M10	Attapulgite (Ga.)	14.50	18.50	4.20	37.20	24.5	90	49	50	200	16.5	102	40.0	3.5
S07	Dune sand (N.Y.)	18.50	5.75	37.00	61.25	12.5	40	56	50	590	16.5	160	n.a.	n.a.
S17	Muck (N.Y.)	13.00	5.10	0.40	18.50	3.50	30	225	55	200	20.0	79	145.0	19.0
M03	Adobe (N.M.)	4.00	5.70	2.30	12.00	5.80	20	n.a.	50	500	15.5	140	13.0	n.a.
S16	Montmorillonite (Wyo.)	23.00	18.50	5.60	47.10	2.55	5	n.a.	50	55	6.5	n.a.	2.5	4.0
M11	Glauconite (N.J.)	60.00	16.50	1.70	78.20	11.5	155	n.a.	30	70	16.5	105	3.3	4.0
S11	Genesee silt loam (N.Y.)	19.00	11.00	1.75	31.75	3.55	10	185	40	310	15.0	81	147.5	5.8

Table 3 (continued)
ADDITIONAL CHEMICAL PROPERTIES OF THE REFERENCE SUBSTRATES

		Nitric acid-extractable metals (mg/g)					Nitric acid extraction (µg/g)						Oxal. extr.	
No.	Name of substrate	Fe	Al	Ca	Fe+Al+Ca	Mg	Cr	Cu	Pb	Mn	Ni	Zn	Cu	Zn
M02	Montmorillonite (Ariz.)	16.00	19.50	2.55	38.05	4.0	5	8	0	70	9.5	65	4.3	1.2

Note: Nitric acid-extractable metals, in milligrams or micrograms per gram, were determined after digestion in nitric acid-hydrogen peroxide (Krishnamurty et al.[54]); Cu and Zn also by extraction of the residue from extraction with 0.1 *M* hydroxylamine in 0.01 *M* nitric acid (1 hr, room temperature) with 0.175 *M* ammonium oxalate in 0.10 *M* oxalic acid, 4 hr at room temperature (K. V. Krishnamurty, personal communication); results in micrograms per gram; indication was by atomic absorption spectrometry; n.a., not analyzed.

Table 4
SIGNIFICANT CORRELATIONS AMONG CHEMICAL PROPERTIES OF REFERENCE SUBSTRATES

A. Organic Content and Associated Fractions

	Protein	Org. C	Ign.	LTA	Cu,ox	Zn,ox	Total N
Protein content, μg/g	1.000						
Organic carbon, %	0.974***[a]	1.000					
Weight loss on ignition, %	0.894***	0.904***	1.000				
Weight loss on LTA, %	0.964***	0.980***	0.940***	1.000			
Cu, NH_4-oxalate, μg/g	0.808***	0.701***	0.619***	0.713***	1.000		
Zn, NH_4-oxalate, μg/g	0.539**	0.514**	0.395*	0.484**	0.401*	1.000	
Total N, %	0.982***	0.996***	0.920***	0.986**	0.724***	0.502**	1.000

B. HNO_3-H_2O_2-Extractable Metals (all in μg/g)

	Al	Ca	Cr	Cu	Fe	Pb	Mg	Mn	Ni	Zn
Aluminum (Al)	1.000									
Calcium (Ca)		1.000								
Chromium (Cr)			1.000							
Copper (Cu)				1.000						
Iron (Fe)			0.957***		1.000					
Lead (Pb)	0.340(*)					1.000				
Magnesium (Mg)	0.573**	0.412*				0.419*	1.000			
Manganese (Mn)			0.341(*)	0.752***	0.436*	0.313(*)		1.000		
Nickel (Ni)			0.748***		0.765***			0.398*	1.000	
Zinc (Zn)					0.559**	0.415*	0.538**	0.527**	0.476**	1.000

[a] Levels of significance: (*) 0.1; *0.05; **0.01; ***0.0001. All variables were entered directly (arithmetic scale).

Table 5
VIRUS ADSORPTION TO REFERENCE SUBSTRATES

No.	Name of substrate	Log distribution coefficient						Free energy of adsorption (−ΔG)					
		ECHO-1	ECHO-7	Cox	Polio-2	Reo-3	Ave.	ECHO-1	ECHO-7	Cox B-5	Polio-2	Reo-3	Ave.
M12	Magnetite sand (Calif.)	3.60	3.19	5.53	5.01	3.67	4.20	49.32	46.50	49.46	48.59	62.82	51.34
S13	Brown sandstone (N.Y.)	4.52	3.62	3.27	5.00	4.52	4.18	53.98	29.99	53.88	47.29	61.49	49.33
S02	Fossilif. limey shale (N.Y.)	3.95	2.70	3.26	4.74	3.14	3.56	30.30	43.91	46.59	47.22	59.16	45.44
S14	Quartz conglomerate (N.Y.)	3.04	2.61	3.28	3.69	5.07	3.54	46.56	43.42	57.02	47.52	57.45	50.39
M04	Quartz (N.Y.)	2.46	3.81	3.84	3.47	3.35	3.39	43.17	49.58	47.76	50.35	56.66	49.50
S15	Kame conglomerate (N.Y.)	3.52	2.47	3.40	3.37	3.49	3.25	48.71	42.67	48.46	47.94	51.91	47.94
S01	Bituminous shale (N.Y.)	3.52	3.28	1.93	4.59	2.82	3.22	48.60	47.00	44.80	40.10	57.90	47.68
S08	Lake Ontario beach sand (N.Y.)	1.82	3.25	4.47	3.39	3.01	3.19	39.90	46.84	46.06	53.79	57.23	48.76
M09	Halloysite (Utah)	3.50	2.75	1.95	4.80	2.87	3.17	46.89	42.65	43.55	38.66	58.40	46.03
M08	Dickite (Utah)	3.75	2.40	3.07	4.20	2.45	3.17	49.75	42.28	42.81	46.14	55.05	47.61
M01	Dolomite (N.Y.)	3.45	2.48	2.61	3.23	3.97	3.15	48.38	42.72	43.82	47.05	56.59	47.71
S05	Arenaceous shale (N.Y.)	2.96	2.38	3.10	4.14	2.84	3.08	45.65	42.23	44.95	46.33	56.27	47.09
S06	Calcareous shale (N.Y.)	3.49	2.83	2.01	3.73	3.22	3.06	48.38	26.47	46.94	40.48	52.95	43.04
M05	Calcite (Ill.)	3.66	3.23	2.52	3.31	2.23	2.99	49.36	46.71	41.70	43.20	51.25	46.44
S03	Illite-bearing shale (N.Y.)	2.45	2.78	2.27	3.64	2.80	2.79	25.61	44.34	44.71	41.89	52.64	41.84
S18	Loess (Kan.)	2.61	1.97	2.61	3.80	2.80	2.76	43.80	40.01	44.73	43.71	54.66	45.38
S09	Dune sand (Mich.)	2.48	3.18	2.87	3.10	1.93	2.71	43.26	46.44	40.20	45.17	51.05	45.22
M07	Kaolinite (S.C.)	3.27	2.54	2.20	2.55	2.82	2.67	47.17	24.02	44.83	41.48	46.40	40.78
M06	Kaolin (Ga.)	2.26	2.61	1.89	3.41	3.17	2.67	41.86	43.41	46.61	39.85	51.12	44.57
S12	Genesee silt (N.Y.)	2.38	1.89	2.36	3.46	3.21	2.66	42.54	39.59	46.93	42.40	51.94	44.68
S10	Colonie sandy loam (N.Y.)	1.98	2.04	2.09	3.74	3.07	2.58	40.48	40.43	46.18	40.19	54.43	44.34
S04	Argillaceous shale (N.Y.)	2.11	1.78	2.38	3.37	2.30	2.39	41.03	39.00	42.03	42.46	50.96	43.10
M10	Attapulgite (Ga.)	2.28	1.74	1.41	3.96	1.98	2.27	41.92	38.80	40.35	37.31	54.77	42.63
S07	Dune sand (N.Y.)	1.25	1.69	3.22	3.04	1.76	2.19	36.52	38.49	39.26	46.93	50.49	42.34
S17	Muck (N.Y.)	2.81	0.41	1.90	2.97	1.93	2.00	45.10	31.80	40.19	40.03	51.32	41.69
M03	Montmorillonite (Wyo.)	1.14	2.03	1.43	2.85	2.18	1.93	35.80	40.36	41.42	37.44	50.15	41.03
S16	Adobe (N.M.)	0.82	1.62	1.88	3.74	1.44	1.90	34.17	38.15	37.52	39.82	31.91	36.31
M11	Glauconite (N.J.)	0.95	1.60	1.40	2.56	1.96	1.70	34.97	40.55	40.26	37.29	29.02	36.42

Table 5 (continued)
VIRUS ADSORPTION TO REFERENCE SUBSTRATES

		Log distribution coefficient						Free energy of adsorption ($-\Delta G$)					
No.	Name of substrate	ECHO-1	ECHO-7	Cox	Polio-2	Reo-3	Ave.	ECHO-1	ECHO-7	Cox B-5	Polio-2	Reo-3	Ave.
S11	Genesee silt loam (N.Y.)	0.48	1.62	1.14	2.62	1.29	1.43	32.37	38.15	36.73	35.92	47.62	38.16
M02	Montmorillonite (Ariz.)	0.58	0.76	1.39	2.42	0.58	1.15	32.96	33.55	32.96	37.21	45.42	36.42
	Average	2.86	2.50	2.56	3.60	2.80	2.76	42.28	40.34	44.09	43.13	52.30	44.44

Note: As in Tables 2 and 3, the substrates are arranged by decreasing overall suitability as virus adsorbents (Column Log Distribution Coefficient — Average).

mercury porosimetry, this difficulty is avoided. Therefore, measurements by this method were used for the calculations of both the distribution coefficient and the calculations of the free energy of adsorption according to Murray.[82] For the calculation of PDADMA adsorption to substrates, a surface measurement intermediate between BET and porosimetry (30 nm pores) may be most appropriate. Calculation of the results per unit surface by porosimetry gave slightly better correlations than BET surface. Furthermore, it appeared appropriate to use log transformations of a variable when the data set consisted mostly of very low values and when there were few high values that then would carry undue weight in the analyses. The results are presented in Table 5.

Table 5 also shows the arithmetic averages of the adsorption results for the five viruses. These averages have no scientific basis except to illustrate that there was a common tendency in the behavior of the five viruses in that the regressions for the averages show a better fit than the regressions for any single virus. (The adoption of sanitary standards based on a hypothetical "average virus" has been tempting to some practitioners in the field; it is not the purpose of this presentation to promote such a concept. Sanitary standards should be conservative and should be based on the most hazardous or least readily adsorbed virus.)

Table 5 shows the results both as $-\Delta G$ and as the distribution coefficients (as base 10 logarithms) which are the coefficients (a) of Equation 11, assuming that the exponent of the Freundlich isotherm is unity. This assumption is often but not always correct. The Freundlich coefficients for poliovirus on Ottawa sand and for some coxsackievirus and echovirus 7 were calculated for several substrates in synthetic freshwater medium. Coefficients of 1.2 to 1.3 were found for poliovirus on Ottawa sand and for coxsackievirus on quartz and kaolinite. The coefficients for echovirus 7 and for the other viruses on the remaining selected substrates were not significantly different from unity. These substrates were Ottawa sand, Arizona montmorillonite, Lake Ontario beach sand, and magnetite. The distribution coefficients equal the Henry coefficient. The effect of a higher exponent would be to overestimate adsorption of viruses at concentrations lower than those used in our experiments (10^6 to 10^8 PFU/mℓ) but at environmental concentrations ($<10^{-1}$/mℓ), the error would be negligible.

The first porosimetry measurements were carried out on 10 substrates selected to cover the range from weak to strong adsorption of viruses. These were the seven minerals, dolomite (M01), montmorillonite (M02), South Carolina kaolinite (M06), halloysite (M08), dickite (M09), magnetite sand (M12), and glauconite (M11), and the three soils, New York dune sand (S07), New York beach sand (S08), and muck (S17). The free energies of adsorption were calculated. More than 90% of the variance in free energy of adsorption was explained by three variables, i.e., the content in organic carbon, and electrophoretic mobility, with some additional improvement in the prediction by acid-extractable iron.[81] For the ten substrates, both organic carbon and acid-extractable iron were considered predictors of van der Waals attraction. It was clear that for a broader selection of substrates, other metal oxides would have to be considered also.

Subsequent analyses were carried out on all 30 substrates using multiple regression procedures and tests for the normal distribution of the residuals by normal distribution plots and the Kolmorgorov-Smirnov test for goodness of fit. The variables included in the analyses were essentially those that from theory and experimental manipulations were likely to be important. Trial runs were used to examine the sensitivity of the tests to various representations and transformations of the variables. These regression analyses therefore were not primarily intended to detect additional important variables but to see whether and how the individual correlations would join to provide a predictive capability for viral adsorption to a substrate characterized by measurements of these properties. The assumption of a linear relationship between variables is not always justified, e.g., a sigmoid relationship was established for the effect of pH.[57]

Calculations for all 30 substrates with the same variables did not produce as good a correlation as for the 10 selected substrates. Most of these substrates were soils, i.e., mixtures of minerals that owe adsorbent properties to components which do not necessarily dominate the bulk composition of the material. Therefore, it needed to be determined whether better correlations were obtained for the substrates that are relatively pure minerals. However, here too, certain relevant properties may not have been adequately represented by the measurements. The 12 minerals among the reference substrates are dolomite, calcite, quartz, magnetite, attapulgite, halloysite, dickite, glauconite, and two each of the clays montmorillonite and kaolinite (Table 1). For these, the distribution coefficients, a, predicted from mass density (spec. gravity, ρ); equilibrium pH in synthetic freshwater (pH); electrophoretic mobility (mob); organic C (C); the sum of HNO_3-extractable iron, aluminum, and calcium (met); and PDADMA (pdp) adsorption are

$$\log(a) = 2.19 + 0.25\,\rho - 0.01\,\text{pH} + 0.75\,\text{mob} - 0.32\log(C) + 0.57\log(\text{met}) + 0.17\log(\text{pdp}) \quad (r^2 = 0.806;\ 5\ \text{d.f.})$$

or, in this particular sample, with $r^2 = 0.790$:

$$\log(a) = 3.00 + 0.154\,\rho + 0.601\,\text{mob} + 0.24\log(\text{met}) - 0.144\log(\text{pdp})$$

where ρ is in g cm^{-3}, mob in ($\mu m\ sec^{-1}$)/(V cm^{-1}), C in %, met in $\mu g\ g^{-1}$, and pdp in mg m^{-2}. For the free energy of adsorption, the regression results are

$$-\Delta G = 32.87 + 1.35\,\rho + 1.02\,\text{pH} + 6.1\,\text{mob} - 3.2\log(C) + 4.7\log(\text{met}) + 3.2\log(\text{pdp}) \quad (r^2 = 0.826;\ 5\ \text{d.f.})$$

This can be reduced, without much loss in accuracy ($r^2 = 0.806$), to

$$-\Delta G = 34.2 + 3.0\,\rho + 7.3\,\text{mob} - 4.2\log(C) + 5.9\log(\text{met}) + 6.4\log(\text{pdp})$$

In both cases, the greatest reduction in sums of squares is from mob followed by approximately equal contributions from met and pdp.

Of the 12 minerals, the greatest deviations were found with Arizona montmorillonite, which performed worse than predicted, possibly as the result of its crystal structure, and Wyoming montmorillonite and quartz, which performed better than predicted, the quartz possibly because of surface contamination with iron oxide (the sample was ground "Herkimer diamond" known for small embedded crystals of iron oxide).

The single-variable correlation coefficients are shown in Table 6. The strong influence of electrophoretic mobility is in part due to its very strong negative correlation with PDADMA adsorption and vice versa. Both measure surface charge development. For the predictors of van der Waals attractive forces, the situation is less clear. Specific gravity and metal content would be positively, organic carbon negatively correlated. The first two behave as expected. Organic carbon, which alone, on a basis of substrate weight (rather than surface), had shown a strong negative correlation with adsorption, does not stand out, but is conspicuously and positively correlated with metal content. This resembles the strong correlation of organic carbon with certain chelated metals described earlier. It represents a confounding factor in regression analysis.

The predictions of adsorptive behavior as logarithms of the distribution coefficient for each of the five viruses and for their average for the 30 substrates are shown in Table 7 together with the individual residuals. The reduction of sums of squares here is no less significant than with the 12 minerals; in fact, it is slightly more so and just significant at

Table 6
SINGLE-VARIABLE CORRELATION MATRIX FOR VIRUS ADSORPTION TO 12 MINERALS[a]

Variable	Sp. gr.	Equ. pH	Mobil.	Log C	Mog Me	Lg PDADMA	ΔG
Spec. gravity	1	−0.051	0.326	0.029	0.137	−0.468	0.320
Equilibrium pH	−0.051	1	−0.316	0.296	0.136	0.625	0.535
El. mobilities	0.326	−0.316	1	−0.035	−0.036	−0.801	0.809
Log organic C	0.029	0.296	−0.035	1	0.797	0.257	0.103
Log (Fe + Al + Ca)	0.137	0.135	−0.036	0.797	1	−0.009	0.178
Log PDADMA ads.	−0.468	0.625	−0.801	0.257	−0.009	1	−0.557
Log dist coeff	0.381	−0.261	0.859	0.045	0.174	−0.749	

[a] Coefficients significant at or about 0.576 ($p = 0.05$) and 0.708 ($p = 0.01$). Units of measurement are the same as in the remainder of this report. PDADMA adsorption is on the basis of surface area as determined by Hg porosimetry. Log distribution (Henry) coefficients and $-\Delta G$ are averages for five viruses.

the $p = 0.01$ level, associated with an increase in degrees of freedom. The contributions of the variables are evident from the variance/covariance matrix in Table 8. By scanning the variances on the diagonal one notices that mobility again makes the greatest contribution to the regression, whereas specific gravity and equilibrium pH make the smallest. Mobility and PDADMA adsorption owe part of their respective contributions to their correlation with each other. For $-\Delta G$ the results are similar in principle:

$$-\Delta G = 26.9 + 0.70\,\rho + 1.75\,\text{pH} + 3.8\,\text{mob} - 2.5\log(\text{C}) + 3.4\log(\text{met}) + 0.256\log(\text{pdp}) \quad (r^2 = 0.531,\ 23\ \text{d.f.})$$

For unknown reasons mobility makes the greatest, pdp and density (as ρ) make the smallest contribution to regression.

Table 9 shows the individual correlations for the individual independent and dependent variables. $-\Delta G$ and the distribution coefficients are very closely correlated as would be expected.

The principal conclusion that can be drawn from this example is that the variables behave as expected. The analysis also confirms that the present choice of variables is to some extent redundant. The variables substitute for each other to varying degrees in the several substrates. Compared with our earlier calculations, it appears, e.g., that the introduction of mass density (as ρ) has consistently lessened the relative contribution of organic carbon. Results are most consistent when all of the variables shown in the equations are retained. With some viruses or groups of substrates, the contribution of the one or other variable may not be significant.

The predictive power of multiple linear regression with a linear model does not exceed that of an interpretation of relevant measurements on a substrate in terms of the DLVO theory. The remaining uncertainties about the properties of the substrates are presumably in the presence of highly adsorbent components on the surface and generally in differences between surface vs. bulk composition. The uncertainties in the IEPs of the picornavirusesare equally significant, together with uncertainties relating to changes in state of aggregation and relative infectivity. Table 7 illustrates the uncertainties of the individual predictions. It mirrors earlier presentations[81] on log virus removal on a weight basis for regression on important single variables. For "average virus" behavior, about 1 to 1.5 log units in each direction should be allowed for substrate-to-substrate variation unexplained by regression. Virus-to-virus variation appears to vary from substrate to substrate; some substrates adsorb viruses more consistently than others.

Table 7
MULTIPLE REGRESSION, LOG DISTRIBUTION COEFFICIENTS

No.	Name of substrate	Echo-1	Echo-7	Cox-B5	Polio-2	Reo-3	(Ave.)	Ave. virus
	Solution Vectors							
	Specific gravity	−0.187	0.092	1.040	0.051	−0.045		0.194
	Equilibrium pH	0.016	0.160	0.312	0.093	0.120		0.134
	Electroph. mobility	0.727	0.566	0.578	0.262	0.698		0.632
	Log organic carbon	−0.293	−0.516	−0.317	−0.393	−0.584		−0.393
	Log Fe + Al + Ca (HNO3)	0.967	0.408	0.172	0.656	0.976		0.614
	Log PDADMA Adsorption	−0.415	−0.249	0.262	0.115	0.051		−0.014
	Constant	2.470	0.805	−2.340	2.250	0.920		0.919
	Coeff of distribution	0.366	0.453	0.601	0.372	0.376		0.532
	Residuals (Better-Than-Predicted Adsorption)							
S13	Brown sandstone (N.Y.)	1.91	1.15	0.79	1.37	1.75	1.39	1.350
S02	Fossilif. limey shale (N.Y.)	1.12	0.09	0.66	0.93	0.11	0.58	0.384
S15	Kame conglomerate (N.Y.)	0.85	0.15	0.91	0.00	0.96	0.57	0.602
S08	Lake Ontario beach sand (N.Y.)	−0.25	1.03	1.23	−0.18	0.48	0.46	0.541
S19	Muck (N.Y.)	1.31	−0.46	0.59	0.09	0.40	0.39	0.478
M08	Halloysite (Utah)	0.31	0.34	0.13	1.10	−0.11	0.35	0.402
M02	Montmorillonite (Wyo.)	0.42	0.75	−0.40	0.22	0.74	0.35	0.018
M04	Quartz (N.Y.)	−0.31	0.77	0.95	−0.37	0.25	0.26	0.331
M12	Magnetite sand (Calif.)	0.71	−0.07	−0.03	0.50	−0.05	0.21	0.298
S04	Argillaceous shale (N.Y.)	0.09	−0.21	0.60	0.35	0.17	0.20	0.239
S03	Illite-bearing shale (N.Y.)	−0.35	0.70	0.21	0.07	0.20	0.17	0.101
S05	Arenaceous shale (N.Y.)	0.24	−0.21	0.39	0.42	−0.09	0.15	0.115
S06	Calcareous shale (N.Y.)	0.60	0.29	−0.65	0.02	0.30	0.11	−0.015
M06	Kaolinite (S.C.)	0.59	−0.01	0.32	−0.80	0.38	0.09	−0.059
M05	Calcite (Ill.)	0.70	0.84	0.11	−0.45	−0.74	0.09	0.018
S01	Bituminous shale (N.Y.)	0.27	0.63	−0.95	0.66	−0.38	0.05	0.311
S18	Loess (Kan.)	0.13	−0.41	0.06	0.19	0.12	0.02	0.037
M09	Dickite (Utah)	0.64	−0.50	0.12	0.29	−0.72	−0.03	−0.100
S14	Quartz conglomerate (N.Y.)	−0.50	−0.27	0.12	−0.87	0.98	−0.11	−0.201
M11	Glauconite (N.J.)	−0.64	0.09	−0.06	−0.23	0.12	−0.14	−0.128
S10	Colonie sandy loam (N.Y.)	−0.51	−0.28	−0.61	0.17	0.46	−0.15	0.067
M01	Dolomite (N.Y.)	−0.15	−0.28	−0.30	−0.82	0.63	−0.18	−0.991
S09	Dune sand (Mich.)	−0.29	0.49	0.11	−0.72	−1.07	−0.30	−0.347
S12	Genesee silt (N.Y.)	−0.38	−0.68	−0.46	−0.29	0.33	−0.30	−0.454
M07	Kaolin (Ga.)	−0.27	−0.44	−0.99	−0.23	0.38	−0.31	−0.184
M10	Attapulgite (Ga.)	−0.50	−0.81	−0.60	0.29	−0.94	−0.51	−0.371
M02	Montmorillonite (Ariz.)	−0.46	−0.76	−0.69	−0.44	−1.27	−0.72	−0.514
S07	Dune sand (N.Y.)	−1.47	−0.88	0.21	−0.60	−1.03	−0.75	−0.706
S11	Genesee silt loam (N.Y.)	−1.85	−0.29	−1.03	−0.69	−0.95	−0.96	−0.948
S16	Adobe (N.M.)	−1.96	−0.76	−0.76	−0.01	−1.40	−0.98	−0.991
	(Poorer-Than-Predicted Adsorption)							
	SD	0.964	0.656	0.692	0.641	0.834		−1.5

Note: Units of measurement for the variables are the same as given throughout this report (see Tables 2 and 3).

Table 8
VARIANCE/COVARIANCE IN % OF TOTAL VARIANCE, "AVERAGE VIRUS", 30 SUBSTRATES[a]

Variable	Sp. gr.	Equ. pH	Mobil.	Log C	Log Me	Lg PDADMA
Log distribution coefficient						
Spec. gravity	1.995					
Equilibrium pH	0.441	2.070				
El. mobilities	0.557	−2.620	31.745			
Log organic C	1.668	−0.322	−4.292	14.184		
Log (Fe + Al + Ca)	1.891	1.816	−1.761	−4.289	16.349	
Log PDADMA ads.	0.030	−0.031	0.126	0.166	0.027	0.010
−ΔG						
Spec. gravity	0.758					
Equilibrium pH	0.607	10.338				
El. mobilities	0.355	−6.108	33.534			
Log organic C	1.135	−0.795	−4.872	17.300		
Log (Fe + Al + Ca)	1.100	3.830	−1.709	−4.472	14.572	
Log PDADMA ads.	−0.058	0.217	−0.403	−0.572	−0.081	0.101

[a] Units of measurement are the same as in the remainder of this report.

Table 9
SINGLE-VARIABLE CORRELATION MATRIX FOR VIRUS ADSORPTION TO REFERENCE SUBSTRATES

Variable	Sp. gr.	Equ. pH	Mobil.	Log C	Log Me	Lg PDADMA	−ΔG
Spec. gravity	1	0.262	0.002	0.110	0.275	−0.043	0.274
Equilibrium pH	0.262	1	−0.323	−0.006	0.312	0.213	0.254
El. mobility	0.002	−0.323	1	0.307	−0.077	−0.219	0.360
Log organic C	−0.110	−0.006	0.307	1	0.301	0.316	−0.133
Log (Fe + Al + Ca)	0.275	0.312	−0.077	0.301	1	−0.067	0.347
Log PDADMA abs.	−0.043	0.213	−0.219	0.316	−0.067	1	−0.250
Log dist coeff	0.382	0.097	0.407	−0.156	0.348	−0.348	0.909

Note: Coefficients of 0.361 or higher are significant at $p = 0.05$, coefficients of 0.463 or higher at $p = 0.01$ (28 d.f.) Units of measurement are the same as in the remainder of this report. PDADMA adsorption is on the basis of surface area as determined by Hg porosimetry. Log distribution (Henry) coefficients and −ΔG are average for five viruses.

V. ADSORPTION TO CLASSES OF NATURAL MATERIALS

Numerous isolated experiments on adsorption of viruses to natural substrates have been described in the literature. The conditions of these experiments have been so varied that in only a few instances can results be compared or expressed in some standardized format. In the following discussion, published results on major groups of natural materials are reviewed from the standpoint of how the chemical nature of these materials may explain their behavior. The discussion is limited to studies of animal viruses. Bacteriophages are considered when they resemble animal viruses or when experimentation on animal viruses is lacking.

A. Quartz and Sands

Silica is a relatively inert, poorly ionized material. The IEP of silanol groups is about pH 1 and that of siloxane groups about 9.3. Both are responsible for the IEP of pure quartz. The number of siloxane groups, and with it the IEP, increases upon grinding or acid treatment of the pure material.[83] Murray[82] used a strong acid-cleaned sample of silica (α-SiO_2) in adsorption experiments with poliovirus and measured a ζ potential of -44.5 ± 8.5 mv at an ionic strength of 0.02. This author also gives a Hamaker constant, A_{11}, for this material of 8.0 to 8.8 10^{-13} erg and $A_{123} = 0.3$ to 0.49×10^{-13} erg for the system quartz-water-poliovirus. The free energy for poliovirus adsorption by this material was -32.4 ± 2.4 kJ mol^{-1}. Murray concluded that the significant adsorption which occurred despite unfavorable double-layer interaction was, within experimental error, accounted for by van der Waals attraction. In our experiments with the unpurified mineral, we found much stronger interaction ($-\Delta G = 50.53$ kJ mol^{-1}). The analysis (Table 2) shows a substantial iron content and some alumina, which are likely to be responsible for the increased adsorption. As noted above, our mineral contains iron oxide inclusions, which become a surface contamination of the ground material.

Zerde et al.[83a] adsorbed poliovirus 1, reoviruses 1 and 3, and coliphages MS2 and T2 to colloidal silica modified to carry positive or negative charges. The experiments were done in buffers with an ionic strength of 0.1 at pH levels of 4.0, 6.4, and 8.5. Viruses adsorbed to pelleted silica and viruses in the supernates were assayed. At pH values below their isoelectric points, conditions favoring a positive charge on the viruses, all viruses adsorbed exclusively to negatively charged silica. Conversely, at pH values above their isoelectric points, conditions favoring a negative charge on the viruses, all viruses adsorbed to positively charged silica. In near isoelectric state, the viruses adsorbed to both positively and negatively charged silica, but to a lesser degree.

Ottawa sand (ASTM C190, 20 to 30 mesh) as used in water filtration plants was used as a reference material in our studies. It was a relatively poor adsorbent on a weight basis because of its small surface area. The other sands used in our study showed great variation, i.e., they were either much better or much poorer adsorbents than predicted by regressions. The fact that brown sandstone showed the greatest deviation (better-than-predicted performance) also suggests that surface components are of critical importance for adsorption by relatively inert materials. On the basis of surface area, the sandstone performed almost as well as magnetite, which contained five times as much iron oxide. Sand filters and columns have been evaluated for virus removal, with varying results.[35,39,45,84-86] The performance of sands was greatly dependent on the presence of divalent cations in the percolating medium. Adsorption was reduced at low ionic strengths and in the presence of organic matter or detergents. Lo and Sproul[87] may have observed artificially high adsorption because they used a steel-ball mill to grind the silicic minerals. In our experience, the use of such a mill introduces considerable contamination of the substrate with iron oxides. The use of a tungsten carbide mill eliminates this error.

B. Clay Minerals

The clay minerals are represented by the kaolin group of 1:1 layer crystals, which includes kaolinite, dickite, and halloysite (Table 1). They consist of alternating sheets of tetrahedrally coordinated cations (Si- or Al-O tetrahedra) and of octahedrally coordinated cations (Al^{3+}, Mg^{2+}, Fe^{2+} etc.). The 7.2-AE-thick sheets are held together by hydrogen bonds. In their crystalline form, these clays have a relatively small surface area and low surface charge. The IEP appears to be generally near pH 3.0 but the groups at the edges of the plates are thought to have an IEP near pH 5.0 (see below). Among the clay minerals in our study were dickite, a distorted kaolinite, and halloysite, a hydrated and disordered kaolinite mineral that derives a negative charge from isomorphic replacement.

Among the 2:1 family of clay minerals are montmorillonite, mica, chlorite, talc, vermiculite, and illite. Two sheets of Si– or Al– tetrahedra are bound to one sheet of the

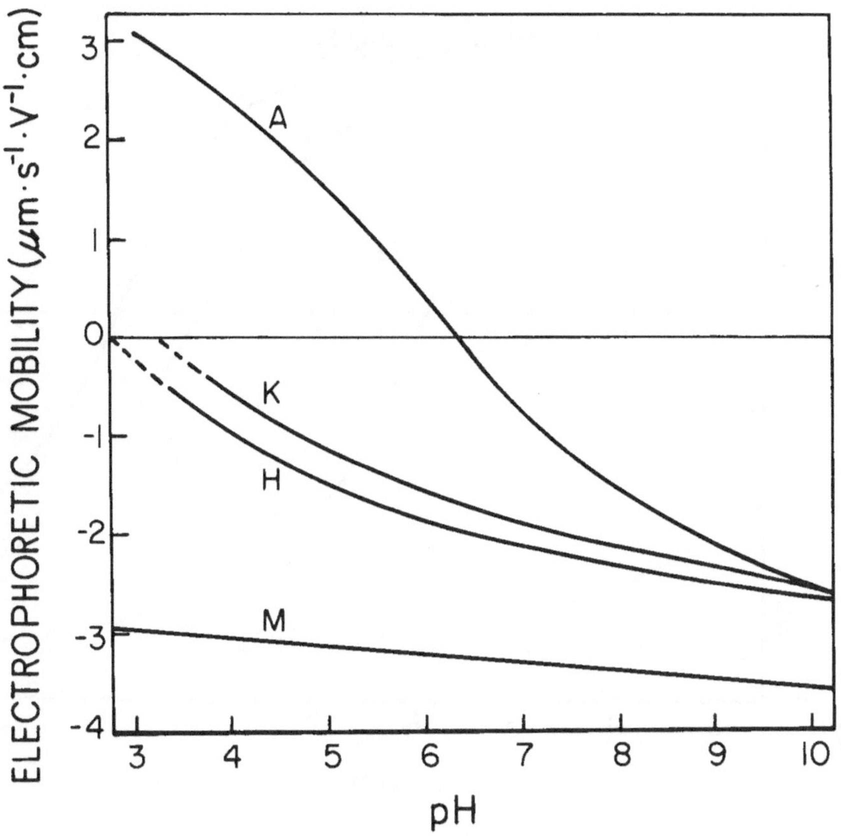

FIGURE 5. Electrophoretic mobilities of clay minerals in dilute electrolytes. A = Ohaupo allophane, New Zealand; K = Mt. Egerton kaolinite; H = Mangawara halloysite; M = Wyoming montmorillonite. A, H, M in 10^{-3} *M* NaCl. (Adapted from Taylor, D. H., Ph.D. thesis, University of Waikato, Hamilton, New Zealand, 1977); K in 10^{-4} *M* KCl (Adapted from Buchanan, A. S. and Oppenheim, R. C., *Aust. J. Chem.*, 21, 2367, 1968.)

octahedra, forming a 10 AE-thick unit. Extensive isomorphic replacement leads to a strong negative charge. Cations adsorbed between the layers neutralize this charge; hydration of the layers causes the layered crystals to swell. Interlayer van der Waals forces are relatively weak. These minerals tend to have large surface areas as measured by the gas adsorption technique and have high cation-binding capacities, including the penetration of small cations between the layers. The IEP are low (usually below three). Montmorillonite is a typical, highly ordered representative. Attapulgite is a fibrous 2:1 clay mineral with the sheets joined by oxygen atoms. Open channels running parallel to the fibers give it some zeolitic properties. A high surface area and a moderate cation-exchange capacity are typical. Glauconite is a hydrous silicate of iron and potassium which precipitates in sand-size crystals in oceanic sediments near continental shores. It readily binds other metal ions.

The adsorptive behavior of the clay minerals for negatively charged ions and colloids, to a large extent, is related to their development of positive charges at the plate edges.[56,88] The IEP of the plate edges of kaolinite is near pH 7.3.[89] Consistent with this view, adsorption increases with loss of order in the crystal structure and parallels increasing pH-dependent charge development, i.e., increasing electropositivity at decreasing pH in combination with an increase in IEP. The increase in adsorption from montmorillonite to halloysite and kaolinite and finally to the volcanic, amorphous allophane illustrates these relationships. Ohaupo allophane (New Zealand) has an IEP of 6.5 (Figure 5). This value is between that of aluminum

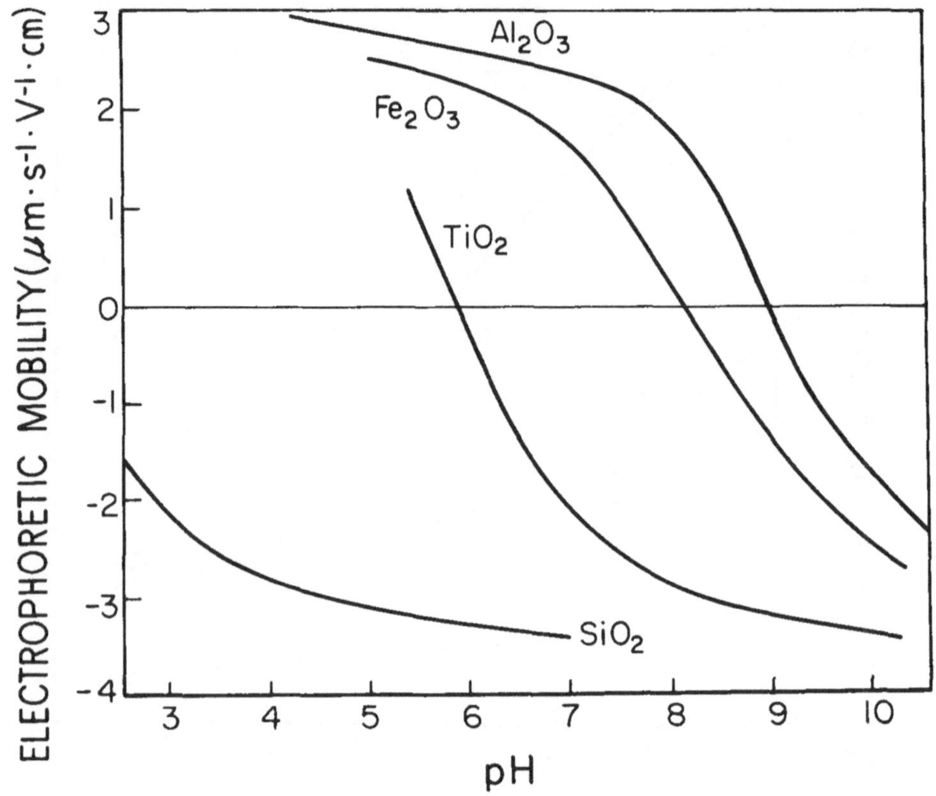

FIGURE 6. pH-dependent electrophoretic mobility of hydrous oxides in 10^{-3} *M* in different electrolytes. (From Fuhs, G. W. and Taylor, D. H., *Water Reuse,* Middlebrooks, E. J., Ed., Ann Arbor Science Publishers, Ann Arbor, Mich., 1982, 603. With permission.)

silicate and those of the oxides of iron and aluminum with IEPs at pH 8.0 to 9.0 (Figure 6).[28,56,88,90] Removal of the oxides decreases the IEP of allophane from pH 9.0 to near 5.0.[91] The same oxides appear during the weathering of feldspars and are instrumental in the ability of soils to remove electronegative ions, e.g., phosphates, from wastewater effluents during recharge. For a more detailed discussion see Dixit.[92]

In our studies, we consistently observed virus removal to increase in the order montmorillonite < kaolinite < halloysite < dickite, the last two sometimes showing equal performance (Table 5). Attapulgite ranked between the montmorillonites and kaolinite. Glauconite was a poor adsorbent of viruses. This correlates well with its reported ability to bind heavy metal cations.[93] The ability to predict virus adsorption was good with Wyoming montmorillonite, dickite, South Carolina kaolinite, and glauconite. Arizona montmorillonite, Georgia kaolin, and attapulgite performed worse, halloysite somewhat better than predicted.

The adsorptive behavior of clay minerals as a function of the ionic environment and pH has been reviewed by several authors.[2,94,95] Adsorption of the low-IEP virus, coliphage T2 by montmorillonite, illite, and kaolinite at pH 7.0 increased with ionic strength; adsorption was strongest upon addition of $Al_2(SO_4)_3$.[43] In our experiments, the high (i.e., unfavorable) equilibrium pH in artificial freshwater of two montmorillonite samples was noteworthy (Table 2). At this pH, where double-layer repulsion is expected to be strong, increases in electrolyte content increased poliovirus adsorption.[57] Cation binding to a montmorillonite enhances adsorption markedly.[61,96]

C. Metal Oxides

Since the early work with aluminum oxide and poliovirus[23,24] and with hematite and myxovirus,[97] there have been many studies showing the great affinity of metal oxides for viruses. In water treatment processes, an artificially produced floc of aluminum or magnesium hydroxide is used to remove viruses.

The physical chemistry of the natural metal oxides will not be discussed here exhaustively. The surface properties of each vary with the composition of the suspending medium as a result of adsorption of counterions and other species. van der Waals attractive forces of metal oxides are high and are exceeded only by those of metals. Murray[82] gives individual Hamaker coefficients for metal oxides (A_{11}) of 10.6 to 15.5 $\times$ 10^{-13} erg and a combined coefficient for the metal oxide-water-virus system (A_{123}) of 0.53 to 1.06 $\times$ 10^{-13} erg. In addition, pH-dependent charge development at ambient pH is favorable for the attraction of negatively charged colloids. To illustrate this point, Figure 6 shows the electrophoretic mobilities of several oxides in 10^{-3} M indifferent electrolyte. Murray[82] measured the ζ potential for αFe_2O_3 as -28.8 ± 2.5 mv at 0.02 ionic strength. Corresponding values for MnO_2 and CuO were -20.7 ± 5.9 and -17.6 ± 6.1 mv, respectively.

The free energy of adsorption ($-\Delta G$) of poliovirus to αFe_2O_3 at 0.02 ionic strength was 52.9 ± 4.0 kJ mol^{-1}.[82] This is in very good agreement with our free energy of 48.59 kJ mol^{-1} for poliovirus on magnetite sand, predominantly Fe_3O_4, at 0.003 ionic strength. The values for the five viruses studied ranged from 62.82 for the most strongly bound, reovirus, to 46.50 for the least strongly bound, echovirus 7. The use of magnetite as a virus adsorbent on a technical scale was recommended by Bitton et al.[26]

Copper oxide adsorbs viruses well and causes the disintegration of adsorbed poliovirus. Some loss of infectivity also occurs upon adsorption to manganese dioxide and aluminum oxide, whereas iron oxide and silica produce no such effect.[100] Virus disintegration by several mechanisms has also been observed in other soils.[101]

D. Coal

Anthracite coal is used as natural filter material in water and advanced wastewater treatment plants. A bituminous coal from India adsorbed T4 and MS2 bacteriophages moderately well.[37,38] MS2 phage adsorption followed a Langmuir isotherm, with saturation at 0.1% surface coverage. Adsorption increased with decreasing pH, with an abrupt increase below pH 7.0, and with increasing ionic strength. Formation of hydrogen bonds between the virus and the coal is suspected as adsorption was correlated with the hydrogen content of the coal and was markedly reduced by treating the coal with acetyl chloride to convert surface hydrogen to acetate and anhydride groups.[37] Because the coal did not contain macro- or micropores accessible to the virus, the adsorption effectiveness per unit surface area appeared to be much greater in coal than in activated carbon. The IEP of Upper Silesian coals varied between 4.5 and 8 as a function of composition. The IEP increased with the degree of coalification and decreased as a result of surface oxidation.[98,99]

E. Organic Material

As noted above, natural surfaces are not clean but carry adsorbed organic materials, at least as monomolecular films. The nature of these microlayers varies with the nature of the surface and with the composition of the aqueous medium. Polar lipids liberated from decaying microorganisms and other living matter are avid seekers of surfaces and interfaces. The nature of these microlayers has not been studied thoroughly, but amino acids, oligo- and polypeptides and other surface-active materials, and nucleotides and bases are also likely to be significant components. The constituents of the microlayers in turn are utilized by bacteria which excrete exopolymer (glycocalyx) and which form a microfouling community by encapsulating themselves and other microorganisms. The presence of amino acids in microlayers raises the question of randomly arranged yet specific adsorption sites for viruses. Specific amino acid residues on biologic surfaces are instrumental in the binding of viruses to the surfaces of susceptible animal tissue during infection.[102,103] The low surface coverage of viruses on many natural surfaces at saturation may suggest the involvement of specific adsorption sites.

Bacterial exopolymers have been characterized to some extent. They contain mostly sugars and some dissociating carboxyl groups and, for this reason and their high degree of hydration, are poor adsorbents for viruses. Fuhs et al.[64a] produced bacterial overgrowth on several of the natural sands and magnetite sand, with a consequent dramatic decrease of virus adsorption. Retention of viruses from suspension on magnetite decreased from 99.9 to 40%.
to 40%.

Humic matter in various forms of transformation (to peat and coal) has an IEP of pH 3.0 or lower and upon adsorption confers this property to mineral surfaces.[79] Organic matter from humified plant material gave a more negative charge to amorphous aluminum silicate. The charge varied with the molar ratio Al/(Al + Si) of the clay and was largest at low ratios.[104] Organic soil extract was prepared by extraction with 0.1 *M* Na-pyrophosphate and separated into fractions of MW $>10^5$ and 10^4 to 10^5 by ultrafiltration. Isoelectric focusing of the first fraction produced several bands at pH 6.5 to 5; the second fraction had three main bands; and mixtures of the two fractions simulated the pattern of the soil extract.[105] Reduction of virus adsorption in the presence of humic matter has been documented extensively.[27,76,77] This humic material confers a more negative charge to the mineral surfaces or competes directly with the virus for adsorption sites.

Increased organic content makes topsoils less suitable as virus adsorbents than the corresponding subsoils. However, dosing with sewage effluents over several decades does not exhaust the virus-binding capacity of a soil, probably because the virus is degraded and its constituents are either metabolized or become part of the natural background of trace organic constituents in the soil (Fuhs et al.[64a]).

Association of viruses in sewage to allochthonous or activated sludge solids has been noted by several workers, but no information on the mechanisms has been given.[106-109]

VI. SUMMARY AND CONCLUSIONS

Virus adsorption to natural materials is a well-studied phenomenon, yet there are still errors of approximately one order of magnitude, in each direction, in the mathematical predictions of viral adsorption. This is understandable if the complexity of natural surfaces and the variability of the aqueous environment are considered. As is shown in this review, we find most if not all observations to support the DLVO theory of interaction of charged colloids. The predictions are good with pure (and practically pure) materials. Discrepancies between the bulk and surface compositions of complex materials such as soils require a careful selection of analytical techniques for their characterization. The additional complexity introduced by adsorbed microlayers, particularly of organic substances, has not yet been adequately investigated.

Incomplete characterization of viral surface properties in dilute aqueous media introduces major uncertainties into the prediction of virus-adsorptive behavior. The effect of conformation changes on IEP in the picornaviruses as a function of the composition of the medium to which the virus was previously exposed is certainly of major consequence. Other uncertainties arise from the formation of aggregates, viral association with other solids, and possible binding of other proteins, e.g., fecal antibodies.

Nevertheless, some conclusions on the control of environmental viruses can be drawn and certain recommendations can be made:

1. The predictions of the DVLO theory are qualitatively useful despite uncertainties remaining in quantitative predictions.
2. The predictions of the theory apply mostly to monodisperse virus particles. Different kinetics apply to clumps of viruses and particle-associated viruses which typically are removed by sedimentation or filtration rather than by adsorption. Upon decay or other dissociation of particulate matter, however, liberated virus particles will obey the theory. The theory, therefore, is a valuable tool under any circumstances.
3. London-van der Waals forces and double-layer compression may both be maximized by the use of appropriate substrates and by chemically conditioning the aqueous phase that contains the virus particles.[5] Such treatment may consist of pH adjustments and/or additions of calcium[85] or aluminum[110] ions. Both treatments have greatly improved removal of viruses in sand columns. If wastewater treatment is combined with groundwater recharge, these treatment steps may be combined with the removal of organic particulates which then further assists the removal of viruses.
4. Control of viruses is best practiced during treatment of wastewater, rather than potable water. Chemical conditioning and adsorption-flocculation-filtration processes at present are more common in potable water treatment. Even though chemical additions for optimizing adsorption processes need not be higher than they are in potable water treatment, the operating parameters in potable water treatment are more critical because treatment times and flow paths are limited. When considered a part of wastewater treatment, soil formations in recharge areas or adsorption-sedimentation processes in receiving waters always provide substantial additional protection. In virus removal, disinfection should not be the principal protective barrier, but rather a means for destroying monodisperse virus particles that have escaped other treatment. If applied in this manner, doses of disinfectants required may be low.
5. Removal of viruses by treatment processes is no more difficult than the removal of bacteria; theory even suggests that viruses may be more easily removed by adsorption than larger particles. In addition, viruses do not grow outside their hosts. Malina[111] discussed the concept of allocating removal of viruses to unit treatment processes as a viable concept for the control of viral contamination of water for reuse.[5] Uncertainties much greater than one or two orders of magnitude are encountered in the control of many environmental contaminants and are effectively dealt with by adopting appropriate safety margins and measures. If all these facts are considered, it appears that in many locations removal of viruses is not difficult. In these locations, the need to remove viruses should not be a major impediment to water reuse.

REFERENCES

1. **Drewry, W. A. and Eliassen, R.,** Virus movement in groundwater, *J. Water Pollut. Control Fed.*, 40, R257, 1964.
2. **Bitton, G.,** Adsorption of viruses onto surfaces in soil and water, *Water Res.*, 9, 473, 1975.
3. **Gerba, C. P., Wallis, C., and Melnick, J. L.,** Fate of wastewater bacteria and viruses in soil, *J. Irrig. Drain. Div. Am. Soc. Civ. Eng.*, 101, 157, 1975.
4. **DuBoise, S. M.,** Poliovirus Survival and Movements in Soil Following Application of Wastewater Effluent, M. A. thesis, University of Texas, Austin, 1977.
5. **Fuhs, G. W. and Taylor, D. H.,** Virus uptake by minerals and soils, in *Water Reuse*, Middlebrooks, E. J., Ed., Ann Arbor Science Publishers, Ann Arbor, Mich., 1982, 603.
6. **Murray, J. P. and Parks, G. A.,** Poliovirus adsorption on oxide surfaces — correspondence with DLVO-Lifshitz theory of colloid stability, *Adv. Chem.*, 189, 97, 1980.

7. **Renn, C. E.,** The bacteriology of interfaces; principles and applications in aquatic microbiology, in *Proc. 1963 Rudolfs Res. Conf. Rutgers Univ.*, Heukelekian, H. and Dondero, N. C., Eds., John Wiley & Sons, New York, 1964, 193.
8. **Derjaguin, B. V. and Landau, L. D.,** Theory of the stability of strongly charged lyophobic soils and of the adhesion of strongly charged particles in solutions of electrolytes, *Acta Physicochim. U.S.S.R.*, 14, 633, 1941.
9. **Verwey, E. J. W. and Overbeek, F. T. G.,** *Theory of the Stability of Lyophobic Colloids*, Elsevier, Amsterdam, 1948.
10. **Prieve, D. C. and Ruckenstein, E.,** Role of surface chemistry in particle deposition, *J. Colloid Interface Sci.*, 60, 337, 1977.
11. **Gregory, J.,** The calculation of Hamaker constants, *Adv. Colloid Interface Sci.*, 2, 396, 1969.
12. **Weiss, L., Nir, S., Harlos, J. P., and Subjeck, J. R.,** Long distance interactions between Ehrlich ascites tumor cells, *J. Theor. Biol.*, 51, 439, 1975.
13. **Lyklema, J.,** Principles of the stability of lyophobic colloidal dispersions in non-aqueous media, *Adv. Colloid Interface Sci.*, 2, 65, 1969.
14. **Weiss, L. and Harlos, J. P.,** Cell contact phenomena and their implication in cell communication, in *Intercellular Communications*, deMello, W. C., Ed., Plenum Press, New York, 1977, 33.
15. **Rogers, H. J.,** Adhesion of microorganisms to surfaces: some general considerations of the role of the envelope, in *Adhesion of Microorganisms to Surfaces*, Ellwood, D. C., Melling, J., and Rutter, P., Eds., Soc. Gen. Microbiol. and Academic Press, London, 1979, 29.
16. **Marshall, K. C., Stout, R., and Mitchell, R.,** Mechanism of the initial events in the sorption of marine bacteria to surfaces, *J. Gen. Microbiol.*, 68, 337, 1971.
17. **Vold, M. J.,** The effect of adsorption on the van der Waals interaction of spherical colloidal particles, *J. Colloid Sci.*, 16, 1, 1961.
18. **van Oss, C. J., Good, R. J., Neumann, A. W., Wieser, J. D., and Rosenberg, P.,** Comparison between attachment and detachment approaches to the quantitative study of cell adhesion to low-energy solids, *J. Colloid Interface Sci.*, 59, 505, 1977.
19. **Fletcher, M.,** The attachment of bacteria to surfaces in aquatic environments, in *Adhesion of Microorganisms to Surfaces*, Ellwood, D. C., Melling, J., and Rutter, P., Eds., Soc. Gen. Microbiol. and Academic Press, London, 1979, 87.
20. **Lips, A. and Jessup, N. E.,** Colloidal aspects of bacterial adhesion, in *Adhesion of Microorganisms to Surfaces*, Ellwood, D. C., Melling, J., and Rutter, P., Eds., Soc. Gen. Microbiol. and Academic Press, London, 1979, 5.
21. **Zisman, W. A.,** Relation of the equilibrium contact angle in liquid and solid constitution, *Adv. Chem. Ser.*, 43, 1, 1964.
22. **Becher, P.,** Interaction parameter calculations from contact angle data, *J. Colloid Interface Sci.*, 59, 429, 1977.
23. **Drescher, J.,** Adsorptiansvorgänge bei Virusarten. I. Mitt. Aufnahme der Adsorptionsisotherme und Prüfung des pH-Wertes und der Molarität für das System Para-Influenzavirus-γ-Aluminiumoxyd, *Zentralbl. Bakteriol. I. Orig.*, 168, 217, 1957.
24. **Drescher, J.,** Comparison of the adsorption of influenza virus strain B/Berlin/2/55 on aluminum oxide and on aluminum hydroxide, *Am. J. Hyg.*, 74, 104, 1961.
25. **Gerba, C. P. and Lance, J. C.,** Poliovirus removal from primary and secondary sewage effluent by soil filtration, *Appl. Environ. Microbiol.*, 36, 247, 1978.
26. **Bitton, G., Pancorbo, O., and Gifford, G. E.,** Factors affecting the adsorption of polio virus to magnetite in water and wastewater, *Water Res.*, 10, 973, 1976.
27. **Burge, W. D. and Enkiri, N. K.,** Virus adsorption by five soils, *J. Environ. Qual.*, 7, 73, 1978.
28. **Taylor, D. H., Bellamy, A. R., and Wilson, A. T.,** The interaction of bacteriophage R17 and reovirus type 3 with the clay mineral allophane, *Water Res.*, 14, 229, 1980.
29. **Cerofolini, G. F., Jaroniec, M., and Sokolowski, S.,** A theoretical isotherm for the adsorption on heterogeneous surface, *Colloid Polymer Sci.*, 256, 471, 1978.
30. **Delbrück, M.,** Adsorption of bacteriophage under various physiological conditions of the host, *J. Gen. Physiol.*, 23, 631, 1940.
31. **Lance, J. C. and Gerba, C. P.,** Poliovirus movement during high-rate land filtration of sewage water, *J. Environ. Qual.*, 9, 31, 1980.
32. **Drescher, J.,** Über Adsorptionsvorgänge bei Virusarten. V. Untersuchungen über die Adsorption von A/2/Sing/1/57-Influenzavirus on γ-Aluminiumoxyd, *Zenstralbl. Bakteriol. Parasitenk. I. Abt. Orig.*, 179, 17, 1960.
33. **Moore, R. S., Taylor, D. H., Sturman, L. S., Reddy, M. M., and Fuhs, G. W.,** Poliovirus adsorption by 34 minerals and soils, *Appl. Environ. Microbiol.*, 42, 963, 1981.
34. **Vilker, V. L. and Burge, W. D.,** Adsorption mass transfer model for virus transfer in soils, *Water Res.*, 14, 783, 1980.

35. **Vilker, V. L., Fong, J. C., and Seyyed-Hoseyni, M.,** Poliovirus adsorption in narrow particle size fractions of sand and montmorillonite clay, *J. Colloid Interface Sci.*, 92, 422, 1983.
36. **Cookson, J. T., Jr. and North, W. J.,** Adsorption of viruses on activated carbon; equilibria and kinetics of the attachment of *Escherichia coli* bacteriophage T4 on activated carbon, *Environ. Sci. Technol.*, 1, 46, 1967.
37. **Oza, P. P. and Chaudhuri, M.,** Removal of viruses from water by sorption on coal, *Water Res.*, 9, 707, 1975.
38. **Oza, P. P. and Chaudhuri, M.,** Virus-coal sorption interaction, *J. Environ. Eng. Div. Am. Soc. Civ. Eng.*, 102, 1255, 1976.
39. **Chaudhuri, M., Koya, K. V. A., and Sriramulu, N.,** Some notes on virus retention by sand, *J. Gen. Appl. Microbiol.*, 23, 337, 1977.
40. **Valentine, R. C. and Allison, A. C.,** Virus particle adsorption. I. Theory of adsorption and experiments on the attachment of particles to non-biological surfaces, *Biochim. Biophys. Acta*, 34, 10, 1959.
41. **Ruckenstein, E. and Prieve, D. C.,** Dynamics of cell deposition on surfaces, *J. Theor. Biol.*, 51, 429, 1975.
42. **Fuhs, G. W., Moore, R. S., Reddy, M. M., Sturman, L. S., and Taylor, D. H.,** A laboratory study of virus uptake by minerals and soils, *Proc. Water Reuse Symp.*, American Water Works Association Research Foundation, Denver, Colo., 3, 2274—2281, 1979.
43. **Carlson, J. F., Jr., Woodard, F. E., Wentworth, D. F., and Sproul, O. J.,** Virus inactivation on clay particles in natural waters, *J. Water Pollut. Control Fed.*, 40, R89, 1968.
44. **LaBelle, R. L. and Gerba, C. P.,** Influence of pH, salinity, and organic matter on the absorption of enteric viruses to esturaine sediment, *Appl. Environ. Microbiol.*, 38, 93, 1979.
45. **Duboise, S. M., Moore, B. E., and Sagik, B. P.,** Poliovirus survival and movement in a sandy forest soil, *Appl. Environ. Microbiol.*, 31, 536, 1976.
46. **Lance, J. C., Gerba, C. P., and Melnick, J. L.,** Virus movement in soil columns flooded with secondary sewage effluent, *Appl. Environ. Microbiol.*, 32, 520, 1976.
47. **Young, D. C. and Sharp, D. G.,** Poliovirus aggregates and their survival in water, *Appl. Environ. Microbiol.*, 33, 168, 1977.
48. **Floyd, R.,** Viral aggregation: mixed suspensions of poliovirus and reovirus, *Appl. Environ. Microbiol.*, 38, 980, 1979.
49. **Floyd, R. and Sharp, D. G.,** Viral aggregation: quantitation and kinetics of the aggregation of poliovirus and reovirus, *Appl. Environ. Microbiol.*, 35, 1079, 1978.
50. **Floyd, R. and Sharp, D. G.,** Viral aggregation: effects of salts on the aggregation of poliovirus and reovirus at low pH, *Appl. Environ. Microbiol.*, 35, 1084, 1978.
51. **Gerba, C. P. and Schaiberger, G. E.,** Aggregation as a factor in loss of viral titer in seawater, *Water Res.*, 9, 567, 1975.
52. **Ritter, H. L. and Drake, L. C.,** Pore-size distribution in porous materials, *Ind. Eng. Chem., Anal. Ed.*, 17, 782, 1945.
53. **Drake, L. C. and Ritter, H. L.,** Macropore-size distribution in some typical porous substances, *Ind. Eng. Chem., Anal. Ed.*, 17, 787, 1945.
54. **Krishnamurty, K. V., Shpirt, E., and Reddy, M. M.,** Trace metal extraction of soils by nitric acid-hydrogen peroxide, *At. Absorpt. Newsl.*, 15, 68, 1976.
55. **Peech, M.,** Hydrogen ion activity, in *Methods of Soil Analysis*, Black, C. S., Ed., American Society of Agronomy, Madison, Wis., 1965, 914.
56. **Taylor, D. H.,** Interpretation of the adsorption of viruses from their electrokinetic properties, in *Chemistry in Water Reuse*, Cooper, W. J., Ed., Ann Arbor Science Publishers, Ann Arbor, Mich., 1981, 595.
57. **Taylor, D. H., Moore, R. S., and Sturman, L. S.,** Influence on pH and electrolyte composition on adsorption of poliovirus by soils and minerals, *Appl. Environ. Microbiol.*, 42, 976, 1981.
58. **Moore, R. S., Taylor, D. H., Reddy, M. M., and Sturman, L. S.,** Adsorption of reovirus to minerals and soils, *Appl. Environ, Microbiol.*, 44, 852, 1982.
59. **Ostle, A. G. and Holt, J. G.,** Elution and inactivation of bacteriophage on soil and cation–exchange resin, *Appl. Environ. Microbiol.*, 38, 59, 1979.
60. **Stotzky, G.,** Population dynamics of microorganisms in soil, *Crit. Rev. Microbiol.*, 2, 59, 1972.
61. **Lipson, S. M. and Stotzky, G.,** Adsorption of reovirus to clay minerals: effect of cation exchange capacity, cation saturation, and surface area, *Appl. Environ. Microbiol.*, 46, 673, 1983.
62. **Goyal, S. M. and Gerba, C. P.,** Comparative adsorption of human enteroviruses, simian rotavirus, and selected bacteriophages to soils, *Appl. Environ. Microbiol.*, 38, 241, 1979.
63. **Koya, K. V. A. and Chaudhuri, M.,** Virus retention by soil, *Prog. Water Technol.*, 9, 43, 1977.
64. **Hukuhara, T. and Wada, H.,** Adsorption of polyhedra of a cytoplasmic polyhedrosis virus by soil particles, *J. Invertebr. Pathol.*, 20, 309, 1972.

64a. **Fuhs, G. W., Chen, M., Sturman, L.S., and Moore, R.S.**, Virus absorption to mineral surfaces is reduced by microbial overgrowth and organic coatings, *Microbial. Ecol.*, 11, 25, 1985.
65. **Pollard, E. C.**, The surface of viruses, in *The Physics of Viruses*, Pollard, E. C., Ed., Academic Press, New York, 1953, 122.
66. **Taylor, D. H. and Bosmann, H. B.**, Measurement of the electrokinetic properties of vaccinia and reovirus by laser-illuminated whole-particle microelectrophoresis, *J. Virol. Methods*, 2, 251, 1981.
67. **Mandel, B.**, Characterization of type 1 poliovirus by electrophoretic analysis, *Virology*, 41, 554, 1971.
68. **La Colla, P., Marcialis, M. A., Mereu, G. P., and Loddo, B.**, Behavior of a guanidine-dependent strain of poliovirus 1 in sucrose density and pH gradients, *Experientia*, 28, 1115, 1972.
69. **Ward, R. L.**, Mechanism of poliovirus inactivation by ammonia, *J. Virol.*, 26, 299, 1978.
70. **Young, D. C. and Sharp, D. G.**, Evidence for two infectious conformational states for echovirus during chlorine disinfection, Abstr. Q(H)8, ASM Annu. Meet., Honolulu, 1979.
71. **Taylor, D. H. and Bosmann, H. B.**, The electrokinetic properties of reovirus type 3: Electrophoretic mobility and zeta potential in dilute electrolytes, *J. Colloid Interface Sci.*, 83, 153, 1981.
72. **Mitchell, R.**, Distribution of bacteria and viruses in seawater, *J. Sanit. Eng. Div. Am. Soc. Civ. Eng.*, 97, 425, 1971.
73. **Bitton, G. and Mitchell, R.**, Effect of colloids on the survival of bacteriophages in seawater, *Water Res.*, 8, 227, 1974.
74. **Lo, S., Gilbert, J., and Hetrick, F.**, Stability of human enteroviruses in estuarine and marine waters, *Appl. Environ. Microbiol.*, 32, 245, 1976.
75. **Smith, E. M., Gerba, C. P., and Melnick, J. L.**, Role of sediment in the persistence of enteroviruses in the estuarine environment, *Appl. Environ. Microbiol.*, 35, 685, 1978.
76. **Bitton, G., Masterson, N., and Gifford, G. E.**, Effect of a secondary treated effluent on the movement of virus through a cypress dome soil, *J. Environ. Qual.*, 5, 370, 1976.
77. **Bixby, R. L. and O'Brien, D. J.**, Influence of fulvic acid on bacteriophage adsorption and complexation in soil, *Appl. Environ. Microbiol.*, 38, 840, 1979.
78. **Scheuerman, P. R., Bitton, G., Overman, A. R., and Gifford, G. E.**, Transport of viruses through organic soils and sediments, *J. Environ. Eng. Div. Am. Soc. Civ. Eng.*, 105, 629, 1979.
79. **DeSilva, J. A. and Toth, S. J.**, Cation-exchange reactions, electrokinetics, and viscometric behavior of clay-organic complexes, *Soil Sci.*, 97, 63, 1964.
80. **Wilson, D. E. and Kinney, P.**, Effects of polymeric charge variations on the proton-metal ion equilibria of humic materials, *Limnol. Oceanogr.*, 22, 281, 1976.
81. **Fuhs, G. W., Sturman, L. S., Taylor, D. H., and Moore, R. S.**, Evaluation of soils for suitability for virus adsorption, *Proc. Water Reuse Symp.*, American Water Works Association Research Foundation, Denver, Colo., 3, 2241—2250, 1981.
82. **Murray, J. P.**, Physical Chemistry of Virus Adsorption and Degradation in Inorganic Surfaces, Ph.D. thesis, Stanford University, Calif. and Environmental Research Laboratory, Environmental Protection Agency, Cincinnati, Ohio, EPA-600/2-80-134, 1980.
83. **Lekki, J.**, A model of the surface of silica in aqueous solutions (in Polish), *Fizykochem. Probl. Mineralurgii*, no.12, 87, 1980; *Chem. Abstr.*, 93,226200x.
83a. **Zerda, K. S., Gerba, C. P., Hou, K. C., and Goyal, S. M.**, Adsorption of viruses to charge-modified silica, *Appl. Environ. Microbiol.*, 49, 91, 1985.
84. **Poynter, S. F. B. and Slade, J. S.**, The removal of viruses by slow sand filtration, *Prog. Water Technol.*, 9(1), 75, 1977.
85. **Jenkins, S. R., Copperthite, C. A., Barton, J. L., and Jordan, C. G.**, Removal of bacteriophage-escherichia coli T1 by sand columns using calcium^{2+} ions as a filter aid, *Water Res.*, 14, 437, 1980.
86. **Dizer, H., Nasser, A., and Lopez, J. M.**, Penetration of different human pathogenic viruses into sand columns percolated with distilled water, groundwater, or wastewater, *Appl. Environ. Microbiol.*, 47, 409, 1984.
87. **Lo, S. H. and Sproul, O. J.**, Polio-virus adsorption from water onto silicate minerals, *Water Res.*, 11, 653, 1977.
88. **Taylor, D. H.**, The Colloid and Surface Chemistry of Allophane, Ph.D. thesis, University of Waikato, Hamilton, N.Z., 1977.
89. **Rand, B. and Melton, I. E.**, Particle interactions in aqueous kaolinite suspension. I. Effect of pH and electrolyte upon the mode of particle interaction in homoionic sodium kaolinite suspensions, *J. Colloid Interface Sci.*, 60, 308, 1977.
90. **Buchanan, A. S. and Oppenheim, R. C.**, The surface chemistry of kaolinite, *Aust. J. Chem.*, 21, 2367, 1968.
91. **Kitano, M.**, Effect of sericin on allophane (in Japanese), *Nippon Sanshigaku Zasshi*, 48, 429, 1979; *Chem. Abstr.*, 92, 204117c.

92. **Dixit, S. P.,** Influences of pH on electrophoretic mobility of some soil colloids, *Soil Sci.*, 133, 144, 1982.
93. **Spoljaric, N. and Crawford, W. A.,** Glauconitic greensand: a possible filter of heavy metal cations from polluted waters, *Environ. Geol.*, 2, 215, 1978.
94. **Daniels, S. L.,** The adsorption of microorganisms onto solid surfaces: a review, *Dev. Ind. Microbiol.*, 13, 211, 1972.
95. **Filip, Z.,** Relationships between soil microorganisms and clay minerals — a review, *Z. Pflanzenernaehr. Bodenkd.*, 142, 375, 1979.
96. **Sykes, I. K. and Williams, S. T.,** Interactions of actinophage and clays, *J. Gen. Microbiol.*, 108, 97, 1978.
97. **Warren, J., Neal, A., and Rennels, D.,** Adsorption of myxoviruses on magnetite iron oxides, *Proc. Soc. Exp. Biol. Med.*, 121, 1250, 1966.
98. **Sobieraj, S. and Majka-Myrcha, B.,** Study of elektrokinetic properties and floatability of Upper Silesian coals with various petrographic compositions (in Polish), *Fizykochem. Probl. Mineralurgii*, no. 12, 165, 1980; *Chem. Abstr.*, 93,24235x.
99. **Fijal, T., Malysa, E., and Stachurski, J.,** Effect of pH of the medium on the ekectrokinetic potential of coals of different degrees of coalification (in Polish), *Fizykochem. Probl. Mineralurgii*, no. 12, 175, 1980; *Chem. Abstr.*, 93,222892b.
100. **Murray, J. P. and Laband, S. J.,** Degradation of poliovirus by adsorption to inorganic surfaces, *Appl. Envir. Microbiol.*, 37, 480, 1979.
101. **Yeager, J. G. and O'Brien, R. T.,** Structural changes associated with poliovirus inactivation in soil, *Appl. Environ. Microbiol.*, 38, 702, 1979.
102. **Lonberg-Holm, K. and Philipson, L.,** Molecular aspects of virus receptors and cell surfaces, in *Cell Membranes and Viral Envelopes*, Blough, H. A. and Tiffany, J. M., Eds., Academic Press, New York, 1980, 789.
103. **Lonberg-Holm, K., and Philipson, L., Eds.,** *Receptors and Recognition* (Series B), Vol. 8 (Part 2), Chapman and Hall, London, 1981, 277.
104. **Perrott, K. W.,** The influence of organic matter extracted from humified clover on the properties of amorphous aluminum silicates. 1. Surface charge, *Austr. J. Soil Res.*, 16, 327, 1978.
105. **Ceccanti, B., Bertolucci, M. T., and Nannipieri, P.,** Characterization of soil organic matter and derivative functions by isoelectric focussing, *Anal. Chem., Symp. Ser.*, 3, 75, 1980.
106. **Malina, J. F., Jr., Raganathan, K. R., Sagik, B. P., and Moore, B. E. D.,** Poliovirus inactivation by activated sludge, *J. Water Pollut. Control Fed.*, 47, 2178, 1975.
107. **Wellings, F. M., Lewis, A. L., and Mountain, C. W.,** Demonstration of solids-associated virus in wastewater and sludge, *Appl. Environ. Microbiol.*, 31, 354, 1976.
108. **Boardman, G. D. and Sproul, O. J.,** Protection of viruses during disinfection by adsorption to particulate matter, *J. Water Pollut. Control Fed.*, 49, 1857, 1977.
109. **Farrah, S. R., Goyal, S. M., Gerba, C. P., Conklin, R. H., and Smith, E. M.,** Comparison between adsorption of poliovirus and rotavirus by aluminum hydroxide and activated sludge flocs, *Appl. Environ. Microbiol.*, 35, 360, 1978.
110. **Lance, J. C. and Gerba, C. P.,** Effect of ionic composition of suspending solution on virus adsorption by a soil column, *Appl. Environ. Microbiol.*, 47, 484, 1984.
111. **Malina, J. E.,** Effect of unit processes of water and wastewater treatment on virus removal, in *Viruses and Trace Contaminants in Water and Wastewater*, Borchardt, J. A., Clelland, J. K., Redman, W. J., and Olivier, G., Eds., Ann Arbor Science Publishers, Ann Arbor, Mich., 1977, 33.

Chapter 9

METHODS FOR RAPID DETECTION AND RAPID IDENTIFICATION OF VIRUSES

John E. Herrmann

TABLE OF CONTENTS

I. INTRODUCTION

The need for rapid viral diagnosis in clinical situations is well recognized in that a diagnosis of viral infection will influence patient management. This includes both appropriate treatment and institution of infection control measures in hospital environments. Although the availability of antiviral agents is limited at present, specific diagnosis can eliminate unnecessary treatment. Rapid diagnosis in this situation usually means the ability to obtain results the same day a patient specimen is submitted, or the next day at the latest.

For environmental samples, there has long been a need for rapid methods to determine if viruses are present in samples such as food and drinking water. The recent findings that viruses, especially rotavirus, can be recovered from treated, municipal drinking water[1-3] makes it more important that a technology for rapid detection of viruses be developed. Rotavirus is the major cause of pediatric hospitalization for acute diarrheal disease, and, in developing nations, rotavirus infections are a major cause of infant mortality.

There are reports of waterborne outbreaks of gastroenteritis with suspected rotavirus etiology,[4-6] and the knowledge that rotaviruses survive water treatment[7] and have been recovered from drinking water supports the likelihood that some rotavirus infections are waterborne.

Monitoring for viruses in environmental samples such as sewage effluents, sludge, and source waters is also becoming more necessary, but the need for a rapid result is usually less important. The major interest in these samples is virus survival, which can be measured by infectivity assays.

The major problems in developing rapid techniques for monitoring any environmental sample for viruses is the sensitivity required and the large number of viruses, in terms of varieties, that potentially require monitoring. The types include those in the enterovirus group, adenoviruses, hepatitis viruses, rotavirus, Norwalk virus, Norwalk-like viruses, and possibly astroviruses and caliciviruses.

Many of the viruses within the virus groups do not share common antigens and there is little or no cross-reactions among the groups. Thus, it would not be practical to monitor for all of these viruses with immunochemical methods. Rather, selection of an indicator virus would be necessary if a monitoring technique based on antigen detection by immunoassay was to be developed. The same would be true if other type-specific detection methods, such as nucleic acid hybridization, were used.

In addition to the technical difficulties above, techniques for virus detection that are not based on infectivity do not distinguish between infective and noninfective viruses. This is important for environmental samples which have undergone treatment that inactivates virus, but retain inactivated virus particles that will react in an immunochemical test. For this reason, for any types of monitoring test developed, infectivity data would need to be supplied.

Even with these limitations, the progress made in the last few years in methods for direct detection of viruses in clinical samples, without the need for culture in vitro, suggests that the methods will be applicable to at least some types of environmental samples. The current status of these methods and potential means for improving them will be the emphasis given in this chapter.

II. VIRAL ANTIGEN DETECTION

A. Radioimmunoassay

Radioimmunoassay (RIA) techniques for detecting viral antigens, especially hepatitis B surface antigen, have been used for some time. Although RIA has been replaced, for most assays, by enzyme-immunoassay (EIA) for detection of viral antigens, not all RIA tests have been converted to EIA.

RIA tests can be performed in two general formats: (1) competitive binding in solution, and (2) solid-phase RIA. The competitive assay was the one originally described by Yalow and Berson[8] and was used in a number of assays before Catt and Tregear[9] described solid-phase assays in 1967. The advantage of solid-phase assays is that they provide an easily controlled means for separating antigen-antibody complexes.

B. Enzyme-Immunoassay

EIA evolved as a result of the findings by Nakane and Pierce[10] that antibodies can be labeled with enzymes for use in histochemical staining procedures, and by Catt and Tregear,[9] who described solid-phase RIA. The substitution of enzyme labels for radioactive ones in the solid-phase RIA resulted in solid-phase EIA tests for human chorionic gonadotropin by van Weeman and Schuurs,[12] and for IgG detection by Engvall and Perlmann.[11] The latter authors coined the term "enzyme-linked immunosorbent assay (ELISA)" for solid-phase EIA tests.

Initial solid-phase EIA tests were not as sensitive as the corresponding RIA, but improvement in enzyme labeling techniques have made the two types of assays comparable for detecting several antigens and antibodies. In some systems where RIA and EIA have been directly compared for viral antigen detections, there is little difference in the sensitivity or specificity of the two assays.[13]

The advantages of enzyme labels over radioactive ones are mainly convenience in use, in that the labeled immunoreagents are stable for long periods, and the precautions and disposal procedures required for radioisotopes are unnecessary. In addition, the use of chromogenic substrates for the enzyme labels permits visual interpretation of test results in some cases. The only real disadvantages of EIA tests are the loss of antibody reactivity that may result from conjugation to enzymes, and the limits of substrate detection. For example, enzymes that have molecular weights higher than that of IgG molecules such as B-D-galactosidase (mol wt 540,000 daltons) can cause steric interference with antibody activity.[14] With regard to limits of substrate detection, improvement of enzyme detection by use of fluorogenic, luminescent, or radioactive substrates (reviewed by Yolken[15]) has been proposed.

The general principles of EIA tests and details of earlier studies have been reviewed[15-18] and will not be reviewed in detail here. Rather, the major emphasis here will be on current developments in EIA methodology and the application of EIA to detection of viral antigens in clinical and environmental samples.

C. Agglutination Tests

Because there is no need for a separation step, agglutination tests, based on specific antibody immobilized on polystyrene (latex) particles, offer an attractive alternative to EIA or RIA tests. In addition, antibody-coated latex particles are stable for long periods, and the tests require only 15 to 30 min for completion. The major problem with these tests to date for direct detection of viruses is lack of sensitivity. A latex agglutination test for diagnosis of rotavirus infection is available and useful for rapid screening of fecal samples, but is less sensitive than EIA for detecting rotaviruses in the same samples.[19] A test such as this may be useful for monitoring viruses in sewage and for detecting concentrated from water, if the test sensitivity can be improved. The judicious application of monoclonal antibodies and better techniques for immobilizing antibody on the particles may give latex agglutination techniques the sensitivity required for detecting a variety of virus types. At this time, however, agglutination techniques have not been sufficiently developed for general use in rapid viral diagnosis.

D. Fluorescent Antibody Techniques

For rapid diagnosis of some virus infections, direct detection of viruses in clinical specimens with fluorescent antibodies (FA) remains a highly useful technique. The technique is particularly useful for diagnosis of respiratory virus infections and permits diagnosis within 2 to 3 hr after the specimen is received. In comparison to isolation in cell culture, the FA technique is approximately 96% sensitive and 98% specific for diagnosis of respiratory syncytial virus, influenza virus type A, and parainfluenza viruses types 1 to 3.[20] The FA technique is less sensitive for diagnosis of herpes simplex in skin lesions (88.7%), and specificity may also be a problem.[20] Use of monoclonal antibodies for herpes simplex diagnosis by FA does not increase its sensitivity.[21]

The FA technique requires cells containing viral antigen and this constitutes a limitation on the method. Moreover, the technique is not useful for detecting viruses in samples such as serum or feces. For the same reasons, the FA technique is not useful for direct detection of viruses in environmental samples.

In combination with cell culture, the FA technique can be used to shorten the time required for virus identification. This technique has been used for identifying polioviruses in experimentally contaminated water[22] and for identifying naturally occurring rotaviruses in sewage.[23]

E. Electron Microscopy

Electron microscopy has been used for direct observation of virus particles in clinical samples for some time. The advantage of the technique is that all types of viruses can be recognized. Specific identification can usually be made by morphological characteristics alone, but immune electron microscopy can be used, if necessary. The limitation of the technique is sensitivity, in that a minimum of 10^6 viral particles are required. The technique has been used to detect adenoviruses seeded into water,[24] but has not been used effectively for detecting naturally occurring viruses in water.

III. DETECTION OF VIRAL NUCLEIC ACID

An alternative to antigen detection for rapid viral diagnosis is the detection of viral nucleic acid. The advantages of this technique are high specificity due to the requirement of complementarity for nucleic acid strand reassociation with the sample and DNA probe, and high sensitivity because of the high avidity of complementary strands for each other. The disadvantages of the current procedures for viral nucleic acid detection by hybridization are the use of radioactive probes, the time required to obtain results (18 to 24 hr or more) for many of the assays, and the complexity of the procedure relative to EIA or RIA tests.

The radioactive isotope most often used is ^{32}P, which is hazardous to handle and has a short half-life. To eliminate this problem, biotinylated probes have been developed, which, when combined with biotinavidin-enzyme complexes, provide stable reagents and also reduce assay time to approximately 5 hr. The general technique for preparation and use of biotinylated probes has been published[25] and applied to probes for herpes simplex virus.[26] However, a comparison of ^{32}P-labeled probes and biotinylated probes for detecting viral nucleic acid in clinical specimens has not yet been reported.

Probes to detect RNA viruses have also been developed, including probes for polioviruses.[27] Whether the sensitivity of the probes will be sufficient to detect viruses in environmental samples remains to be demonstrated. The potential high specificity of nucleic acid probes would appear to limit their use as a broad spectrum method for monitoring virus contamination of water or other samples, as discussed above for antigen immunoassay techniques. However, as discussed more fully below, the probes described for polioviruses[27] appear to be broadly reactive with other enterovirus types. This might make them more suitable as a general monitoring technique.

IV. METHODOLOGY

A. Enzyme-Immunoassay

1. Design of Enzyme-Immunoassays

a. Solid-Phase Assays

The solid-phase or heterogeneous EIA requires immobilization of antigens or antibodies on a solid surface as a means for separating antigen-antibody complexes. Solid-phase surfaces used to immobilize antigens or antibodies have for most applications been polystyrene beads, tubes, and wells of microtiter plates, or wells of polyvinyl chloride microtiter plates. Coupling of proteins to these surfaces is usually achieved by passive adsorption. More recently, adsorption of antigens or antibodies to nitrocellulose membranes has been used for detection of viruses by EIA.[28] Covalent linkage of antibodies or antigens to a variety of surfaces, including porous glass,[29] nylon,[30] cellulose,[31] agarose,[32] and polyacrylamide,[33] has been described.

The majority of solid-phase EIA tests useful for detecting viral antigens and antibodies utilize plastic plates or beads, with antigen or antibody passively adsorbed to the solid phase. In situations where a given antigen does not readily attach, antibody to the antigen is applied first. For some antigens, nonspecific adsorbents such as poly-L-lysine have been used to enhance antigen adsorption, or the C1q component of complement has been used to capture antigen-antibody complexes.[15] Most of the assays described are of the noncompetitive type, although a number of competitive assays have been described. The disadvantage of many competitive assays for antigen detection is that labeled antigen is used which is usually more difficult to prepare than labeled antibody. With the advent of monoclonal antibodies, competitive assays for specific antibodies are becoming more common.

The lower limit of sensitivity for detecting antigen by EIA in most studies is approximately 100 pg to 1 ng, although lower limits have been described. This level of sensitivity is sufficient to detect virtually all culture-propagated viruses, but is not always sensitive enough to detect antigen directly in clinical specimens.

The types of formats most often used are shown in Table 1.

b. Homogeneous Assays

To avoid the need for separation of antigen-antibody complexes, homogeneous EIA tests were developed.[34] Homogeneous assays are based on reaction of antigen with an antibody-enzyme complex. This results in steric hindrance of the enzyme, which causes a decrease in product after reaction with enzyme substrate. The major advantage of homogeneous assays is that they do not require the separation and washing steps required in heterogeneous assays. The major disadvantage of this type of EIA is that it has been difficult to apply it to detection of high molecular weight antigens with the degree of sensitivity required. Thus, the homogeneous assay has been used mainly for detection of hormones, drugs, and other low molecular weight substances. Means to improve the sensitivity of homogeneous assays have been devised[15] but have not been widely used. Thus, the applications discussed here will be limited to EIA tests of the solid-phase type.

2. Enzymes and Substrates

Most EIA formats require covalent coupling of enzymes to antibody or antigen. A number of enzymes and coupling techniques have been tried.[15] The most consistent results have been obtained with horseradish peroxidase coupled with the aid of periodate,[35] and alkaline phosphatase coupled with the aid of glutaraldehyde.[36] Most of the assays clinically useful in diagnosing infectious diseases use chromogenic substrates, although fluorogenic and radioactive substrates have been described,[15,37] as have luminescent ones (review by Seitz[38]). Other enzymes, such as β-D -galactosidose and glucose oxidase have also been used. How-

Table 1
ENZYME IMMUNOASSAYS FOR DETECTING VIRAL ANTIGENS

EIA Type/Procedure

Noncompetitive EIA

Direct EIA		**Indirect EIA**	
1.	Antigen-specific antibody is attached to solid phase	1.	Same as direct EIA
2.	Test specimen containing antigen is added	2.	Same as direct EIA
3.	Enzyme-labeled specific antibody is added	3.	Specific antibody (prepared in species different from that used in step 1) is added
4.	Enzyme substrate is added	4.	Enzyme-labeled antiglobulin specific for antibody used in step 3 is added
5.	Substrate hydrolysis product is proportional to the amount of antigen present	5, 6.	Same as 4 and 5 in direct EIA

Competitive EIA

Enzyme-labeled antigen method		**Enzyme-labeled antibody method**	
1.	Antigen-specific antibody is attached to solid phase	1.	Antigen is attached to solid phase (may require use of specific antibody)
2.	Test specimen containing antigen is added	2.	Test specimen containing antigen plus enzyme-labeled specific antibody is added
3.	Enzyme-labeled antigen is added	3, 4.	Same as 4 and 5 in enzyme-labeled antigen method
4.	Enzyme substrate is added		
5.	Substrate hydrolysis product is inversely proportional to the amount of antigen present		

ever, the high carbohydrate content of peroxidase permits use of the periodate method for coupling, a method that is more efficient than methods that require bifunctional reagents. Several chromogenic substances have been commonly used in combination with the substrate for peroxidase, hydrogen peroxide. These include a *O*-phenylenediamine (OPD), 5-aminosalicylic acid, and 2,2′-azinodi (3-ethylbenzothiazoline-6-sulfonate) (ABTS). All give comparable sensitivity in detecting peroxidase. Several other enzymes have been used less frequently for heterogeneous EIA, including acetylcholinesterase, adenosine deaminase, catalase, and urease. Catalase, which has a specific activity of 40,000 units per milligram would appear to be preferable to peroxidase, which has a maximum specific activity of 4500 units per milligram. However, the product of catalase activity absorbs in the ultraviolet range, which is not always satisfactory for biological materials. Further details on the use of various enzyme and substrate systems for EIA are available in a recent review.[39]

A more recent development in EIA which has found application in diagnostic microbiology is the use of avidin and biotin.[40] The test is based on the high affinity complex formed between biotin and avidin. The most common method utilizes specific antibody labeled with biotin; the indicator system is enzyme-labeled avidin. A recent adaptation of this method using biotinyllated beta-lactamase in combination with avidin was effective for detecting rotavirus antigen.[41]

3. Solid-Phase Supports

In addition to antibody affinity and the sensitivity of the indicator system used, the sensitivity of many antigen detection systems depends on the amount of antibody that can be effectively immobilized on a solid phase. Methods for immobilization of antibody on plastic surfaces are usually based on simple adsorption, although covalent-linking methods have been utilized. The amount of immunoglobulin that can be immobilized by passive adsorption on different plastics varies and has been reported to be from 0.9 to 5.7 ng/mm^2.[42]

With covalent coupling, the quantity of immunoglobulin immobilized may be increased from 100 to 590 ng/mm^2. Increasing the amount of antibody immobilized with covalent techniques may increase the EIA sensitivity to some degree. However, increasing the concentration of antibody for coating surfaces by passive adsorption beyond 10 μg/mℓ does not increase immunoassay sensitivity, probably because of desorption of antibody from the plastic surface and because of steric hindrance of antibody that is adsorbed.

Solid-phase surfaces other than plastic tubes, beads, or particles, such as porous glass,[29] have been used to immobilize antibody. However, the majority of EIA tests for microbial infections are done on either plastic plates or beads. Another development noted above that may apply to a variety of viral antigens and antibodies is the use of nitrocellulose membrane discs as a solid phase. This technique was successfully developed as a visual-readout method for detecting adenovirus antigens and antibodies.[28]

4. Immunoreagents

The use of antibodies that are highly specific and have high affinity is the most critical aspect of most EIA techniques.

For example, the sensitivity of the EIA for detecting *Clostridium deficile* toxin was increased from 58.6 to 95% by changing the immunoreagents used.[43] The diluents used for antigenpreparation can also alter sensitivity. Disrupting viruses with detergents or other chemicals may increase the sensitivity of some assays but decrease the sensitivity of others.[15] A number of diluents not usually used in EIA tests were tested by Conroy and Esen[44] for adsorbing a plant protein to polystyrene. These included detergents, acids, alcohols, and urea. Use of alcohols or urea increased the EIA sensitivity substantially. Whether these diluents would be useful with microbial antigens remains to be determined.

The large number of monoclonal antibodies being developed for viral antigens may also improve immunoassay reliability. The use of monoclonal antibodies in EIA tests offers two potential advantages: (1) improved specificity due to the nature of monoclonal antibodies: and, (2) improved sensitivity by allowing for clearer EIA cut-off values. Sensitivity couldalso be increased by increasing the amount of detector antibody used in an EIA. However, because monoclonal antibodies react with only one epitope of a given antigen, more than one monoclonal antibody may be needed to achieve the desired sensitivity. In practice, monoclonal antibodies have been used successfully in latex agglutination tests and also in immunofluorescence techniques.[45] Their use in EIA has been limited, but the number of such antibodies described for microbial antigens to date suggests that applications in clinical diagnosis will increase; the EIA tests that have been developed look promising. For EIA detection of adenovirus group antigens in stools, monoclonal antibodies were as sensitive as polyclonal antibodies and were more sensitive for detecting noncultivatable adenoviruses (presumably enteric serotypes). All 12 stool samples positive by electron microscopy (EM) were monoclonal EIA positive.[46] For diagnosis of rotavirus infection, we have found that a monoclonal EIA was100% sensitive and specific for samples from adults, neonates, and young children.[46a] This sensitivity was possible because of the high affinity and broad group specificity of the monoclonal antibody used.[47]

Monoclonal antibodies to hepatitis B surface antigen have been described and were more sensitive than polyclonal antibodies for detecting viral antigen in serum by RIA.[48]

B. Nucleic Acid Hybridization

The principle of nucleic acid hybridization is that a single-stranded DNA probe specific

Table 2
NUCLEIC ACID HYBRIDIZATION ASSAYS FOR VIRAL DNA

Assay Type/Procedure

Single probe filter hybridization	Sandwich hybridization
1. Treat sample to release viral DNA (heat, organic solvent, or protease-salt)	1. Immobilize virus-specific probe No. 1 on nitrocellulose filter
2. Dry sample on nitrocellulose filter	2. Treat sample to release DNA; denature DNA by boiling
3. Heat for 2 hr at 80°C to denature DNA on filter	3. Mix with labeled probe No. 2, react on filter containing probe No. 1 for 18 hr
4. Hybridize with viral DNA-specific radiolabeled probe, 4—24 hr	4. Detect by isotope counting or autoradiography
5. Detect by isotope counting or autoradiography	

for a given viral DNA (or RNA) recognizes and binds in a complimentary manner to denatured (single-stranded) DNA or RNA from the target virus. In practice, this reaction is usually demonstrated with radiolabeled probes and nitrocellulose filters as a solid phase for the hybridization. The procedures, though tedious and time-consuming compared with EIA, have been sufficiently developed so that they can be routinely done in many laboratories.

The general procedures, condensed from several methods reported,[26,49-52] are shown in Table 2. Note that the "sandwich" method requires that two separate probes (which need to be derived from two nonoverlapping regions of the genome) be used.

For single-stranded RNA viruses, molecular probes from DNA clones of the viral RNA genome are used. However, single-stranded RNA is inactivated by some of the procedures used to extract DNA and many clinical materials contain high levels of ribonuclease activity. Ribonuclease inhibitors may be used to circumvent the latter problem.[26]

The major problem at the moment for application of nucleic acid hybridization to routine virus detection is the relative complexity of the procedure. Complete details may be found elsewhere[26,49-52] and will not be repeated here for the various viruses. However, one example may be useful. The procedure for detecting cytomegalovirus in urine specimens requires centrifugation of the sample for 1 hr at 20,800 rpm, treatment of the pellet with detergent and proteinase, extraction of the mixture once with phenol-chloroform-isoamyl alcohol and twice with ethylether, and then precipitation with ethanol.[52] This procedure is for extraction of the DNA only prior to hybridization procedures. Such complexity is not a problem for research laboratories but is not practical in clinical laboratories for routine screening.

V. APPLICATIONS IN RAPID VIRAL DIAGNOSIS

A. Virus Detection in Clinical Samples

1. Enzyme-Immunoassay

The interest in EIA methods for rapid diagnosis of viral infections has been high because of the time and expense required for isolation of the agents in cell culture. Moreover, someviruses cannot be cultivated or are difficult to cultivate. The latter include enterovirus 72 (hepatitis A virus), hepatitis B virus, and rotaviruses. Tests for hepatitis B surface antigen and e antigen have been commercially available for some time, have been extensively evaluated, and need not be elaborated upon here. Detection of enterovirus 72 antigen byEIA has also been reported[53,54] but is not at this time commercially available. The sensitivity of the EIA developed by Mathiesen et al.[53] was 77% (10/13) compared with immune electron microscopy (IEM). Locarini et al.[54] detected enterovirus 72 in 85% (17/20) of samples

Table 3
ENZYME IMMUNOASSAYS FOR DETECTING VIRAL ANTIGENS IN CLINICAL SPECIMENS

Virus	Specimen	Ref.
Adenovirus	Respiratory secretions	13, 59, 86
	Stools	74
Coronavirus	Respiratory secretions	87
Coxsackievirus	Stools	75, 76
Cytomegalovirus	Urine	70
Enterovirus 72 (hepatitis A virus)	Stools	53, 54
Hepatitis B virus	Serum	88
Herpes simplex virus	Dermal lesions	63—68, 89
Influenza A virus	Respiratory secretions	59—62, 90
Parainfluenza virus	Respiratory secretions	13, 91
Respiratory syncytial virus	Respiratory secretions	13, 55—58, 92
Rotavirus	Stools	71, 86
Varicella-Zoster virus	Dermal lesions	69

positive by IEM and found no false-positive EIA reactions in fecal samples from patients with hepatitis B or non-A-non-B hepatitis virus infections.

Besides detection by EIA of hepatitis viruses, the EIA tests most frequently developed have been for respiratory viruses, herpes viruses, and gastroenteritis viruses. These and other viruses that have been detected in clinical samples by EIA are shown in Table 3.

a. Respiratory Viruses

Because antiviral agents are becoming available for some of the respiratory virus infections, rapid methods for diagnosis are essential for prompt treatment. Rapid diagnosis by EIA has been proposed for a number of respiratory viruses. Several methods have been described for diagnosis of respiratory syncytial virus. Specificity does not appear to be a problem with any EIA reported, but the sensitivity of EIA for respiratory syncytial virus is less than that found by culture. Compared with culture, sensitivities have been 79.3% (23/29),[55] 60.9% (25/41),[56] 78.7% (37/47),[57] and 82.8% (77/93).[58]

Diagnosis of viral influenza by EIA has also been reported, with variable results. Compared with culture, Harmon and Pawlik[59] reported a sensitivity of 53% (21/40). A later report by Harmon et al.[60] on an EIA that employed fluorogenic substrates gave a sensitivity of 87% (27/31). By use of a radioactive substrate, Coonrod et al.[61] detected influenza virus in nasal washes, but the maximum sensitivity on any given day of infection was 48% (12/25). A similar assay tested on samples from 12 volunteers gave sensitivities of 78 to 100%, depending on the day tested.[62] Tests for adenoviruses in respiratory secretions have also been developed. Harmon and Pawlik[59] compared an EIA with tissue culture isolation and detected antigen by EIA in 62% (13/21) of adenovirus-positive specimens.

b. Herpesviruses

The interest in sexually-transmitted herpes virus and the availability of treatment have led to the development of a number of EIA tests for rapid diagnosis of herpes infection. Most of these tests lack sufficient sensitivity to be used as a substitute for culture. Two evaluations of a commercial EIA for herpes genital infection (Ortho Diagnostic Systems, Inc.) have been reported. Morgan and Smith[63] found the test to be 71.9% (105/146) sensitive and 100% specific in 366 control specimens. Warford et al.,[64] however, found the test to be only 52.5% (155/295) sensitive and 96.9% (834/860) specific. Some of the other EIA tests developed have given similar results. Lawrence et al.[65] developed an EIA which was 50.5%

Table 4
DIAGNOSIS OF VIRAL INFECTIONS BY NUCLEIC ACID HYBRIDIZATION

Virus	Specimen	Ref.
Adenovirus	Respiratory secretions	50
Cytomegalovirus	Urine	52, 80
	Buffy-coat cells	52
Enteroviruses[a]	Cerebrospinal fluid	27
Hepatitis B virus	Serum	51, 82, 83
	Liver	83
Herpes simplex virus	Dermal lesions	84
Influenza virus	Respiratory secretions	26
Rotavirus	Stools	81

[a] Experimentally inoculated samples.

(94/186) sensitive and 99.1% (423/427) specific. An EIA reported by Grillner and Landqvist[66] was 75.9% (44/58) sensitive and 100% specific. Two assays reported appear to have higher efficiencies. Nerurkar et al.,[67] with a biotin-avidin EIA, obtained a 95.6% sensitivity and a 91.4% specificity; Land et al.,[68] with a detergent-treated specimen, obtained a sensitivity and specificity of 94%.

Development of EIA tests for other viruses in the herpesvirus group have also been reported. Ziegler[69] developed an EIA for varicella-zoster virus that detected antigens in 8/8 culture-positive specimens, and Pronovost et al.[70] developed a chemiluminescent EIA for cytomegalovirus that detected antigens in 9/11 culture-positive specimens.

c. *Gastroenteritis and Enteric Viruses*

The two most important gastroenteritis viral agents for which EIA tests have been developed are rotaviruses and enteric adenoviruses. An EIA test for rotaviruses was first developed by Yolken et al.,[71] and commercial assays are now available. A recent evaluation of two commercial products (Rotazyme, Abbott Laboratories, and Enzygnost, Behring) showed the sensitivity of Rotazyme to be 88% and Enzygnost, 98%.[19] The standard for comparison was EM. Both EIA methods appear suitable for use if EM is not available, although Rotazyme is known to cause false-positive results in samplesfrom neonates[72,73] and is insensitive in samples from adults.[46a]

Enteric adenoviruses (types 40 and 41) are difficult to isolate and thus EIA methods would be useful if sensitivity were satisfactory. Preliminary results from Johansson et al.[74] suggest that development of an EIA specific for enteric adenoviruses is possible.

Because of the large number of types within the enterovirus group, EIA tests have not been widely applied to these viruses. Type-specific tests have been used to identify coxsackievirus A and B in stool specimens.[75,76] The test had an overall sensitivity of 70% (14/20) compared with tissue culture isolation. The detection limit was approximately 10^3 infectious units.

Another virus in the enteric group which is important in environmental transmission is Norwalk virus. We have recently developed an EIA for this virus which is as sensitive as RIA for detecting Norwalk virus in stools of volunteers.[76a]

2. *Nucleic Acid Hybridization*

The goal of nucleic acid hybridization studies in rapid viral diagnosis is to provide a more reliable means than immunoassays for detecting viruses in clinical specimens. This goal has not yet been achieved, but a number of attempts have been made. The applications in diagnostic virology are summarized in Table 4.

For a given virus, the assays generally have approximately the same sensitivity as EIA or RIA. A nucleic acid hybridization assay for adenovirus in nasopharyngeal swabs gave the same specificity as the RIA, and detected virus in 88% (22/25) of the RIA positive samples.[50] Culture was not done on all samples, so the true sensitivity of the method is unknown.

In one study on detection of cytomegalovirus (CMV) in urine and buffy-coat specimens, the DNA-probe assay was 92% (22/24) as sensitive as culture and 88% (23/26) specific for urine specimens. In buffy-coat specimens, the assay was 93% (13/14) sensitive and 60% (32/53) specific.[52] In another study on CMV detection in urine specimens, DNA hybridization was 100% sensitive and specific for all specimens from 9 patients positive and 21 patients negative for CMV.[80]

An interesting finding was reported by Rotbart et al.[27] for detecting enteroviruses in cerebrospinal fluid. The probe used was cloned cDNA to poliovirus 1, which reacted with coxsackieviruses A9 and B1 and echovirus 11 as well as poliovirus. The assay detected 10^3 infectious units of poliovirus and 10^4 infectious units of the other enteroviruses. Viruses were detected in experimentally inoculated cerebrospinal fluid, but tests on actual clinical specimens were not done. The broad specificity of the probe, however, suggests it might be useful for generic identification of enteroviruses.

Another RNA virus for which probes have been developed is influenza virus.[26] Specimens from naturally occurring human infections have not yet been tested, but viruses were detected in nasal washings from infected monkeys.

Probes for a double-stranded RNA virus, rotavirus, have also been described.[81] The sensitivity of the assay was 100% on 91 samples positive by EM and correlated well with EIA on a number of other samples. By serial dilution of positive stool samples, the probe technique was 10 to 100 times as sensitive as the EIA method used.

Use of probes for diagnosis of hepatitis B infection[82,83] has the advantage of quantifying viruses in sera of carriers by direct detection of viral DNA, but does not offer any advantages over currently available immunological tests for hepatitis B antigens as a purely diagnostic test. DNA probes have not been useful, in immunological tests, for detecting viral DNA integration in cellular DNA of liver specimens.[83]

Current EIA tests have not yet been clinically useful in diagnosing herpes simplex infections because these tests lack sensitivity. Reports on DNA probes, that can detect about 1 pg of viral DNA or 10^4 infectious units, indicate that the sensitivity and specificity of the probes is approximately that for EIA tests. In a comparison with culture isolation for detecting HSV infections, DNA detection was 80% sensitive for HSV 2, and 90% for HSV 1. The specificity was 100%.[84]

B. Virus Detection in Environmental Samples

Direct detection of viruses in environmental samples by any nonculture method has been limited. Immunochemical methods for detecting viruses in water and sewage have been described for hepatitis B antigen,[77] but application of immunochemical methods to field samples have been limited to rotavirus detection in sewage. Steinmann[78] tested a commercial rotavirus EIA (Behringwerke) on domestic sewage concentrated by adsorption to filters followed by ultracentrifugation, or by chemical precipitation. In samples concentrated by the first method, six samples positive by EM were also positive by the EIA. The chemical precipitation method was ineffective. In another study on rotavirus in sewage, the Rotazyme assay (Abbott Laboratories) was compared with a cell culture-immunofluorescence technique.[79] The EIA technique was not effective, giving a 23% false-positive rate.

VI. SUMMARY AND CONCLUSIONS

Although many of the current EIA tests for viral antigens have not realized their potential, there are reasons to believe this situation will improve. For detection of antigen, which offers a rapid and direct means of diagnosing microbial infections, the major problem has been lack of sensitivity. Increasing the sensitivity of polyclonal EIA tests by using more concentrated immunoreagents or more sensitive enzyme substrates has often resulted in a loss of specificity. From the reports available, it appears that appropriate monoclonal antibodies may solve the problem of sensitivity for detecting many infectious agents in clinical samples. With the increasing number of monoclonal antibodies available for various viral antigens, we can expect that more of them will be utilized for EIA detection systems. If the affinities of monoclonal antibodies can be increased, the EIA tests may be sufficiently sensitive.

A similar problem with sensitivity exists with nucleic acid hybridization techniques. Current tests require 10^3 to 10^5 infectious units of virus for detection, which is similar to that required for EIA. For direct detection of viruses in environmental samples, concentration of the samples is needed with the present level of sensitivity of the nonculture methods. Even with concentration, the ability to detect virus would depend on the sample tested. For example, the number of infectious viruses present in sewage may be in the range of 100 to 1000 or more per liter, whereas river water may contain only 10 or less infectious viruses per liter in wastewater.[85] For identification of specific virus types, such as rotaviruses, a monoclonal antibody-based EIA test might be suitable. For a more broad-spectrum assay, probes that react with more than one enterovirus type, such as the one for poliovirus[27] described above, may also be applicable to the same kind of samples. Further development and simplification of nucleic acid detection methods are required, however, to be practical for routine use.

REFERENCES

1. **Sekla, L., Stackiw, W., Kay, C., and Van Buckenhout, L.,** Enteric viruses in renovated water in Manitoba, *Can. J. Microbiol.*, 26, 518, 1966.
2. **Payment, P.,** Isolation of viruses from drinking water at the Pont-Viau water treatment plant, *Can. J. Microbiol.*, 27, 417, 1981.
3. **Keswick, B. H., Gerba, C. P., DuPont, H. L., and Rose, J. B.,** Detection of enteric viruses in treated drinking water, *Appl. Environ. Microbiol.*, 47, 1290, 1984.
4. **Lycke, E., Blomberg, J., Berg, G., Erikson, A., and Madsen, L.,** Epidemic acute diarrhea in adults associated with infantile gastroenteritis virus, *Lancet*, 2, 1056, 1978.
5. **Linhares, A. C., Pinheiro, F. P., Freitas, R. B., Gabbay, Y. B., Shirley, J. A., and Beards, G. M.,** An outbreak of rotavirus diarrhea among a nonimmune, isolated South American Indian community, *Am. J. Epidemiol.*, 113, 703, 1981.
6. **Sutmoller, F., Azeredo, R. S., Lacerda, M. D., Barth, O. M., Pereira, H. G., Hoffer, E., and Schatzmayr, H. G.,** An outbreak of gastroenteritis caused by both rotavirus and *Shigella sonnei* in a private school in Rio de Janeiro, *J. Hyg.*, 88, 285, 1982.
7. **Satar, S. A., Raphael, R. A., and Springthorpe, V. S.,** Rotavirus survival in conventionally treated drinking water, *Can. J. Microbiol.*, 30, 653, 1984.
8. **Yalow, R. S. and Berson, S. A.,** Immunoassay of endogenous plasma insulin in man, *J. Clin. Invest.*, 39, 1157, 1960.
9. **Catt, K. J. and Tregear, G. W.,** Solid phase radioimmunoassay in antibody-coated tubes, *Science*, 158, 1570, 1967.
10. **Nakane, P. K. and Pierce, G. B., Jr.,** Enzyme-labeled antibodies: preparation and application for the localization of antigens, *J. Histochem. Cytochem.*, 14, 929, 1966.
11. **Engvall, E. and Perlmann, P.,** Enzyme-linked immunosorbent assay (ELISA), quantitative assay of immunoglobulin G, *Immunochemistry*, 8, 871, 1971.
12. **van Weeman, B. K. and Schuurs, A. H. W. M.,** Immunoassay using antigen-enzyme conjugates, *FEBS Lett.*, 15, 232, 1971.

13. **Sarkkinen, H. K., Halonen, P. E., Arstila, P. P., and Salmi, A. A.**, Detection of respiratory syncytial, parainfluenza type 2 and adenovirus antigens by radioimmunoassay and enzyme immunoassay from nasopharyngeal specimens of children with acute respiratory infection, *J. Clin. Microbiol.*, 13, 258, 1981.
14. **Herrmann, J. E. and Morse, S. A.**, Conjugation of enzymes to anti-poliovirus globulin: effect of enzyme molecular weight on virus neutralization capacity, *Immunochemistry*, 11, 79, 1974.
15. **Yolken, R. H.**, Enzyme immunoassays for the detection of infectious antigens in body fluids: current limitations and future prospects, *Rev. Infect. Dis.*, 4, 35, 1982.
16. **Yolken, R. H.**, Enzyme-linked immunosorbent assay (ELISA): a practical tool for rapid diagnosis of viruses and other infectious agents, *Yale J. Biol. Med.*, 53, 85, 1980.
17. **Hilderbrand, R. L.**, ELISA (enzyme-linked immunosorbent assay), in *Rapid Diagnosis in Infectious Diseases*, Rytel, M. W., Ed., CRC Press, Boca Raton, Fla., 1979.
18. **Voller, A., Bartlett, A., and Bidwell, D., Eds.**, *Immunoassays for the 80s*, University Park Press, Baltimore, Md., 1981.
19. **Morinet, F., Ferchal, F., Colimon, R., and Perol, Y.**, Comparison of six methods for detecting human rotavirus in stools, *Eur. J. Clin. Microbiol.*, 3, 136, 1984.
20. **Gardner, P. S.**, Rapid virus diagnosis, in *Immunoassays for the 80s*, Voller, A., Bartlett, A., and Bidwell, D., Eds., University Park Press, Baltimore, Md., 1981.
21. **Goldstein, L. C., Corey, L., McDougall, J. K., Tolentino, E., and Nowinski, R. C.**, Monoclonal antibodies to herpes simplex viruses: use in antigenic typing and rapid diagnosis, *J. Infect. Dis.*, 147, 829, 1983.
22. **Katzenelson, E.**, A rapid method for quantitative assay of poliovirus from water with the aid of the fluorescent antibody technique, *Arch. Virol.*, 50, 197, 1976.
23. **Smith, E. M. and Gerba, C. P.**, Development of a method for detection of human rotavirus in water and sewage, *Appl. Environ. Microbiol.*, 43, 1440, 1982.
24. **Fournier, J. G., Rousset, S., and Bouteille, M.**, Application of immune electron microscopy to the detection of virus in water, *C.R. Acad. Sci. Paris*, 286, 1637, 1978.
25. **Leary, J. J., Brigati, D. J., and Ward, D. C.**, Rapid and sensitive method for visualizing biotin labeled DNA probes hybridized to DNA or RNA immobilized on nitrocellulose: bio-blots, *Proc. Natl. Acad. Sci. U.S.A.*, 80, 4045, 1983.
26. **Richman, D. D., Cleveland, P. H., Redfield, D. C., Oxman, M. N., and Wahl, G. M.**, Rapid viral diagnosis, *J. Infect. Dis.*, 149, 298, 1984.
27. **Rotbart, H. A., Levin, M. J., and Villarreal, L. P.**, Detection of enteroviruses in cerebrospinal fluid with a poliovirus 1-derived molecular probe, *Abstr. 1984 ICAAC*, 426, 166, 1984.
28. **Bode, L., Beutin, L., and Kohler, H.**, Nitrocellulose-enzyme-linked immunosorbent assay (NC-ELISA) — a sensitive technique for the rapid visual detection of both viral antigens and antibodies, *J. Virol. Methods*, 8, 111, 1984.
29. **Lynn, M.**, Inorganic support intermediates: covalent coupling of enzymes on inorganic supports, in *Immobilized Enzymes, Antigens, Antibodies, and Peptides*, Weetall, H. H., Ed., Marcel Dekker, New York, 1975.
30. **Hendry, R. M. and Herrmann, J. E.**, Immobilization of antibodies on nylon for use in enzyme-linked immunoassay, *J. Immunol. Methods*, 35, 285, 1980.
31. **Ferrua, B., Maiolini, R., and Masseyeff, R.**, Coupling of gamma globulin to microcrystalline cellulose by periodate oxidation, *J. Immunol. Methods*, 25, 49, 1979.
32. **Streefkerk, J. G. and Deelder, A. M.**, Serodiagnostic application of immunohistoperoxidase reactions on antigen-coupled agarose beads, *J. Immunol. Methods*, 7, 225, 1975.
33. **Aurameas, S. and Guilbert, B.**, Dosage enzyme-immunologique de proteines a l'aide d'immunoadsorbants et d'antigens marque aux enzymes, *C. R. Acad. Sci. Paris*, 273, 2705, 1971.
34. **Rubinstein, K. E., Schneider, R. S., and Ullman, E. F.**, Homogenous enzyme immunoassay: a new immunochemical technique, *Biochem. Biophys. Res. Commun.*, 47, 846, 1972.
35. **Nakane, P. K. and Kawaoi, A.**, Peroxidase-labelled antibody, a new method of conjugation, *J. Histochem. Cytochem.*, 22, 1084, 1974.
36. **Avrameas, S., Ternynck, T., and Guesdon, J. L.**, Coupling of enzymes to antibodies and antigens, *Scand. J. Immunol.*, 8, 7, 1978.
37. **Avrameas, S. and Guesdon, J. L.**, Immunoenzyme techniques: principles and recent developments, in *Medical Virology*, de la Maza, L. and Peterson, E. M., Eds., Elsevier, New York, 1982, 33.
38. **Seitz, R.**, Immunoassay labels based on chemiluminescence and bioluminescence, *Clin. Biochem.*, 17, 120, 1984.
39. **Blake, C. and Gould, B. J.**, Use of enzymes in immunoassay techniques: a review, *Analyst*, 109, 533, 1984.
40. **Guesdon, J. L., Ternynck, T., and Avrameas, S.**, The use of avidin-biotin interaction in immunoenzymatic techniques, *J. Histochem. Cytochem.*, 27, 1131, 1979.

41. **Yolken, R. H. and Wee, S. B.,** Enzyme immunoassays in which biotinillated β-lactamase is used for the detection of microbial antigens, *J. Clin. Microbiol.*, 19, 356, 1984.
42. **Herrmann, J. E.,** Quantitation of antibodies immobilized on plastics, in *Methods in Enzymology*, Langone, J. J. and Van Vainakis, H., Eds., Academic Press, New York, 1981, 239.
43. **Laughon, B. E., Viscidi, R. P., Gdovin, S. L., Yolken, R. H., and Bartlett, J. G.,** Enzyme immunoassays for detection of *Clostridium deficile* toxins A and B in fecal specimens, *J. Infect. Dis.*, 149, 781, 1984.
44. **Conroy, J. M. and Esen, A.,** An enzyme-linked immunosorbent assay for zein and other proteins using unconventional solvents for antigen adsorption, *Anal. Biochem.*, 137, 182, 1984.
45. **Nowinski, R. C., Tam, M. R., Goldstein, L. C., Stong, L., Kuo, C. C., Corey, L., Stamm, W. E., Handsfield, H. H., Knapp, J. S., and Holmes, K. K.,** Monoclonal antibodies for diagnosis of infectious diseases in humans, *Science*, 219, 637, 1983.
46. **Anderson, L. J., Godfrey, E., McIntosh, K., and Hierholzer, J. C.,** Comparison of a monoclonal antibody with a polyclonal serum in an enzyme-linked immunosorbent assay for detecting adenovirus, *J. Clin. Microbiol.*, 18, 463, 1983.

46a. **Herrmann, J. E., Blacklow, N. R., Perron, D. M., Cukor, G., Krause, P. J., Hyams, J. S., Barrett, H. J., and Ogra, P. L.,** Enzyme immunoassay with monoclonal antibodies for the detection of rotavirus, *J. Infect. Dis.*, 152, 830, 1985.

47. **Cukor, G., Perron, D. M., Hudson, R., and Blacklow, N. R.,** Detection of rotavirus in human stools by using monoclonal antibody, *J. Clin. Microbiol.*, 19, 248, 1984.
48. **Wands, J. R., Carlson, R. I., Shoemaker, H., Isselbacher, K. J., and Zurawski, V. R., Jr.,** Immunodiagnosis of hepatitis B with high-affinity IgM monoclonal antibodies, *Proc. Natl. Acad. Sci. U.S.A.*, 78, 1214, 1981.
49. **Engleberg, N. and Eisenstein, B. I.,** The impact of new cloning techniques on the diagnosis and treatment of infectious diseases, *N. Engl. J. Med.*, 311, 892, 1984.
50. **Virtanen, M., Laaksonen, M., Soderlund, H., Palva, A., Halonen, P., and Ranki, M.,** Novel test for rapid viral diagnosis: detection of adenovirus in nasopharyngeal mucus aspirates by means of nucleic-acid sandwich hybridization, *Lancet*, 1, 381, 1983.
51. **Berninger, M., Hammer, M., Hoyer, B., and Gerin, J. L.,** An assay for the detection of the DNA genome of hepatitis B virus in serum, *J. Med. Virol.*, 9, 57, 1982.
52. **Spector, S. A., Rua, J. A., Spector, D. H., and McMillan, R.,** Detection of human cytomegalovirus in clinical specimens by DNA-DNA hybridization, *J. Infect. Dis.*, 150, 121, 1984.
53. **Mathiesen, L. R., Feinstone, S. M., Wong, D. C., Skinhoej, P., and Purcell, R. H.,** Enzyme-linked immunosorbent assay for detection of hepatitis A antigen in stool and antibody to hepatitis A antigen in sera: comparison with solid-phase radioimmunoassay, immune electron microscopy, and immune adherence hemagglutination assay, *J. Clin. Microbiol.*, 7, 184, 1978.
54. **Locarini, S. A., Garland, S. M., Lehmann, N. I., Pringle, R. C., and Gust, I. D.,** Solid-phase enzyme-linked immunosorbent assay for detection of hepatitis A virus, *J. Clin. Microbiol.*, 8, 277, 1978.
55. **Chao, R. K., Fishaut, M., Schwartzman, J. D., and McIntosh, K.,** Detection of respiratory syncytial virus in nasal secretions from infants by enzyme-linked immunosorbent assay, *J. Infect. Dis.*, 139, 483, 1979.
56. **Hornsleth, A., Brenoe, E., Friis, B., Knudsen, F. U., and Uldell, P.,** Detection of respiratory syncytial virus in nasopharyngeal secretions by inhibition of enzyme-linked immunosorbent assay, *J. Clin. Microbiol.*, 14, 510, 1981.
57. **Hornsleth, A., Friis, B., Anderson, P., and Brenoe, E.,** Detection of respiratory syncytial virus in nasopharyngeal secretions by ELISA: comparison with fluorescent antibody technique, *J. Med. Virol.*, 10, 273, 1982.
58. **McIntosh, K., Hendry, R. M., Fahnestock, M. L., and Pierik, L. T.,** Enzymal-linked immunosorbent assay for detection of respiratory syncytial virus infection: application to clinical samples, *J. Clin. Microbiol.*, 16, 329, 1982.
59. **Harmon, M. W. and Pawlik, K. M.,** Enzyme immunoassay for direct detection of influenza type A and adenovirus antigens in clinical specimens, *J. Clin. Microbiol.*, 15, 5, 1982.
60. **Harmon, M. W., Russo, L. L., and Wilson, S. Z.,** Sensitive enzyme immunoassay with β-D -galactosidase-Fab conjugate for detection of type A influenza virus antigen in clinical specimens, *J. Clin. Microbiol.*, 17, 305, 1983.
61. **Coonrod, J. D., Betts, R. F., Linnemann, C. C., Jr., and Hsu, L. C.,** Etiological diagnosis of influenza A virus by enzymatic radioimmunoassay, *J. Clin. Microbiol.*, 19, 361, 1984.
62. **Berg, R. A., Rennard, S. I., Murphy, B. R., Yolken, R. H., Dolin, R., and Straus, S. E.,** New enzyme immunoassays for measurement of influenza A/Victoria/3/75 virus in nasal washes, *Lancet*, 1, 851, 1980.

63. **Morgan, M. A. and Smith, T. F.,** Evaluation of an enzyme-linked immunosorbent assay for the detection of herpes simplex virus antigen, *J. Clin. Microbiol.,* 19, 730, 1984.
64. **Warford, A. L., Levy, R. A., and Rekrut, K. A.,** Evaluation of a commercial enzyme-linked immunosorbent assay for detection of herpes simplex virus antigen, *J. Clin. Microbiol.,* 20, 490, 1984.
65. **Lawrence, T. G., Budzko, D. B., and Wilcke, B. W., Jr.,** Detection of herpes simplex virus in clinical specimens by an enzyme-linked immunosorbent assay, *Am. J. Clin. Pathol.,* 81, 339, 1984.
66. **Grillner, L. and Landqvist, M.,** Enzyme-linked immunosorbent assay for detection and typing of herpes simplex virus, *Eur. J. Clin. Microbiol.,* 2, 39, 1983.
67. **Nerurkar, L. S., Namba, M., Brashears, G., Jacob, A. J., Lee, Y. J., and Sever, J. L.,** Rapid detection of herpes simplex virus in clinical specimen by use of a capture biotin-streptovidin enzyme-linked immunosorbent assay, *J. Clin. Microbiol.,* 20, 109, 1984.
68. **Land, S. A., Skurrie, I. J., and Gilbert, G. L.,** Rapid diagnosis of herpes simplex virus infections by enzyme-linked immunosorbent assay, *J. Clin. Microbiol.,* 19, 865, 1984.
69. **Ziegler, T.,** Detection of varicella-zoster viral antigens in clinical specimens by solid-phase enzyme immunoassay, *J. Infect. Dis.,* 150, 149, 1984.
70. **Pronovost, A. D., Baumgarten, A., and Andiman, W. A.,** Chemiluminescent immunoenzymatic assay for rapid diagnosis of viral infections, *J. Clin. Microbiol.,* 16, 345, 1982.
71. **Yolken, R. H., Kim, H. W., Clem, T., Wyatt, R. G., Chanock, R. M., Kalica, A. R., and Kapikian, A. Z.,** Enzyme-linked immunosorbent assay (ELISA) for detection of human reovirus-like agent of infantile gastroenteritis, *Lancet,* 2, 263, 1977.
72. **Krause, P. J., Hyams, J. S., Middleton, P. J., Herson, V. C., and Flores, J.,** Unreliability of rotazyme ELISA test in neonates, *J. Pediatr.,* 10, 259, 1983.
73. **Chrystie, I. L., Totterdell, B. M., and Banatvala, J. E.,** False positive rotazyme tests on faecal samples from babies, *Lancet,* 1, 1028, 1983.
74. **Johansson, M. E., Uhnoo, I., Kidd, A. H., Madeley, C. R., and Wadell, G.,** Direct identification of enteric adenovirus, a candidate new serotype associated with infantile gastroenteritis, *J. Clin. Microbiol.,* 12, 95, 1980.
75. **Yolken, R. H. and Torsch, V.,** Enzyme-linked immunosorbent assay for the detection and identification of coxsackie B antigen in tissue cultures and clinical specimens, *J. Med. Virol.,* 6, 45, 1980.
76. **Yolken, R. H. and Torsch, V. M.,** Enzyme-linked immunosorbent assay for detection and identification of coxsackieviruses A, *Infect. Immun.,* 31, 742, 1981.

76a. **Herrmann, J. E., Nowak, N. A., and Blacklow, N. R.,** Detection of Norwalk virus in stools by enzyme immunoassay, *J. Med. Virol.,* 17, 127, 1985.

77. **Horst, H.,** Isolation of hepatitis B antigen from water and sewage by means of immunoadsorption in a continuous flow procedure, *Zentralbl. Bakteriol. Hyg. Abt. I. Orig. B,* 158, 578, 1974.
78. **Steinmann, J.,** Detection of rotavirus in sewage, *Appl. Environ. Microbiol.,* 41, 1043, 1981.
79. **Hejkal, T. W., Smith, E. M., and Gerba, C. P.,** Seasonal occurrence of rotavirus in sewage, *Appl. Environ. Microbiol.,* 47, 588, 1984.
80. **Chou, S. and Merigan, T. C.,** Rapid detection and quantitation of human cytomegalovirus in urine through DNA hybridization, *N. Engl. J. Med.,* 308, 921, 1983.
81. **Flores, J., Purcell, R. H., Perez, I., Wyatt, R. G., Boeggeman, E., White, L., Chanock, R. M., and Kapikian, A. Z.,** A dot hybridization assay for detection of rotavirus, *Lancet,* 555, 1983.
82. **Shimizu, Y., Ida, S., Matsukura, T., and Yuasa, T.,** Determination of hepatitis B virus DNA in serum by molecular hybridization, *Microbiol. Immunol.,* 28, 1117, 1984.
83. **Scotto, J., Hadchouel, M., Hery, C., Alvarez, F., Yvart, J., Tiollais, P., Bernard, O., and Brechot, C.,** Hepatitis B virus DNA in children's liver diseases: detection by blot hybridization in liver and serum, *Gut,* 24, 618, 1983.
84. **Redfield, D. C., Richman, D. D., Albanil, S., Oxman, M. N., and Wahl, G. M.,** Detection of herpes simplex virus in clinical specimens by DNA hybridization, *Diagnostic Microbiol. Infect. Dis.,* 1, 117, 1983.
85. **Berg, G.,** Reassessment of the virus problem in sewage and in surface and renovated waters, *Proc. 6th Int. Water Pollution Res.,* B-14, 28, 1972.
86. **Sarkkinen, H. K., Tuokko, H., and Halonen, P. E.,** Comparison of enzyme-immunoassay and radioimmunoassay for detection of human rotaviruses and adenoviruses from stool specimens, *J. Virol. Methods,* 1, 331, 1980.
87. **Macnaughton, M. R., Flowers, D., and Isaacs, D.,** Diagnosis of human coronavirus in children using enzyme-linked immunosorbent assay, *J. Med. Virol.,* 11, 319, 1983.
88. **Wolters, G., Kuijpers, L., Kacaki, J., and Schuurs, A.,** Solid-phase enzyme-immunoassay for detection of hepatitis B surface antigen, *J. Clin. Pathol.,* 29, 873, 1976.
89. **Miranda, Q. R., Bailey, G. D., Fraser, A. S., and Tenoso, H. J.,** Solid-phase enzyme-immunoassay for herpes simplex virus, *J. Infect. Dis.,* Suppl. 136, S304, 1977.

90. **Sarkkinen, H. K., Halonen, P. E., and Salmi, A. A.,** Detection of influenza A virus by radioimmunoassay and enzyme-immunoassay from nasopharyngeal specimens, *J. Med. Virol.*, 7, 213, 1981.
91. **Sarkkinen, H. K., Halonen, P. E., and Salmi, A.,** Type-specific detection of parainfluenza viruses by enzyme-immunoassay and radioimmunoassay in nasopharyngeal specimens of patients with acute respiratory disease, *J. Gen. Virol.*, 56, 49, 1981.
92. **Meurman, O., Sarkkinen, H., Ruuskanen, O., Hanninen, P., and Halonen, P.,** Diagnosis of respiratory syncytial virus infection in children: comparison of viral antigen detection and serology, *J. Med. Virol.*, 14, 61, 1984.

Chapter 10

CELL CULTURES AND OTHER HOST SYSTEMS FOR DETECTING AND QUANTIFYING VIRUSES IN THE ENVIRONMENT

Naomi Guttman-Bass

TABLE OF CONTENTS

I. INTRODUCTION

The exposure of susceptible individuals or populations to water contaminated by human pathogenic viruses may lead to sudden and widespread outbreaks of disease. To prevent epidemics of waterborne viral disease, the degree of viral contamination of our waters must be minimized. Our knowledge of the extent of viral pollution of the water environment is limited by the sensitivity of virus detection methods. The detection of viruses in natural, waste, recreational, and drinking waters and the limitation of their spread by appropriate water treatment requires the development of rapid and reliable monitoring systems appropriate for waterborne viruses. It is the aim of this chapter to evaluate current knowledge of host systems available for detecting viruses in water, with recommendations for the most appropriate systems and areas for future research.

The viruses that are potential waterborne disease agents enter the water environment in a viable form, are stable in that environment, and may infect individuals through water contact. These include the enteric viruses which infect orally and are excreted fecally into sewage and viruses with contact routes of infection and transmission that are stable in water.[1] There are more than 120 virus types that are potential agents for waterborne spread including the enteroviruses, adenoviruses, reoviruses, and the agents of viral gastroenteritis. Not all candidate viruses for water transmission have been detected in water or have been shown to cause waterborne epidemics, in part because of methodological problems associated with their detection physically or epidemiologically.[2]

The detection of viruses in water may be divided into several stages. A representative water sample is obtained, and if necessary, the virus particles may be concentrated from large volumes of water. The water sample may be treated to reduce microbial or chemical contamination and is then assayed for viruses in suitable laboratory systems. This chapter is concerned with the last stage, the detection of viruses in the laboratory by viable host systems. Since viruses can replicate only in viable hosts, this is the only way to demonstrate the presence of infective virus particles. For viruses that are not readily cultivated in simple laboratory host systems, or only partially so, it is possible to use immunochemical techniques for their detection. These methods are rapid and can be used either alone or in combination with a viable host system for detection of viruses. Generally, these techniques are less sensitive for detecting viruses than the more traditional techniques and may detect nonviable viral components also. These techniques will be discussed for viruses for which viable laboratory hosts do not exist or are difficult to work with.

II. HOST SYSTEMS FOR DETECTING VIRUSES

The choice of a host system for detecting viruses in water must consider not only the sensitivity of the system to the viruses of interest, but also the availability, ease of handling, and cost of such a system. Of great importance to water virology is the range of susceptibility of the system to a wide variety of viruses, each present in small numbers. Detecting viruses in water in a viable laboratory host system implies viral viability, but not necessarily pathogenicity for humans.

The traditional laboratory host systems used for detecting viruses are described.

A. Animal Host Systems

The classical means for detecting viruses is to infect animal hosts and observe the development of disease or death. Experimental animals are chosen on the basis of susceptibility to the suspected virus and availability of the host. Most viruses are infective for their natural hosts, but sometimes disease or death may be induced in a heterologous species. The viruses of man are often pathogenic for other primates and sometimes for other animals such as mice. Susceptibility of animals may depend on age, route of infection, species, and genetic background.

The ideal animal host system requires easily managed laboratory animals under controlled environmental and genetic conditions. Where exotic and rare animals are the only host apart from man, supply and maintenance are problems. Even convenient animal hosts may be difficult to standardize and costly to maintain. In general, the use of animal hosts is to be avoided for water virology and are only recommended for viruses that are not readily detected in cell cultures, such as certain group A coxsackieviruses. As an alternative, immunochemical methods may be used in place of animal hosts for detecting specific viral antigens.

B. Organ Cultures

Organ cultures are animal tissue explants whose components retain their in vivo structure and function. These cultures may be used for the recovery of respiratory viruses, but have not been reported to be useful for the recovery of viruses from water. Limitations on availability, reproducibility and sensitivity of these cultures, and complicated detection procedures[3] restrict their use to viruses with no other detection system.

C. Embryonated Eggs

Chick embryos are sensitive to a wide variety of human agents that induce pocks on embryonic membranes, hemagglutination, or death of the embryo.[4] Waterborne enteric viruses do not generally produce characteristic changes in embryonated eggs and do not grow well, if at all, in the chick egg system. This system, although fairly easy to obtain and use, is therefore not particularly useful for water virology.

D. Cell Culture Systems

Cell culture systems consist of cells removed from their parent animal and cultivated or maintained in the laboratory. These systems are preferred over other viable systems for viruses that can readily be detected in cell culture systems. Such systems are relatively easy to use and obtain, inexpensive, and usually sensitive to a variety of viruses. No single system can support the growth of all enteric viruses,[5] but it is possible to choose a combination of systems to provide a wide range of susceptibility, or to choose the most sensitive system available for the viruses most easily found in water and to use these viruses as indicators of the presence of other undetected viruses.

The three categories of all cultures in common use for detecting viruses are primary, diploid, and continuous cell cultures. In general, simian or human cell cultures are the most useful cultures for viruses of interest in the water environment.

1. Primary Cell Cultures

Primary cultures (1°) consist of cells removed from their animal source, dispersed, and plated directly on tissue culture plates or bottles. The cells are maintained in growth medium and spread and multiply to form a confluent monolayer culture that contains cells similar to the original source tissue. Secondary cultures may be made by subpassaging primary cultures. Secondary cultures contain cells morphologically similar to the primary culture,[6] although they may be more uniform. Primary and secondary cultures generally have the same range of virus sensitivities, and will not be considered as independent systems. More than one cell type may be present in the cultures, allowing for a wide range of virus susceptibility.

Sources of cells are usually human or other primate tissues from embryonic, young, or adult animals. Embryonic or young tissues tend to produce primary cultures more sensitive to viruses than those derived from adult tissues. Kidney and lung are common source tissues. The types of primary cultures frequently used for growing and recovering enteric viruses are listed in Table 1.

Table 1
CELL CULTURE SYSTEMS FOR RECOVERING ENTERIC VIRUSES

Primary cultures
- 1° RhMK: Rhesus monkey kidney
- 1° CMK: Cynomolgus monkey kidney
- 1° AGMK: African green monkey kidney
- 1° patas MK: Patas monkey kidney
- 1° VMK: Vervet monkey kidney
- 1° BaK: Baboon kidney
- 1° HEL: Human embryonic lung
- 1° HEK: Human embryonic kidney
- 1° HuAm: Human amnion

Diploid strains
- HL-8: Fetal rhesus monkey kidney
- HFDL: Human fetal lung
- HEL: Human embryonic lung
- IMR-90: Human embryonic lung
- MRC-5: Human embryonic lung
- WI-38: Human embryonic lung
- HFDK: Human fetal kidney

Continuous Lines
- LLC-MK_2: Rhesus monkey kidney
- MA-104: Fetal rhesus monkey kidney
- MFK: Fetal rhesus monkey kidney
- BGM: African green monkey kidney
- BSC-1: African green monkey kidney
- Vero: Aftrican green monkey kidney
- HeLa: Human cervical carcinoma
- KB: Human carcinoma of the nasopharynx
- RD: Human rhabdomyosarcoma
- HEp-2: Human carcinoma of the larynx
- Detroit-6: Human amnion
- FL: Human amnion
- U: Human amnion
- Borrie: Human embroynic lung

The advantages of primary cultures for recovering viruses are their sensitivity to a relatively wide range of agents and the large numbers of cell cultures that can be prepared from each animal. The cultures generally last for several weeks before degenerating,[7] long enough for recovering slow-growing viruses.

Disadvantages of primary cell culture systems include problems of availability of the source animal tissues, whether simian or human. Moreover, the lifetime of each culture is finite since the amount of growth in culture is limited. Also, batches of cell cultures vary in virus susceptibility,[8] and this may lead to inconsistent results. A further disadvantage is the possible presence of adventitious agents in tissues derived from infected animals, especially from monkeys. The result of such presence may be interference with virus growth or premature degeneration of cultures.[9] Some of these agents are extremely hazardous to workers handling the cultures.

Primary cultures are sensitive substrates for recovering viruses from water, but there is a continuing search for equally sensitive systems that can be maintained in the laboratory indefinitely.

2. Diploid Cell Strains

Diploid cell strains or semicontinuous cell lines consist of cells usually derived from primary cells that can be maintained in culture by subcultivation; such cells form a confluent monolayer upon each subpassage. These cultures may retain some characteristics of their source tissue and retain a diploid chromosome number. The early passages consist of rapidly multiplying cells, but usually after 40 to 50 passages, the cells cease multiplication and the cultures degenerate.[10] Strains vary with respect to the maximum number of passages they can undergo before degenerating. Diploid cells have a finite lifetime in culture, but that lifetime is more extensive than that of primary cells. Early passage cells produce stable, uniform monolayers, and may be stored by freezing. Common sources of cells are human embryonic organs such as kidney, lung, skin, muscle, intestine, tonsil, and foreskin.

Cell strains are relatively uniform and stable until the latter stages of degeneration, although sensitivity to viruses may vary with passage number. Diploid cell cultures are easier to obtain and maintain than primary cultures, but may have a more limited range of virus susceptibility. The supply of each strain is ultimately limited since serial passage cannot be carried out indefinitely.[10] Passages can be maintained for two weeks or more, sufficient for the detection of slow-growing isolates.[11] The strains in common use for enteric virus isolation are listed in Table 1.

3. Continuous Cell Lines

Continuous cell lines are cultures that can be subpassaged for a potentially unlimited number of passages and are considered to be immortal. Characteristic of these lines is a heteroploid karyotype and the lack of differentiated characteristics associated with their tissue of origin. The lines are often initiated from malignant tissue or from cells that have undergone transformation in culture. Cells may grow attached to plastic or glass surfaces or in suspended cultures in liquid medium. Some lines form monolayers; others are not restricted to forming a single cell layer.

The viral susceptibilities of these cultures may vary greatly because of their geneticinstability. Cloned lines, or lines maintained in different laboratories, may differ with respect to chromosome number and phenotypic expression.

Continuous lines have the advantage of ready availability and easy growth in the laboratory for indefinite periods of time. These cultures are often used for the preparation of virus stocks. Laboratory strains of viruses may often be adapted to cell lines and grow to high titer in those lines. Continuous cell lines may be relatively insensitive for recovering viruses but be sensitive to laboratory grown viruses. Common lines in use are listed in Table 1.

The cell culture systems preferred for recovering viruses from water would have the wide host range of primary cultures, the uniformity and stability of diploid cell strains, and the virtually unlimited availability of continuous lines.

III. METHODS FOR ASSAYING VIRUSES

The assay of viruses in viable host systems requires the induction of changes in the host, whether morphological or biochemical. Since water virology is concerned with detecting and measuring prevalence of many virus types present in small numbers, the assay method must be sensitive, quantitative, and reliable. There are two major methods for detecting enteric viruses in cell culture hosts: the cytopathic effect (CPE) and plaque assays. For virus recovery studies, a stable available host system is required that may be maintained under assay conditions for several weeks after sample inoculation. The presence of microbial contaminants in the water sample or host culture may interfere with the assays, causing premature culture degeneration or interfering with virus replication and detection. The inoculum may require pretreatment to reduce microbial contamination. Assays are usually performed in the presence of antibiotics.

For purposes of confirmation and identification, unless primary isolation involves direct identification of viral antigens, the virus assay systems should allow recovery of the virus after the assay.

A. Cytopathic Effect

CPEs are the pathological (morphological and physiological) changes induced in a susceptible host cell by infection with cytopathogenic viruses. These changes may be characteristic of the infecting virus type and often lead to the death of the host cell.[12] The assay is generally performed as follows: the growth medium is removed from a monolayer and replaced by an inoculum. The cultures are incubated at a suitable temperature, usually 35 to 37°C for fecal viruses. The virus replicates and spreads throughout the culture until most or all of the cells have been infected. The CPE is observed by microscopic examinationof the cultures. Each positive culture indicates that one ore more infective virus units were present in the inoculum.

The number of viruses present is quantified statistically by dilution assays with replicate samples. The virus dilution where 50% of the replicate samples are positive is the $TCID_{50}$ (50% tissue culture infective dose)[13] and is the most common measure of virus concentration. A second method, calculating the most probable number (MPN) from serial dilutions, can also be used.[14]

The capability of the CPE assay to detect viruses depends on three major factors: the virus, the host cell culture, and the conditions of the assay. Viruses vary in the efficiency and speed with which they produce CPE in susceptible cultures. Detection may take from 12 hr to 21 days or more, depending on the time course of viral replication and release, and the concentration of the virus in the inoculum. Virus strains grown in the laboratory may induce CPE more rapidly than isolates that have not been adapted to or been selected for rapid growth in the laboratory culture system.[15] High viral concentration may cause CPE in relatively resistant cell cultures and a high virus to cell ratio may increase the rapidity of the assay.

Enteroviruses generally produce rapid and extensive CPE in susceptible hosts while other viruses such as reoviruses and adenoviruses may take longer. CPE morphology may be used as a preliminary means to identify viruses, although further passage and serological confirmation are necessary for final identification of viruses.

The host cell culture used for detecting viruses is an important factor in the sensitivity of the assay. The inherent infectibility of the cells by the virus is of primary concern, and depends in part on the type and source of the cells. Other variables that may influence CPE development include cell passage history, yield of virions per cell, and rate of CPE development in the culture. CPE induction in a culture is not always accompanied by complete virus replication, and there may be multiplication of the virus in the absence of CPE.[16]

Treatment of the cell cultures before inoculation of a sample may influence the assay. Maintenance of storage of cultures at 25°C before inoculation may affect the yield of laboratory virus strains, depending on cell and virus type.[17] Time in culture of the cell passage and date of last fluid change before inoculation of the virus may influence CPE development, with poliovirus titers decreasing with more time in culture or time since the previous fluid change.[18] The state of the cells, whether stationary or growing, influence sensitivity to viruses.[19]

Conditions under which an assay is performed influence CPE production. Culture vessels used include tubes, bottles, and petri plates, which may be glass or plastic, sealed or open. Rolling of cultures and temperature of incubation influence the rate of CPE development although the efficiency of enterovirus and adenovirus isolation are not affected.

The maintenance medium used on cultures should be able to maintain the monolayer without degeneration for the length of the assay. The concentration and type of serum in the liquid medium are important because serum may inhibit multiplication of the virus due to the presence of neutralizing antibodies, protease inhibitors, or other factors.[20,21] Atmospheric and pH control may be critical for some viruses.[22] Although an adsorption period prior to medium addition is not generally used, it is necessary for the induction of CPE by certain enterovirus strains.[22] The volume of the inoculum influences CPE induction because the larger the volume, the less chance for virus-cell contact.[23]

CPE production by individual viruses may be enhanced or inhibited by the addition of chemicals to the medium. For example, treatment of cells with IdU 95 (iododeoxyuridine) potentiates the development of CPE by adenoviruses, increasing yield and rapidity of CPE occurrence.[24,25]

To recover viruses, it may be necessary to perform blind passage, where infected cells and fluids are used to infect fresh cells. This may be necessary if the longevity of a single passage is not sufficient for detecting slow growing viruses, or where the concentration of viruses in the inoculum is too small to produce detectable CPE.[26] This is particularly a problem with viruses that do not release efficiently from infected cells and cannot spread efficiently through the culture to produce CPE. Furthermore, with passage, the ability of a given isolate to cause CPE in culture may increase by selection or adaptation to the cells in use.[27] For clinical isolations, up to three passages are often performed. For example, with adenoviruses the CPE is slow to appear when the virus is initially present in small numbers. For adenovirus isolates, however, the total length of incubation on monolayers is more important than the number of passages; the number of isolates in tube cultures of HeLa cells increases if incubation time is greater than 14 days, up to 45 days.[28]

The CPE assay in tube cultures is a sensitive, semiquantitative method for detecting viruses. It is easy to perform, and is the most commonly used method for detecting viruses.

B. Plaque Assay

The plaque assay of animal viruses has been in use since 1952[29,30] for the quantification and detection of viruses. The assay measures the number of infective virions and clumps of virions in an inoculum in plaque-forming units (PFU). The most common method uses a monolayer of cells attached to a solid surface as the host system. The growth medium is removed and viruses are inoculated in a small volume of fluid onto the monolayer and allowed time to adsorb to the host cells. An overlay of agar and medium is added andallowed to solidify. This confines the viral progeny to a localized area of infection, and allows cell destruction in a limited area known as plaque. After several cycles of infection, the plaque is large enough to be seen by eye. A vital stain is usually added to enhance the contrast between the surrounding live cells and the dead cells of the plaque.

The agar suspension method of plaque assay is an alternative to the standard technique.[31,32] Monolayer cells separated from their solid substrate or cells growing in suspension are infected with a virus inoculum. The cell-virus mixture is combined with agar, and remains suspended as this agar solidifies. The surface area available for adsorption of viruses is larger than in the monolayer method and prior adsorption is not necessary. It is a particularly good method for small viruses such as enteroviruses which diffuse easily through agar. More cells are required than in the standard assay, and cells may be less stable for long-term incubation. Mixed cell populations can be used in the cell suspension method, which might be useful in picking up a wide range of viruses.

The efficiency and rate of virus recovery by plaque assay depend on the infecting virus, the host culture system, and the assay conditions. Virus types and strains differ in their efficiencies of plaque production, time course of infection, and yield of progeny per cycle. Plaque morphology and rate of development depend on the type of virus isolated, although identification cannot be made on this basis alone. Isolates of the same type of virus from a single epidemic have produced plaques of more than one type.[27]

The host cell system used must be susceptible to and destroyed by the infecting virus. The virus must successfully adsorb to the host cell, complete the stages of macromolecular synthesis and assembly, and produce infective progeny that are released to infect neighboring host cells. The type of cell and metabolic state, passage history, and interferon production may affect virus plaquing efficiency. Lines of cells can change in sensitivity to viruses with passage. This is particularly true of continuous lines such as HeLa cells, which are genetically unstable.[33,34]

The host system must produce a stable confluent monolayer, since holes in the uniform cell sheet may be mistaken for plaques unless examined microscopically. Ideally, one should use a uniform host cell system with a broad range of sensitivity and constant characteristics. The system should be stable enough that no plaques are formed in the absence of a virus, and allow for rapid plaque formation.

Contamination of the cell cultures by mycoplasma may reduce plaquing efficiency without necessarily impairing cell growth. Cultures should be monitored for such contamination.

The sensitivity of the assay is affected by the components and procedures used. The type of vessel used, whether bottles or plates, plastic or glass, stoppered[36] or open (usually maintained in a 5% CO_2 atmosphere[34]), influences the assay. Temperature of incubation is usually 35 to 37°C although polioviruses, for example, can grow from 30 to 41°C. The efficiency of plaque formation by various viruses at different temperatures may be strain-dependent.

Conditions of adsorption of inoculum to the monolayer must be controlled. In general, the smaller the volume of inoculum, the more sensitive the assay because the probability of cell-virus contact increases.[23,37] Time allowed for adsorption usually ranges from 30 min to 3 hr. Shaking during adsorption or suspension of the cells and shaking followed by settling may increase recovery of viruses.[37] If the inoculum is toxic, as water samples sometimes are, toxicity should be removed before the agar overlay is added.[38]

Cells multiply best at pH 7.2 to 7.4, although they do well over a pH range of 6 to 8. Viruses usually multiply best in healthy multiplying cells.[39] Bicarbonate-CO_2 systems are often used for pH control.[40,41] The components of the agar overlay that may affect the development of plaques include medium, serum, agar, antibiotics and other additives, and stain.

The type and concentration of serum are important factors. The serum used should contain no antibodies or other inhibitors to virus growth.[21] Some viruses form plaques only in the absence of serum. Serum is used for cell maintenance during the plaque assay, since the monolayer must last as long as possible without degenerating, and cells vary with respect to their serum requirements. The species from which serum is obtained influences plaque formation,[42] and different lots of serum influence plaque formation differently. Substitutes include skim milk,[43,44] lactalbumin hydrolysate, yeast extract, and bovine albumin-yeast extract.[45]

The solid medium used to localize the virus during a plaque assay is an influential factor in plaque production. Variables include type, brand, batch, concentration, and depth of the overlay medium.[32] The solid overlay medium most commonly used is agar. The concentration of agar used is usually in the range of 0.2 to 1.2%. As the concentration of agar decreases, the mobility of the viruses in the assay increases, as does plaque size, decreasing the time required for detection.

The agar itself may contain substances inhibitory for some viruses. The degree of inhibition varies with brand and lot of agar, although there is no consensus on which type is best.[27,43,46] One inhibitory substance may be the sulfated polysaccharides present in agar[47] which may be counteracted by protamine sulfate[48] or DEAE dextran.[49] Different virus strains react differently to these substances, and the overall effect may be due to the combination of components present in the assay.[50]

Alternatives to agar, such as gels of hydrolyzed starch,[51] tragacanth gum,[52] agarose,[53] and methyl cellulose[32] may be used for plaque assays.

Chemical additives may be used in plaque assays and may enhance, inhibit, or have no effect on plaque development, depending on the virus type and strain, and other plaque assay components. There is no universal additive that enhances plaques of all the enteric viruses. Additives may influence plaque number, size, and speed of formation.

Magnesium chloride[43,54,55] enhances plaquing by many of the enteroviruses tested, although some enteroviruses are inhibited. Other additives that increase detection of some enteric viruses include protamine sulfate,[55] dextran sulfate,[49,55,56] DEAE dextran,[49,55,57] proteases,[55,58] and cysteine.[55,59]

Unfortunately, the effect of a particular additive on the plaquing efficiency of a mixture of virus isolates is unpredictable because enhancement or inhibition varies with virus type and strain. The routine inclusion of additives in water studies should be approached with caution unless the encouragement of certain virus types at the expense of others is desired.

The stain used to visualize plaques is important; variables include type, concentration, and time of addition. There have been several reports that neutral red, the most commonly used stain, may decrease plating efficiency if present in the initial overlay.[60,61]

It is common to add neutral red in a second overlay, applied several days after the first, although this may cause virus mixing due to lateral migration.[62] Filtering the neutral red has also been reported to reduce neutral red toxicity.[43] Reoviruses[58] and adenoviruses[48] are among the viruses most sensitive to neutral red. Alternative stains include crystal violet and trypan blue. Neutral red at high concentrations is cytotoxic.

C. Comparison of Plaque Assay to CPE Assay

The advantages and disadvantages of plaque assay and CPE methods for water virology may be summarized as follows.

In the plaque assay, the progeny from each infecting virus are localized so that many lesions or isolates can be detected in the same vessel. Isolates are already partially purified. With the CPE method, there is diffusion throughout, so that each virus-containing vessel has one or many infecting virus particles. Unless dilution assays are performed, mixtures of viruses are easily masked. Overgrowth may also occur with the plaque assay, with plaques overlapping because of high concentrations of viruses in the inoculum. Here, dilution may be necessary. The addition of antisera against the overgrowing viruses may help isolate less efficient agents.

A larger sample volume can usually be applied in the CPE method. For the plaque assay, more vessels or a larger surface area of cells may be necessary to test the same inoculum volume. The CPE method is simpler to perform and more economical than the plaque assay.

Cells tend to be more stable longer under liquid than under agar medium, so for slow-growing viruses, the CPE method may have an advantage. If too small a fraction of the cells exhibit CPE, the isolate may not be detected.[63] In several cases of enterovirus isolations, viruses were detected by plaque assay without the parallel appearance of CPE.[41,64] This was true even for viruses capable of producing CPE. The reverse may occur as well. In one study[65] comparing the efficiency of virus isolation from clinical specimens by the two methods, both methods were equal in efficiency of isolation, with some isolates picked up by both and some by one of the two methods. The serotype of the enterovirus was not the factor determining which assay was positive for the individual isolate. Both methods require confirmation by transmission.

The relative sensitivity of the methods varies with virus and cell type. Generally, enteroviruses have similar titers by tube dilution and plaque assays,[66,67] although there are exceptions. Adenoviruses have been reported to have a higher plaque than tube dilution titer.[61] Statistically, for the same virus concentration, 1 $TCID_{50} = 0.69$ plaques.[32]

Cytotoxic material in an inoculum may interfere less with plaque assays than with tube dilution assays because in the plaque assay contaminants are localized while in the CPE method they may spread and destroy the entire cell culture. In the plaque assay, local toxic effects may falsely be observed as plaques. Both assay methods require confirmation of the virus by transmission.

D. Immunoassays

Other methods for detecting viruses may be used. Hemadsorption is a method where erythrocytes adhere to monolayer cells infected by viruses that incorporate hemagglutinin into the plasma membrane of infected cells. This method is not useful for most of the enteroviruses, although some enteroviruses, adenoviruses, and the reoviruses have hemagglutinins and can be detected in this way.[52]

Viruses may be detected immunologically with antibodies tagged with markers to assay for the presence of viral antigen in the sample, or after partial virus replication and antigen production in culture. Such assays include radioimmunoassay (RIA), enzyme-linked immunosorbent assay (ELISA), and fluorescent antibody staining of cells. These methods are especially useful for viruses that do not replicate in culture and may be present in high concentrations in the water environment, such as hepatitis A virus (enterovirus 72) and rotaviruses. The assays are rapid and may be used to detect virtually any virus, but are less sensitive than standard tissue culture techniques. Furthermore, specific antibodies for each serologically distinct type of virus must be used, and a prior choice of which viruses to seek out in a particular sample is required. Unless used in combination with some growth in a living host system, these assays do not provide information about the viability of the viral components detected.

Some viruses can be detected only by electron microscopic techniques. Before the advent of cell culture methodology, viruses had to be detected in animal systems. The advantages of cell culture systems, including ease of handling and relative standardization, have allowed such culture systems to replace animal hosts, except where there is no substitute for animal inoculation. The recent developments in immunological technology for detecting viruses may in turn lead to the replacement of cell culture methodology by more rapid methods.

IV. VIRUS HOST RANGE

The choice of the most efficient host cell type for the recovery of viruses from water is based on information gleaned from three types of studies: the host range of laboratory-grown viruses, recoveries of viruses from clinical specimens, and recoveries of viruses from water samples. The first two types of studies will be discussed in this section, and recoveries of viruses from water will be treated separately in the following section. Laboratory virus strains are those that have been passaged in experimental systems, whether animal or cell cultures; isolates are obtained directly from clinical specimens. These studies do not necessarily reflect the host range of individual isolates of the same type, but indicate which hosts are likely to be susceptible to different viruses.

Clinical studies involve testing tissues or body fluids for the presence of viruses. When successful, such studies usually result in the recovery of a single virus type as the possible disease agent. The concentration of viruses in clinical specimens may be low, and there may be inhibitory material in the specimen that prevents detection of viruses. Moreover, such viruses have not adapted to the host system used for isolation, unless several passages have been performed. Growth may be slower than with prototype viruses, and host range may be more limited. More than one virus may be recovered from a specimen, although this is relatively infrequent.[68,69] Where there is a disease, a presumption of the likely etiological agent dictates the host system selected for isolation attempts. Many types of viruses can be recovered from individuals without overt disease, and a combination of cell culture systems must be used to detect the widest range of viruses.

The enteric viruses that infect and multiply in the intestinal tract of man are of primary interest to the water virologist. As a group, these viruses are ether and acid resistant, are nonenveloped, and are relatively stable in the water environment. There exist several morphological groups of enteric viruses, and these viruses may not be limited to the gastrointestinal tract as the site of replication. These groups include the enteroviruses, adenoviruses, and reoviruses, and more recently detected fecal viruses such as rotaviruses, Norwalk and Norwalk-like viruses, and other viruses collectively referred to as small round viruses. Some of these viruses cannot be detected by traditional cell culture systems.

Different groups of human viruses have different and characteristic host specificities, but individual serotypes and strains vary in virulence for particular host systems in which each can be detected with ease. Different laboratories sometimes report contradictory findings with respect to host cell susceptibility to particular viruses. Some variables that may alter apparent host range of virus types include the history and strain of the cell culture system and the virus, and the kind and conditions of the assay.

Common cell culture systems used for enteric virus multiplication and detection are listed in Table 1.

A. Enteroviruses

The enteroviruses are small (28 nm diameter), nonenveloped RNA-containing viruses that are the acid-stable subgroup of the picornaviruses. These viruses are the etiological agents of a wide variety of diseases. They are excreted in the feces and are found in the water environment. As a group, they grow rapidly and well in cell cultures and are among the most common viruses recovered from water. The enteroviruses are divided into several major subgroups: polioviruses, group A coxsackieviruses, group B coxsackieviruses, echoviruses, and most recently, a few high number enteroviruses including hepatitis A virus (enterovirus 72). The groups are differentiated on the basis of pathogenicity for laboratory animals, disease induced, and serology, but are not distinguishable morphologically by electron microscopy.

The mode of infection of these viruses in cell culture has been extensively studied. The viruses replicate in the cytoplasm of infected cells. One of the major determinants of the host range of these viruses is the presence of a specific receptor on the cell surface that adsorbs the infecting particle.[70] Absence of the receptor, however, may not be the only barrier to infection.[71]

1. Polioviruses

There are three serotypes of poliovirus, the agents of paralytic poliomyelitis. All three grow well in susceptible cell cultures, including primary cultures, diploid strains, and continuous lines. These viruses multiply rapidly and in laboratory cell systems may produce plaques or CPE within 2 or 3 days.

The polioviruses are not limited to a cell culture assay; the production of paralysis in susceptible monkeys was the original method for detecting these viruses. Most other experimental animals, such as suckling mice, are not generally susceptible, so that for detection, a culture system is the most practicable.

Simian or human cells are the culture systems of choice for polioviruses, although some adaptation to other cell types may occur with laboratory virus strains. Polioviruses multiply readily under a variety of media and overlays, as long as the cells are susceptible. In general, a host cell sensitive to one poliovirus type is sensitive to all three; the three poliovirus types apparently share the same cell surface receptor.[72] Efficiency of infection and virus yield per cell vary among isolates and strains.[38]

Polioviruses are found in the water environment. Vaccine strains account for the majority of isolates in countries that vaccinate with live attenuated virus. In countries that do not use live vaccine, the detection of live poliovirus in water may represent a significant health hazard.

Laboratory strains of poliovirus types 1, 2, and 3 produce CPE in virtually all of the primary cell cultures listed in Table 1, including 1° RhMK,[73-77] 1°VMK,[78] 1° AGMK,[17] 1° CMK,[79] and 1° HuAm.[73,74]

CPEs are also produced in the following diploid cell strains: WI-38,[76,79,80] HL-8,[76] HFDL,[81] MRC-5,[83] HEL,[82] and cultures of human ovary,[84] HFDK,[82] and human placental fibroblasts.[80] Continuous lines that are sensitive to polioviruses include HeLa,[74] RD,[82] HEp-2,[73,84,85] LLC-MK_2,[75] BSC-1,[74,86] BGM,[87] and Vero.[17,77,83]

In studies comparing the efficiency of titration of laboratory stocks of poliovirus by CPE, continuous and primary human and simian cell cultures usually had similar sensitivities.[17,74,78,86] Sometimes, diploid strains were less insensitive than continuous[83] or primary[79] cultures. Polioviruses may also replicate in cell cultures without the appearance of CPE,[63] perhaps because of the small fraction of cells infected in relatively resistant cell cultures.

Polioviruses are readily titrated by plaque assay. Sensitive cells include human and simian primary cultures, continuous lines, and diploid strains.[8,73] In general, additives to the overlay medium are not required for plaque formation, although some strains of polioviruses may multiply better in the presence of $MgCl_2$,[43] cysteine,[55,59] or other enhancers as noted above.

The relative efficiency of recovery of polioviruses from clinical specimens by a number of cell cultures has been investigated in several studies. Most polioviruses have been recovered from fecal specimens, but polioviruses may also be recovered from throat swabs, saliva, blood, cerebrospinal fluid, and various tissues.

Studies on comparative recoveries are usually performed by parallel inoculation of a specimen on several strains of cell cultures, and scoring the development of cytopathic effect.

Primary rhesus monkey kidney cell cultures are the traditional cultures used for poliovirus recovery and most comparative studies use these cultures as a standard. In comparisons between 1° RhMK cultures and other systems, similar numbers of clinical recoveries were made in 1° HEK,[88,89] 1° BaK,[90] HeLa,[69,91] Vero,[77] LLC-MK_2,[75] BGM,[92] and BSC-1.[74] HeLa cells are less efficient than 1° CMK or HEp-2 for poliovirus recovery.[15]

In one large-scale study, 1° HEK cultures were more sensitive for poliovirus recovery than 1° RhMK cultures,[68,69] in contrast to other studies reporting similar efficiencies for 1° HEK and primary simian cultures.[88,89,93] Polioviruses from stool specimens had a higher titer in 1° HK than in 1° RhMK cultures, although both kinds of cells were susceptible,[94] which is consistent with a bias towards 1° HK cultures in samples with small amounts of virus.

Human diploid cell strains from kidney or lung tissue are less sensitive than primary or continuous cultures for poliovirus recovery.[3,5,68,69,88,91,93,95] The difference in efficiency between diploid strains and other cultures was less than twofold and in some cases was marginal.

It is possible to recover polioviruses from clinical specimens by plaque assay, sometimes without the appearance of CPE.[64] Similar recovery efficiencies by plaque assay were found for 1° RhMK, MFK, and BGM cells, although BGM showed the greatest number of plaques for each positive sample.[95]

In summary, the recovery of polioviruses from clinical specimens by cell culture methods is relatively simple. Most simian and human kidney and lung culture systems seem to be sensitive, with diploid cultures somewhat less sensitive than primary and continuous cultures.

2. Group A Coxsackieviruses

The group A coxsackieviruses consist of 23 serotypes (A1-A24, where A23 is echovirus 9) that cause a wide variety of disease including herpangina and aseptic meningitis.[52] These viruses were originally isolated in suckling mice, and suckling mice remain the most sensitive system for recovering these viruses. The group A coxsackieviruses are distinguished from other enteroviruses by their pathological effect in suckling mice. The disease induced is a flaccid paralysis after intracerebral or intraperitoneal inoculation. Muscle degeneration is induced, and death may occur within a week.[96]

With the exception of a few serotypes, the group A coxsackieviruses are not easily grown in cell cultures. Some strains can be adapted to growth in cell culture by passage, with an increase in titer over time, a decrease in the time for CPE to appear, and a decrease in mouse virulence.[97]

Embryonated eggs, chick cell cultures,[98] and mouse embryonic cell cultures[99] are not sensitive to most of the group A coxsackieviruses.

Of the simian and human cell cultures tested for sensitivity to group A coxsackieviruses (originally isolated in mice), the most sensitive for a wide range of serotypes are 1° HuAm, HFDK, and RD cell cultures. Of the 23 serotypes, only A1, 13, 22, and 24 were negative in RD cells.[82,102,103] Note that these are summaries of the results obtained by several laboratories, with individual differences for particular serotypes. Overall, only types A1 and A22 have not produced CPE in any cell culture system to date.

Many other cell types have been tested for sensitivity to various group A coxsackieviruses passaged either in mice or in cell cultures. HeLa cells were sensitive to types A7, 9, 10, 11, 13, 15, 16, 18, 20, 21;[74,96] 1° RhMK were sensitive to types A7, 9, 10, 14, 16;[74,102] BSC-1 were sensitive to A7, 8, 9, 10, 14, 16;[74] BGM were sensitive to A7, 9, 14, 16;[87] human ovarian cell cultures were sensitive to A4, 6, 7, 14, 15, 16, 18, 21;[84] and WI-38 cells were sensitive to A7, 9, 13, 15, and 18.[79] Generally, human embryonic fibroblasts are more sensitive than primary monkey kidney cultures to group A coxsackieviruses.[104] A few group A coxsackievirus serotypes have wide cell culture ranges, particularly coxsackieviruses A7, 9, and 16.

Some group A coxsackieviruses that grow in cell cultures may also be assayed by the plaque technique. Coxsackievirus A9 produces plaques in 1° AGMK and 1° RhMK cultures,[105] BGM[106] (with a slightly lower titer than 1° RhMK), and HeLa cells.[35]

Different strains of HeLa cells vary in sensitivity to coxsackievirus A9.[35] Titer determined by plaque assay was higher than by $TCID_{50}$. Plaque formation of other group A coxsackieviruses has also been reported.[8,55,103,107,108]

Optimization of plaque assay conditions by enhancers such as $MgCl_2$ or DEAE dextran aided in plaque production for some serotypes.[43,55,108,109]

The recovery of group A coxsackieviruses from clinical specimens by cell culture systems has been reported in a number of studies; the range of serotypes recovered depended on the culture used. In one study where fecal isolates were recovered on 1° HK, WI-38 and 1° RhMK cultures in parallel, no culture picked up all the positive specimens. 1° RhMK cultures were the least sensitive and picked up primarily coxsackievirus A9; the other two cell types recovered larger numbers and a wider range of serotypes.[88] This pattern has not been universal; in other studies, WI-38 cells were less sensitive than primary cultures.[93] In comparisons between 1° RhMK and diploid human fibroblast strains, the latter were somewhat more sensitive than the former,[3,26] particularly for group A coxsackieviruses, except for A9.[26] HL-8 cells have also been used to recover group A coxsackieviruses, with results similar to those with 1° RhMK cultures.[76] Most group A coxsackieviruses recovered by culture systems were types A9 and A16.[3,15,26,76,77,93]

Continuous cell lines such as Vero[77] and HeLa[15,26] were less sensitive than 1° RhMK cultures for group A coxsackievirus recovery. One line of continuous cells, RD, has been reported to be sensitive to a wide range of group A coxsackievirus serotypes.[102] Parallel reisolations from specimens originally positive in suckling mice were made in RD cells and in suckling mice. The RD cells were positive for 8 out of 11 samples including one that was negative in suckling mice, and the mice were positive for 9 samples. The isolates on RD cells included types A2, 4, 5, 6, 10, and 16, a relatively broad range. The RD recoveries, however, required several blind passages for most of the isolates detected.

Coxsackievirus A9 can be recovered by plaque assay on 1° RhMK cultures without the formation of CPE[64] and recoveries can be made by CPE without plaque formation.[95] The relative efficiencies of the plaque assay and the CPE method for recovering group A coxsackieviruses has not been extensively investigated.

In summary, the most useful cell culture systems for recovering group A coxsackieviruses are primary or diploid human or simian cultures, and the RD line of human cells. Suckling mice, however, remain the system most likely to recover the broadest range of serotypes with the highest efficiency.

3. Group B Coxsackieviruses

The group B coxsackieviruses are a group of 6 serotypes that produce spastic but not necessarily fatal paralysis in suckling mice. In humans, these viruses cause a variety of diseases, ranging from mild to serious, including aseptic meningitis and myocarditis.[52]

The group B coxsackieviruses have similar host ranges for cell culture systems, perhaps due to a shared cell surface receptor.[72] There are exceptions to this rule,[82,110] both in host range and degree of cell sensitivity to the viruses.

Human and primate cells are the most susceptible of those available, and may be used to grow and recover these viruses from clinical specimens. Some nonprimate cells, such as primary pig and hamster kidney cultures have been reported to be sensitive to these viruses.[12,16] Growth in embryonated eggs has not been found.[98] In appropriate cells, growth of these viruses is relatively simple. Unlike many group A coxsackieviruses, all group B coxsackieviruses grow well in cell culture systems and are readily recovered in such systems from clinical specimens. Suckling mice may also be used.

Laboratory strains of group B coxsackieviruses produce CPE in a wide range of primary, diploid, and continuous cell cultures. These include 1° RhMK,[73-75,77,111] 1° patas MK,[73] 1° VMK,[78] 1° AGMK,[17] 1° CMK,[79] and 1° HuAm.[73,74,97] Diploid lines for which cell culture sensitivities to group B coxsackieviruses have been reported include HFDL,[81] HFDK,[82] HL-8,[76] and human ovarian[84] cell cultures. Sensitive continuous lines include HeLa,[74,112] HEp-2,[73,85] Vero,[77,82] BGM,[87] LLC-MK_2,[75] BSC-1,[74] KB,[113] and FL cells.[113] In contrast, growth in RD cells is relatively poor.[82,102]

In comparative studies, RD and HEL cells were less sensitive than HFDK and Vero cells to group B coxsackieviruses,[82] and Vero was less sensitive than 1° AGMK for 1° AGMK-grown virus.[17] Primary RhMK cells were more sensitive than 1° BaK cells[90] or human placenta diploid fibroblasts.[80] Primary CMK were more sensitive than WI-38,[79] and FL were somewhat more sensitive than KB, which in turn were more sensitive than HeLa cells.[113] Finally, BGM cells were more sensitive than 1° VMK cells in roller tube assays.[78] Generally, primary and continuous lines were more sensitive than diploid strains, although group B coxsackieviruses grew in all types of cells. CPE production by group B coxsackieviruses was not enhanced by the inclusion of $MgCl_2$ in the medium.[43] Group B coxsackieviruses are easily assayed by plaque assay and produce plaques on 1° AGMK,[105] 1° HuAm,[97] BGM,[106] and other cell cultures. Enhancers aid plaque production of 1° RhMK cultures for some of the B coxsackieviruses,[43,55] but enhancers are not necessary for plaque production and in some cases may be inhibitory.

Cultures of 1° RhMK cells form the common basis for comparison of cell culture recovery efficiencies of group B coxsackieviruses from clinical specimens. Primary RhMK and other primary primate cultures such as 1° CMK[114] and 1° BaK[90] had similar recovery efficiencies. Primary HEK cultures, however, may be less efficient than 1° RhMK[89] or 1° BaK[93] cultures.

Human diploid fibroblast strains, including WI-38[26,69,88] and HFDK,[3,11] are less sensitive to group B coxsackieviruses than 1° RhMK or 1° HEK cultures. HL-8 and 1° RhMK cells pick up small numbers of group B coxsackieviruses with equal frequency.[76]

Similar recovery efficiencies from clinical specimens have been reported for 1° RhMK and for the continuous lines Vero,[7] LLC-MK_2,[75] HeLa,[26] BSC-1,[74] and BGM.[92] HEp-2 cells have also been used for recovering group B coxsackieviruses.[15,85]

Recovery of group B coxsackieviruses have been made by plaque assay on 1° RhMK cultures, without CPE development.[64] In one study, most isolates from clinical specimens that produced CPE on 1° RhMK could be recovered from the same specimens by plaque assay on 1° RhMK and BGM cells.[95] The BGM cells had higher numbers of PFU per sample than did the 1° RhMK cultures.

Overall, continuous lines and primary cultures of human or simian cells are the most efficient for the growth and recovery of group B coxsackieviruses. Human diploid strains are less sensitive than other types of cells for recovery of these viruses.

4. Echoviruses

The echoviruses (enteric cytopathogenic human orphan viruses) consist of a large group of enteroviruses (serotypes 1 to 34, where type 10 is reovirus 1, type 28 is rhinovirus 1A, and type 34 is a variant coxsackievirus A24)[115] originally isolated in cell cultures; serotypes and strains vary in host range. Growth in culture is generally limited to primate cells.[16,116]

Most laboratory grown echovirus types grow in the following cultures: 1° RhMK,[23,73,77] 1° CMK,[79] 1° HuAm,[73,74,80] HFDL,[81,82] diploid HEL,[66] HFDK,[82] HL-8,[76] RD,[82] BGM,[87] HeLa,[74,117] LLC-MK_2,[75] BSC-1,[74] and human ovarian[84] and placental[80] diploid cell cultures. Vero cell cultures are fairly sensitive[77] and resistant[82] to a wide range of echoviruses.

In general, primary monkey kidney cultures tend to be the most sensitive for the widest range of echoviruses. Primary RhMK cultures were the cells on which the prototype echovirus strains were isolated and continue to be the standard cultures against which other cultures are compared. Less sensitive systems include 1° patas MK and HEp-2 cells,[73] although HEp-2 strains varied in their sensitivity. Similar strain-dependent sensitivity has been reported for HeLa cells.[118]

Echoviruses produce plaques on 1° RhMK cultures[73] and in general are easily assayed on other cultures such as BGM[106] cells. Under the most favorable conditions, plaques are detected in 2 to 5 days postinfection in diploid strains of HEL cells.[66] Titers determined by plaque assay and $TCID_{50}$ are similar.[66] Optimal conditions for plaque formation vary with serotype and strain and are influenced by agar concentration and the presence of DEAE dextran.[66] Several other additives may enhance or inhibit plaque formation, depending on the conditions of assay.[43,50,55,57]

Plaque morphology, plaque-forming ability, and relative titer by $TCID_{50}$ and PFU varied among isolates of the same serotype from a single epidemic.[27] Passage in culture exerted a strong selective influence on the characteristics of the virus strain that resulted.[27,119,120]

Echoviruses may be recovered on a variety of cell types. There is not always a correlation between the reported host range of laboratory grown serotypes and serotypes that are recovered from clinical specimens in particular host systems. Relative recovery frequencies and range of types recovered in different cell cultures have been compared in a number of studies. The results will be summarized here, but the reader is referred to the indicated references for more complete information about specific serotypes isolated. In many instances, a few serotypes predominated among the echoviruses recovered and recovery frequencies were biased toward the cell types most sensitive to those serotypes. Furthermore, the most efficient host cell system often did not pick up all of the isolates and could be complemented by a less efficient system.

In comparing the frequencies with which echoviruses were recovered in primary cultures, 1° RhMK and 1° CMK were found to have similar efficiencies,[114] as did 1° HEK and

1° BaK cells.[93] In one study, 1° RhMK and 1° HEK yielded similar recovery rates,[88] but in another, 1° RhMK was more efficient than 1° HEK.[89] In the latter study, however, different temperatures of incubation were used for the two cultures, which may have influenced the results. Among those primary cultures that have been compared, no one culture was obviously better than the others.

Comparisons between diploid and primary cultures sometimes indicated a higher recovery frequency for human diploid strains than for 1° RhMK,[3,11,121] or 1° HEK,[68,69] although in others, the recovery efficiencies were similar.[26,88,91,93]

Continuous lines, such as LLC-MK_2,[75] BGM,[92] HeLa,[15,26,91] and Vero[77] recovered fewer echoviruses than did 1° RhMK cultures. BSC-1 cells,[74] however, were as efficient as 1° RhMK cultures, and HEp-2 cells[15] were more efficient than 1° CMK cultures. In a compraison between 1° HuAm and the Detroit-6 line of human amnion cells, the line cells were less sensitive than the primary cells.[122]

To summarize, primary and diploid cells are generally more sensitive than continuous cells for recovering echoviruses, although line cells support the replication of laboratory strains of echoviruses.

5. Enteroviruses 68 to 72

Five viruses more recently recovered from clinical specimens have been designated enteroviruses 68 to 72. The role of enteroviruses 68 to 71 in waterborne disease has not yet been investigated, although they are potentially spread by water.

At least one of these viruses, enterovirus 70, can be propagated in cultures and forms plaques in human embryonic lung cells.[127]

Enterovirus 72 (hepatitis A virus), the agent of infectious hepatitis, is one of the few viruses that has been shown to cause waterborne epidemics, but only recently have techniques been developed that allow the direct detection of the virus in water. The virus can be propagated in primates, including marmosets, chimpanzees, bush babies, and owl monkeys,[123,123a,123b] but these are not readily accessible systems for routine investigations.

Recent attempts at propagation in cell culture have proved increasingly successful,[124-126a] although a simple, rapid assay system in a tissue culture host does not yet exist. Enterovirus 72 does not produce CPE in most of the cells it impacts in culture and must be detected in these cells by flourescent antibody, RIA, or radioimmunofocus assay techniques. One strain of enterovirus 72, HM 175, has been adapted to cell cultures and can be assayed by immunoflourescence and by other techniques.[126b,126c]

Immunochemical techniques, such as ELISA or IEM (immunoelectron microscopy), are used for the direct detection of hepatitis A viral antigens in clinical specimens or in the environment if the virus is present in high enough concentration.

The choice of the most sensitive cell culture system for the enterovirus group is difficult to make because of variations in the host ranges of the viruses. The selection should take into account the viruses expected to be present in highest concentrations and the systems most efficient for those viruses. Enteroviruses differ in their host range, depending on group, serotype, and strain. In general, primary human and primate kidney cells[41,128] have been the most widely used host systems for recovery attempts. Difficulties in obtaining these systems have led to increased use of cultures that can be propagated and maintained in the laboratory, assuring a continuous and standard supply. Studies of cell culture recovery efficiencies for enteroviruses as a group are summarized below.

Primary RhMK, 1° CMK,[114] and 1° BaK[90] recover enteroviruses from clinical specimens with similar frequencies. In two studies comparing the recovery efficiencies of 1° RhMK and 1° HEK cultures, contradictory results were obtained, with 1° RhMK more efficient in one study[89] and less in the other.[88] These differences may be due to differences in cultures, the viral flora present in the specimens, or the particular conditions of incubation used.

In comparisons between 1° RhMK and diploid strains of cells, HFDK,[3,11,92] WI-38,[88] and HL-8[76] recovered enteroviruses as effectively as 1° RhMK cultures. Three strains of human fetal lung cells (WI-38, MRC-5, and IMR-90) were equally effective for recovering enteroviruses, with time of onset of CPE being similar for all three. IMR-90 had the disadvantage of degenerating after fewer passages than the other two lines.[123]

Continuous lines, including HEp-2,[65] BGM,[92] and Vero,[77] picked up fewer isolates than 1° RhMK cultures. BSC-1 cells, however, recovered almost as many enteroviruses as did 1° RhMK cultures.[74]

In one study comparing plaque and CPE assays, both methods were useful for enterovirus recoveries, each assay picking up isolates that the other missed. Overall, the plaque assay with the addition of $MgCl_2$ was slightly more efficient than plaque assay without additives or the CPE tube assay.[65]

The enteroviruses as a group are best recovered from clinical specimens by simian or human primary or diploid culture systems. This does not exclude the possibility that a particular continuous line or combination of lines may prove to be equally — or more — efficient than the other systems.

B. Adenoviruses

The adenoviruses are a group of DNA-containing viruses that have a larger and more complex structure than the enteroviruses and are distinguishable from them morphologically by electron microscopy. There are more than 41 serotypes[130,130a] that have both shared group-specific antigens and differentiating type-specific antigens. Many of these viruses are associated with respiratory disease or conjunctivitis. Adenoviruses 40 and 41 produce overt gastrointestinal diseases.[130a,131] Adenoviruses are excreted in large quantities in feces and thus enter the water environment. They have been recovered from that environment and have been directly associated with swimming pool outbreaks of conjunctivitis.[2]

Most adenoviruses grow well in cell cultures and are not pathogenic for laboratory animals.[132] Their preferred host systems are human cell cultures, although the induction of a CPE, not necessarily associated with complete replication, can be detected in simian cultures.

Adenovirus infection of monkey cells, such as 1° RhMK, 1°AGMK,[133] or Vero,[134] is semipermissive and is aided by the presence of a helper virus such as SV-40.[135] The block in the production of viable viral progeny is post-transcriptional.[136]

The cytopathic effect produced by adenoviruses is distinct from that produced by enteroviruses and is seen most clearly in continuous lines such as HeLa, KB, and HEp-2, and less rapidly in primary monkey kidney or human amnion cultures.[137,138] Human diploid strains, such as HFDL[81] and human ovarian cell cultures,[84] are also sensitive to adenoviruses although HeLa cells are more productive than WI-38 cultures for growth.[79]

Adenoviruses do not grow well in BSC-1 cells.[86] There has been a report of propagation of several adenovirus types in the HL-8 strain of fetal rhesus monkey kidney diploid cells, but 1° RhMK cultures were negative for growth of these types.[76]

The time of appearance of CPE is influenced by the quantity of viruses inoculated onto the cell culture. Inoculation of a small number of infective particles may produce delayed CPE, taking up to 2 to 4 weeks to appear: larger virus concentrations produce CPE in 2 days, depending on cell culture type.[132,139] One explanation for this difference is that adenoviruses are released from the nucleus of the cell in aggregated form so that dissemination throughout the culture is slow.[139] Subpassage may enhance dissemination by aiding release of the viruses and may decrease the time for CPE appearance. Adenoviruses 40 and 41, major etiological agents of gastrointestinal disease in young children, are fastidious and are detected primarily by electron microscopic techniques. In recent years, these adenoviruses have been grown in Graham 293 cells[139a] and in other cells also.[130a]

Plaque assay can be performed on human cell cultures including HeLa,[140] KB, and HEp-

2 with plaques forming within 8 to 10 days.[141] Titer by plaque assay is higher than by $TCID_{50}$.[61] Arginine may be necessary for adenovirus replication at the level of assembly of progeny virus[141-145] and is required for plaque formation. The concentration of arginine added to the agar overlay in the plaque assay influences plaque formation.[141,146]

The time of addition of the neutral red stain is important[61] because neutral red is inhibitory if added in the first overlay agar. The stain may be added as a second overlay.[61] The type and concentration of serum in the overlay can also influence the efficiency of plaque formation.[146]

Adenoviruses are usually recovered from respiratory or fecal specimens. Best results are usually obtained with human primary cultures such as 1° HEK cultures or continuous cell cultures such as HeLa, LN, KB, or HEp-2.[132]

Recent studies indicate that stool-associated viruses may be replication defective in the cell cultures usually used for recovering them. Partial replication apparently occurs with these viruses; viral antigens are detected in infected cells by staining with fluorescent antibody, but no CPE are detected in a variety of cell types including 1° HEK, HEp-2, MRC-5, 1° RhMK, and 1°AGMK cultures,[147] nor is infective progeny produced.[131] Adenovirus replication is not aided by SV-40 coinfection, centrifugation,[131] or proteolytic enzyme treatment.[147] It is therefore possible that it may be more difficult to detect adenoviruses that enter the water environment with fecal pollution than adenoviruses from nonfecal sources.

Difficulties in recovering adenoviruses from samples that contain mixtures of viruses stem from their relatively slow replication time compared with many other viruses, so that the adenoviruses may be overgrown.

A minimum incubation period of 14 days is recommended for adenovirus isolation efforts; longer times of incubation including subpassages increase the likelihood for recovering adenoviruses.[28]

Primary HEK cells are generally more efficient for recovering adenoviruses than WI-38 cells,[5,69,89,93] HeLa cells,[5,69] and 1° BaK[93] or 1° RhMK[5,69,88,89] cells. HeLa cells are more efficient for recovering adenoviruses than 1° CMK,[15] 1° RhMK, or WI-38[26,91] cells and about equal to HEp-2 cells.[15] Human fetal diploid kidney strains are more efficient than 1° RhMK cells.[11] Primary CMK and 1° RhMK gave similar recovery frequencies,[114] but continuous BGM[92] and BSC-1[74] lines were not able to detect the small number of isolates tested.

In spite of the inefficient replication of adenoviruses in monkey kidney cells, some adenoviruses have been recovered in these cultures.[148] It is possible that adenoviruses may be detected if present in different numbers to produce CPE in the first round of infection. Further dissemination of viruses throughout the culture is probably limited because little, if any, progeny virus is produced.

C. Reoviruses

The reoviruses (respiratory enteric orphan viruses) are a genus of the family Reoviridae that consists of three serotypes of nonenveloped viruses. Reoviridae contain a double-stranded segmented RNA genome. These viruses have diameters of 65 to 75 nm. They are distinguishable from the enteroviruses and adenoviruses morphologically by electron microscopy, and by composition. These viruses may be recovered from the human respiratory and enteric tracts,[149] but are not usually associated with disease. The three serotypes share group specific antigens and are distinguishable by type-specific antibodies. They multiply in the cytoplasm of infected cells and have an associated hemaglutinin.[141] Widespread in nature, they are not as yet associated with a specific disease, but are found in human stools and in the water environment.

The reoviruses are not generally detected in animal systems, although they may have some effect on infant mice.[149]

Reovirus replication and production of CPE is not limited to primate or human cell systems, although these tend to be the most commonly used for virus recovery.[150] Permissive non-primate systems include primary cultures of calf, lamb, pig, hamster, dog, mouse, cat,[16] rabbit, and guinea pig[151] cells. L cells, a line of mouse cells, are commonly used for growing these viruses.[150,152]

The reoviruses produce CPE in a wide range of human and simian continuous cell lines, including LLC-MK_2,[153] BGM,[87] BSC-1,[74,153,154] KB,[150] and the FL line of human amnion cells.[154] Primary cell systems for growing reoviruses include 1° RhMK[74,153] and 1° HuAm.[74] Diploid strains for growing reoviruses include human ovarian cell cultures,[84] WI-38,[153] HFDL,[81] and HL-8[76] cells.

In comparative studies, 1° VMK were more sensitive than BGM cells to two strains of reovirus,[80] and 1° VMK were more sensitive than Vero cells for growing reoviruses, produced in 1°AGMK cells,[17] indicating a trend for higher sensitivity of primary over continuous MK cultures.

Infectivity can sometimes be enhanced by heating reoviruses in the presence of 2 *M* $MgCl_2$[155] or by exposure to proteolytic enzymes.[154,156,157] This increased infectivity is apparently due to the removal of the outer capsid layer of the virus and the activation of the endogenous nucleic acid polymerase in the virion.[158] The increase in infectivity by these methods is strain dependent. Lysine, arginine, and leucine are necessary for reovirus growth[159] DEAE dextran interferes with CPE and virus production in HeLa cells.[160] CPE may be induced in the absence of virus production.[161]

Reoviruses produce plaques but generally need special conditions for successful plaque formation. Pancreatin in the overlay increases plaquing efficiency.[162] Neutral red, if added in a first overlay, may inhibit plaque formation; it is better to add neutral red in a second overlay.[58] Since serum may inhibit plaque formation, serum substitutes such as skim milk, bovine serum albumin, or yeast extract[58] may be used instead or serum may be omitted. Plaques may be detected on monolayers of L cells and 1° RhMK cultures within 4 days of inoculation.[58] Isolates may take longer to produce plaques and under the usual conditions used for enteroviruses, plaques may not be detected.

Reoviruses are generally recovered from fecal swabs or extracts. A variety of cells may be used for recovering these viruses[3] including 1° RhMK, 1° HK,[150] HFDK,[11] HeLa,[149] KB,[149] and BSC-1[74] cells. The slow growth of reoviruses may lead to difficulties in recovering them from a virus mixture unless the sample is diluted enough so that no other viruses remain.

D. Rotaviruses

The human rotaviruses are causal agents of nonbacterial infantile gastroenteritis.[163-165] There are four or more serotypes that infect man[166,166a] and they possess shared antigens. The viruses are members of the family Reoviridae and are therefore reovirus-like in morphology and possess a double protein shell and a double-stranded RNA genome. They are found in large concentrations in stools and were first detected by electron microscopic examination of fecal samples. Rotaviruses have been detected in sewage by immunoassays.[167]

Isolates of human rotaviruses have difficulty in replicating in cell cultures. They may be detected by immunochemical techniques, such as RIA or ELISA, and by immunofluorescent staining for viral antigens after infection of susceptible cells. Centrifugation onto cell monolayers may enhance infectivity, but immunofluorescent techniques have not been reported to be more sensitive than electron microscopic observation. Viral antigens, but not complete viable progeny, may be detected in a pig kidney cell line[168] and in LLC-MK_2 and 1° HEK cell cultures, among others.[169] LLC-MK_2 cells appear to be the most sensitive culture system.[169] Trypsin treatment may enhance infectivity of the virus. This has been known to be the case for nonhuman rotaviruses.[170]

One strain of human rotavirus (Wa) has been adapted for growth in cell cultures by serial passage in newborn gnotobiotic piglets, followed by passage in 1°AGMK.[171] Direct growth and detection of human rotavirus isolates by CPE or plaque formation in cell culture systems has not yet been reported. Rotaviruses have been recovered from human stools in 1°AGMK cells and detected by ELISA (Ward, personal communication, 1985). These isolates may be plaqued in MA104 cells after several passages in 1°AGMK cells. These viruses cannot yet be readily detected with standard host systems and techniques, but until these are developed, more sophisticated, if less sensitive, immunochemical technology can be employed.

E. Norwalk and Norwalk-Like Viruses

The Norwalk and Norwalk-like viruses are viruses recovered from stools of patients with acute infectious nonbacterial gastroenteritis.[172] These viruses have been detected by electron microscopy only.[173] These viruses are 20 to 30 nm in diameter, ether stable, and acid resistant. There are several strains of these viruses, none of which have been grown in cell cultures. All have been identified by immunochemical techniques such as RIA and IEM. The Norwalk agent may or may not be related to other similar agents which have been recovered in various parts of the world.[174] They have not yet been demonstrated directly in sewage. Waterborne transmission appears to be a major path for epidemic dissemination of the Norwalk virus.[174a]

F. Other Small Round Viruses

Other agents, such as coronaviruses, astroviruses, and caliciviruses, have been detectable in fecal samples by electron microscopy and some can replicate in viable host systems in culture.[175-179b]

The role of some of these particles as agents of disease needs to be established. Caliciviruses are known to produce gastroenteritis in young children and others of these viruses have been found in the stools of patients with gastroenteritis.[175,179] These viruses will not be reviewed here in depth because their importance as waterborne agents remains to be established.

G. Non-A, Non-B Hepatitis

Several reports have appeared in the literature in recent years suggesting that a number of outbreaks of waterborne hepatitis in India were produced by an agent that was neither hepatitis A (enterovirus 72), hepatitis B, nor the non-A, non-B hepatitis agent(s) associated with transmission through human blood or blood products.[179c-179f] The agent is now believed to be a 27 nm particle.[179g] The agent has not been grown in cell cultures or in intact animals.

V. HOST SYSTEMS FOR DETECTING VIRUSES IN THE WATER ENVIRONMENT

The detection of viruses in the water environment requires a host system sensitive to a wide range of enteric viruses. The viruses described in the previous section are potentially waterborne, although not all have been detected in water. Viruses that have been found in water include polioviruses, group A and group B coxsackieviruses, echoviruses, adenoviruses, reoviruses, rotaviruses,[180] and recently, hepatitis A virus.[181]

No one host is known that is capable of detecting all of these viruses, but a combination of systems may be selected to maximize the range of sensitivity to diverse viruses.

In processing water samples, the type and source of water influence the choice of detection system and method. For drinking water, where the presence of even one infective unit of any human-pathogenic virus is of concern, the most sensitive system available should be used, with less regard for quantification. Where it is important to have a quantitative picture of virus contamination, as in sewage, or when determining the effectiveness of treatment processes, it is advisable to use a stable, accurate system for detecting viruses. Range and sensitivity are important also, but since viruses are present in larger numbers in wasterwater, their presence is easier to detect.

Several problems are associated with the recovery of waterborne viruses. One problem, shared with clinical virology, is that of recovering viruses that have not been adapted to the host system used to detect them. Host range studies of laboratory-grown viruses may therefore not provide an accurate picture of the spectrum of sensitivity of host cells to virus isolates. Environmental isolates may be less virulent for laboratory host systems than their prototypes. A further consideration is that in water a wide variety of viruses may be present, each in small numbers, unlike clinical samples which usually contain one or at most two virus types. The method of recovery used should minimize interference and overgrowth of one virus type by another that multiplies faster. Since each virus type may not be present in large numbers, the lack of detection of a particular virus by a host system may be due to chance rather than to an insensitivity to infection. Studies that attempt to define the most sensitive system for detecting viruses by the parallel inoculation of cell culture systems with a single sample may well be biased unless based on large sample numbers or on samples with large number of viruses. The water samples may contain viruses inhomogeneoulsy distributed, due to aggregation or adsorption to particulates, so that parallel inoculums may not contain the same viruses. The presence of toxic materials in cell cultures not due to viruses is a special problem of the water environment. Toxicity may be controlled by decontamination of the inoculum with ether[182] or chloroform, or with antibiotics,[106] by removal of the inoculum after virus adsorption and replacement with growth medium,[183] or adsorption in the presence of serum.[95] Such methods should be tested to ensure they do not negatively affect either the host system or the viruses. Confirmation of virus presence should be performed by passage and identification where possible.

Water samples are often subjected to a variety of treatments aimed at concentrating the viruses into a smaller volume. These techniques may, in turn, concentrate certain toxic contaminants,[184] while removing others.[183] Furthermore, the viruses present may be affected by the concentration method and may be differentially concentrated, aggregated or disaggregated, enhanced or inhibited by the treatment.

Finally, it is difficult to predict which individual viruses will be present in a given sample. The viral flora varies with the season and year, depending on the viruses present at a given time in a population, and their likelihood of reaching the water environment in a viable state. A suitable detection system must therefore recover the broadest range of fecal viruses possible, rapidly and consistently. The host systems available are those described earlier in this chapter, and include animal as well as cell culture hosts.

Animals can be appropriate hosts for viruses that are restricted to such hosts for replication, or which are most sensitively detected in such systems. The first reported recovery of a human pathogenic virus from water was performed in 1939 by the inoculation of rhesus monkeys with sewage. The induction of experimental poliomyelitis in the inoculated monkeys demonstrated the presence of poliovirus in the sewage water.[185] Subsequently, with the development of cell culture methods for the growth of polioviruses, it was no longer necessary to use monkeys to detect these viruses.

Coxsackieviruses have been detected by the inoculation of water samples into suckling mice.[186-191] This system is particularly important for group A coxsackieviruses because many of these viruses are not recoverable in cell cultures.

Atypical coxsackieviruses have been recovered from water samples. These viruses produced CPE in cell cultures without producing paralysis in mice. In one report, parallel samples of sewage were tested in cell cultures and in suckling mice. Several isolates of coxsackieviruses B4, B5, and A9 were made in the cell cultures but no coxsackieviruses were recovered in the mice.[192] In another study, samples negative in mice induced paralysis in mice after passage in cell cultures.[190]

Hepatitis A virus, which is difficult to propagate in cell culture systems, may also be detected in water by inoculation into susceptible animals.[123]

The presence of hepatitis A virus antigens is more rapidly and easily determined by immunochemical techniques.

Embryonated eggs are another system for detecting viruses but they are not useful for the enteric viruses. The production of pocks and foci has been reported in eggs inoculated with effluent,[193] but the efficiency of the method and the viruses so detected were not described.

Cell culture systems are the most convenient and widely used hosts for the recovery of viruses from water. The two standard methods for detecting and quantifying viruses by cell culture inoculation are plaque assay and CPE induction in the host cells, as described above. Both methods may be used for recovering viruses. The efficiencies of these techniques for recovering waterborne viruses have been compared in a study with primary monkey kidney epithelial cultures. In general, the efficiency of enterovirus recovery was similar for CPE tube and bottle plaque assay, but variation in efficiency occurred for some virus types. Polioviruses were more frequently detected by plaque assays, and group B coxsackieviruses and echoviruses were more frequently recovered by CPE assay.[194] Another study with three cell substrates showed that polioviruses, group B coxsackieviruses, and echoviruses could be recovered from wastewater by both bottle and plate plaque and CPE tube assays. The relative frequency of recovery by the assays was influenced by the cell and virus types and by the vessels used for plaque assay.[95] The reoviruses in particular were recovered more efficiently by CPE tube assay.[67,95] The inefficency of plaque assay for recovering reoviruses may be due to the need for special conditions, such as the inclusion of protease or omission of serum from the agar overlay to enhance plaque formation.

Overall, virus recovery from wastewater by tube assays may be more efficient than plaque assay.[195] When different cell culture systems have been compared, however, this has not always been the case. In one study, the most efficient general recovery systems, as measured by the number of positive wastewater specimens, were bottle plaque assay and CPE tube assay on 1° RhMK cells; plate plaque assay was less efficient. For MFK and BGM cells, the CPE tube assay was more efficient than bottle or plate plaque assays.[95]

An alternative to the monolayer plaque assay is the agar suspension method of Cooper,[31] which has been used to pick up virus isolates from water.[196,197] In a comparative study of three cell types and three methods for detecting viruses, the best results were obtained with BGM cells in the agar suspension plaque assay, as compared with a CPE tube assay or monolayer plaque assay. Vero cells were best in agar, followed by monolayer plaque assay and tube assay. RD cells gave varied results; the agar suspension plaque assay gave the highest titers, but the monolayer plaque assay produced the most positive samples. The system with the most consistent positive results was the agar suspension assay on BGM cells.[197] It is likely that different viruses were isolated by the assays, but the viruses were not identified.

Plaque and CPE assays are both subject to cytotoxic contaminants in the inoculums. In the CPE assay, the contamination is more likely to spread throughout the vessel, destroying the entire cell monolayer. The plaque assay may localize the contaminant, allowing plaque development in less affected areas of the monolayer. Nonviral elements may, however, induce plaque-like holes in the monolayer, which may be mistaken for viral plaques.

The recovery of more than one virus from a mixed sample is more of a problem with the CPE assay than with the plaque assay because of the dissemination of viruses throughout the liquid medium. Plaque assays result in discrete virus isolates, as long as the plaques have not intermingled. In samples with large numbers of viruses, endpoint dilution of the inoculum allows for the separation of viruses in a mixed sample. After the identification of a dominant virus type, the sample can be replated in the presence of antiserum to that type, allowing other viruses to be detected.[198,199]

The CPE assay has an advantage over the plaque assay for water virology in that in the CPE assay a larger sample volume may be inoculated per monolayer surface area, and the cells tend to be stable for a longer period of time under the liquid medium of the CPE assay.

At present, it is not completely clear whether the plaque or CPE assay method is superior for the recovery of waterborne viruses. Prospective cell culture hosts for these viruses should be tested by both methods with wastewater samples to determine the most efficient one for each cell system. Moreover, the particular viral flora present in the samples influence the relative efficiencies of these assays.

The virus population expected to be present influences the choice of host cell system, because the most sensitive host for that range of viruses, or for a highly abundant indicator virus, should be used. Several studies with several host cell culture systems have indicated that the relative abundance of virus types in a sample is a function of the host range of the cells, the ability of the viruses to induce detectable changes in the host cells, and the actual quantities of each type of virus in the samples.

Several studies have been performed with different cell culture systems for recovering viruses. The relative abundance of the virus types recovered varied with the cell host. In one study, sewage samples were inoculated on 1°AGMK or 1° CMK; primarily echoviruses were recovered with coxsackieviruses and reoviruses recovered in smaller numbers. In contrast, with parallel inocula, U cells picked up coxsackieviruses in the largest numbers, together with a few polioviruses. This study was carried out in Sweden, where the live poliovirus vaccine is not administered, and an attempt was made to selectively recover these viruses in the presence of large numbers of other enteric viruses.[199]

A study of sewage samples in England over a period of several years, with five cell hosts, indicated that echoviruses were the most abundant followed by reoviruses. Next in abundance were polioviruses and group B coxsackieviruses, followed by adenoviruses. Of the cell types studied, 1° HuAm was the most useful, although relative recovery frequencies varied with the year. Other useful cell cultures were 1° RhMK and Vero. Diploid HEL cells were slightly less efficient, and HEp-2 cells were least useful because of premature degeneration of the monolayer.[200]

In the U.S., echoviruses have also been the most abundant viruses recovered.[95,194] Kelly and Sanderson[194] plated processed sewage samples on 1° MK, 1° HuAm, and on Hela cells and found, in order of abundance, echoviruses, coxsackieviruses, polioviruses, reoviruses, and adenoviruses. The 1° HuAm cells were the most efficient system, and 1° MK cells were better than HeLa cells. Schmidt et al.[95] found echoviruses in the greatest abundance in wastewater samples, followed in similar numbers by polioviruses, reoviruses, and group B coxsackieviruses, and then by group A coxsackieviruses. The cell culture systems used, in order of the frequency with which they recovered viruses, were 1° RhMK, RD, MFK, BGM, and HFDK, although by CPE assay the range was no greater than twofold between the most and the least sensitive cultures.

In the studies described above, each using several cell substrates, there is general agreement that echoviruses are the most abundant in wastewater, and adenoviruses the least. This may not always be the case; in South Africa, reoviruses appear to be present in concentrations exceeding those of the enteroviruses.[67,78]

In another study,[201] carried out on raw sewage and effluent in Australia, the reovirus and enterovirus groups were the most prevalent, followed by the adenoviruses. The enterovirus isolates were evenly divided among the polioviruses, group B coxsackieviruses, echoviruses, and untyped groups. The largest number of recoveries was made on HeLa-R cells, followed closely by 1° CMK cells. Lesser numbers were recovered in BGM cells and in the Borrie continuous line of HEL cells.[201] The small numbers of adenoviruses and group A coxsackieviruses usually recovered may reflect the poor efficiencies of the cell culture systems in which they were recovered, competition with other viruses, inefficient concentration from water, or it may be a true picture of the relative paucity of these viruses in wastewaters. The relative efficiencies with which virus groups are recovered in different cell cultures, as reported in comparative studies on water samples, are discussed below:

The polioviruses are relatively easy to detect in water samples by cell culture systems, both by plaque and by CPE assays. Studies on the recovery rates of different host systems have not revealed an absolute advantage for any particular system. In one study of sewage isolates, 2°AGMK cells recovered three times the number of polioviruses as did HFDL cells.[202] In another study, plaque assay on 1° MK cultures was the method that recovered the most polioviruses, followed by 1° MK CPE assay, with few recoveries on HeLa cells and none on 1° HuAm cultures.[194]

BGM cells have been reported to be more efficient for poliovirus recovery from sewage than HEp-2 cells.[203] The frequency of recovery of polioviruses in another study followed the order RD, BGM, 1° RhMK, MFK, HFDK, with very few recoveries on HFDK; plaque and CPE assays had similar sensitivities.[95] In another comparative study, 1° RhMK and 1° HuAm cultures were more sensitive than Vero and HEp-2 cells, and diploid cells were the least sensitive.[200] U cells gave a selective advantage to polioviruses in samples with several virus types as compared with 1° CMK or 1°AGMK cultures.[199] The best systems would seem to be primary and continuous simian or human cell lines; human diploid strains seem to be less sensitive. These same conclusions were reached from clinical studies.

Group A coxsackieviruses may be recovered from sewage or water samples in cell cultures, but many group A serotypes cannot be recovered except in suckling mice. In one study, where a small number of recoveries were made on HFDK and RD cells, none were made on 1° RhMK, MFK, or BGM cultures.[95] In another study, however, RD cells did not increase the range of group A coxsackieviruses detected in other cell lines including 1° HuAm, 1° RhMK, 1°AGMK, Vero, and WI-38.[204] RD cells may be useful for the recovery of group A coxsackieviruses, but more work needs to be done to confirm this. Coxsackievirus A9 may be recovered on 1° MK[192,198,205] cultures.

The group B coxsackieviruses are relatively simple to recover in cell cultures from water samples. In one comparison of recoveries from wastewater, BGM was the most sensitive cell type studied, followed by MFK; 1° RhMK, HFDK, and RD cells did not recover any viruses.[95] In another study, CPE assay on 1° MK cells was more sensitive for recovering group B coxsackieviruses than plaque assay on 1° MK cells or CPE assay on HeLa cells; 1° HuAm cells were relatively inefficient.[194] In a comparison between two primary cells, 1° MK cells were more efficient than 1° HuAm cells for recovering group B coxsackieviruses.[192]

Sellwood and Dadswell[200] recovered about 250 coxsackieviruses (mostly group B) from sewage samples. These authors reported cell sensitivities for recoveries (in descending order of sensitivities) as follow: Vero, 1° MK, HuAm, Hep-2, and diploid HEL cells. Overall, the monkey cell culture systems appear to be more sensitive than the human cell culture systems for recovering group B coxsackieviruses from water; continuous cultures such as Vero and BGM appear to be slightly more sensitive than primary cell cultures for this purpose.

The echoviruses are a large virus group with varying host ranges. Certain echoviruses may be present in large numbers in a given water sample, and this may bias recovery rates in favor of a particular cell type for which the dominant viruses have affinity. In one study, 1° HuAm cells were the most sensitive of those surveyed for echovirus recovery, followed by diploid HEL, 1° RhMK, Vero, and HEp-2 cells.[200] Another study indicated a higher efficiency of recovery, both of numbers and types, in HFDL over 2°AGMK cells.[202] In a third study, the efficiencies of 1° RhMK and RD cells were similar, followed by HFDK, MFK, and BGM cells.[95] In another study, 1° HuAm cells were by far the most efficient recovery system, followed in order by 1° MK CPE assay and 1° MK plaque assay, with no recoveries in HeLa cells.[194]

On a smaller scale, echoviruses were better recovered from mixed virus populations in 1° CMK and 1° GMK cell cultures than in U cells.[199] In a comparison between 1° HuAm and 1° MK, the human amnion cells recovered greater numbers of viruses, although both picked up the same range of types.[192] HeLa cells[183] and BGM cells[196,197] may also be used for recovering echoviruses. In summary, 1° HuAm and 1° MK cell cultures are the most efficient of those tested for recovering echoviruses, but other cell cultures, including human diploid and some continuous lines, may be used.

Adenovirus recoveries are better obtained in human cell cultures than in monkey cell cultures because of the growth characteristics of the viruses. CPE produced by adenoviruses are slow to develop in comparison with enteroviruses,[201] and the presence of adenoviruses may go undetected in samples that contain mixtures of viruses. Recoveries of adenoviruses has been reported to be most efficient in 1° HuAm cultures, followed by diploid HEL cells; HEp-2, Vero, and 1° RhMK were inefficient.[200] Recoveries of adenoviruses were greater in HeLa-R cells than in Borrie cells; BGM and 1° CMK were inefficient.[2]

Reoviruses can be recovered, at times in abundance, from water samples. Plaque assays are less efficient for this purpose than CPE assays,[67] probably because of the need for additives in the agar overlay. The appearance of a reovirus CPE is a slow process, taking approximately 7 to 33 days.[78,201,208] In two studies, reoviruses were recovered most efficiently in primary MK (CMK or RhMK) cultures followed by MFK and BGM cells. Few reovirus recoveries were made in HeLa-R or Borrie cells,[201] and none were made in HFDK or RD cells.[95] One recovery was made by plaque assay on 1° RhMK cells in the presence of serum.[95] In another study, 1° GMK and 1° CMK were more efficient than U cells for recovering a small number of reoviruses in competition with other viruses.[199] In a third study, 1° RhMK cells were more efficient than Vero cells for recovering reoviruses, and in descending order, diploid HEL and HEp-2 cells were less sensitive. No recoveries were made in 1° HuAm.[200] In all of these studies, primary monkey cell cultures were the most sensitive recovery system for reoviruses, human cell cultures the least. Continuous lines of monkey kidney cells may be used for recovering reoviruses, albeit with lower efficiency than primary monkey kidney cell cultures.

The total number of enteric viruses recovered by a given cell system, although in part a function of the viral flora in the water sample, is also a measure of the relative sensitivity of the system. Among primary cultures, 1° HuAm cultures appear to be slightly more sensitive than 1° or 2° RhMK.[194,200] In one study, 18 of 77 isolates were made only in 1° MK, and 19 only in 1° HuAm; the remaining recoveries were made in both.[192] Thus, similar numbers of viruses were recovered in the two cell culture systems, although many of the recoveries were not made in parallel. In this study, the 1° MK cultures were better for recovering polioviruses and group B coxsackieviruses, and the 1° HuAm cells were more efficient for recovering echoviruses. Among the primary MK cultures, 1°AGMK recovered greater numbers of viruses by plaque assay from effluent and sewage than 1° RhMK.[106]

Studies on clinical recovery rates, described in more detail in the previous section, indicated that the primary cultures most efficient for enteric viruses varied among different laboratories. Furthermore, that efficiency may change with time,[208] and it is not feasible at present to recommend a single primary culture as the most useful for recovering these viruses. The cultures commonly used include 1° RhMK, 1° CMK, 1°AGMK, 1° HuAm, 1° HEK, and 1° HEL, and their use is limited by the availability and cost of the cultures. Primary monkey cell cultures are particularly efficient for reoviruses and are useful for recovering polioviruses, group B coxsackieviruses, and echoviruses. A few group A coxsackieviruses, including A9 and A16, are also recovered by these cultures. Primary human cultures are efficient for adenovirus, poliovirus, and echovirus isolations and may recover a wider range of group A coxsackieviruses than the monkey cell cultures.

Diploid human strains appear to be less useful than other cell culture systems for recovering viruses from water. In one study, however, the total number of polioviruses and echoviruses recovered from wastewater in diploid HEL cells was slightly higher than in 2°AGMK cultures.[202] Generally, the diploid human strains are most useful for group A coxsackieviruses and adenoviruses and may recover a broad range of echoviruses,[11] depending on the cell strain. The use of diploid monkey cell strains has not been investigated for water virology, although efficient enterovirus recovery from clinical specimens has been reported for the HL-8 strain.[76]

Continuous lines of primate cells are the most convenient laboratory system for recovering viruses. Genetic variablity of cell lines, however, poses difficulties in evaluating the recovery efficiencies of continuous lines. Contradictory findings of different laboratories may result from the use of continuous lines that derive from the same source, but with passage in culture, have altered virus sensitivities.

One of the most extensively investigated lines for water virology is the BGM line of African green monkey kidney cells. In the first such study, BGM cells recovered greater numbers of viruses in bottle plaque assays from wasterwater samples than did 1° RhMK or 1°AGMK.[106] In a second study, BGM cells detected viruses in water samples consistently more frequently than Vero cells did.[197] In contrast, it has been reported that recoveries of viruses in BGM cells were fewer than in 1° RhMK or RD cultures, about equal to recoveries in MFK cells, and higher than recoveries in HFDK cells by CPE tube assays.[95] In this study, viruses recovered by BGM cells were primarily from the poliovirus, group B coxsackievirus, and the reovirus groups. Assay conditions influenced the recovery efficiencies in that plate plaque assay with BGM cells produced more virus isolates than the same assay with 1° RhMK cultures, although the plate plaque assay was not the most efficient method for either cell culture system. In this study, the number of viruses recovered from the wastewater samples was not high, and chance may account for some of the observed differences in recovery. Moreover, it is known that high passage BGM cells may lose sensitivity to many viruses (Dahling, Berg et al., 1978, unpublished data).

Recovery rates depend both on the viruses in the sample and the ability of cell lines to recover those viruses. BGM cells were reported to be less efficient than 1° CMK cells for recovering enteric viruses by CPE assay because fewer reoviruses were recovered, although larger numbers of enteroviruses were recovered in the BGM cells.[201] In agreement, another wastewater study reported that the relative insensitivity of BGM cells to reoviruses resulted in fewer reoviruses of these viruses in BGM cells than in 1° VMK cultures.[78] Under the plaque assay conditions used by Dahling et al.,[106] it is probable that reoviruses were not recovered efficiently, which might have biased the results. The plaquing efficiency of polioviruses and group B coxsackieviruses is higher in BGM than in 1° RhMK cells[95,106] so that samples that contain large number of these viruses may yield higher counts in BGM cells. In general, BGM cells appear to be most useful for recovering these two virus groups from water.

Vero cells have also been used for recovering viruses from water. These cells appear to be less sensitive than 1° BaK,[195] 1° HuAm,[208] RD, and BGM[197] cell cultures. Reoviruses, coxsackieviruses and polioviruses were the viruses most frequently recovered by Vero cells from wastewater.[200] These cells do not seem to have an advantage over the BGM cell line.

Continuous cell lines of human origin have also been used in water virology investigations.[95,197] RD cells, found to be sensitive to laboratory strains of group A coxsackieviruses,[102] have also been used to recover viruses from water samples. In one study, RD cells recovered as many enteric viruses from effluent as 1° RhMK cells.[95] The cells recovered polioviruses, group A coxsackieviruses and echoviruses, but no group B coxsackieviruses or reoviruses.[95] Thus, this cell line might complement the BGM line for recovering viruses from water samples.

HeLa cells have also been used to recover viruses from environmental waters; in one study, in comparison with 1° HuAm and 1° MK cultures,[194] they were inefficient for recovering enteric viruses.

Recently, however, a line of HeLa-R cells was reported to be more sensitive than 1° CMK, BGM, and Borrie cells to enteric viruses as a whole, and enteroviruses as a group. The HeLa-R line was by far the most efficient for recovering adenoviruses, accounting for 85% of the isolates. Of the cultures tested, the 1° CMK and HeLa-R cells provided the best combination for an efficient two-culture system.[201]

VI. SUMMARY AND CONCLUSIONS

Primary cell cultures are the most sensitive host systems for recovering enteric viruses from environmental waters. It is possible to use continuous lines in place of primary cell cultures, but care must be taken that the subline used is a sensitive one. Careful selection of both a monkey and a human line will maximize the range of viruses detected. Of the lines that have been explored, HeLa-R, BGM, and RD seem to be the best substitutes for primary cultures. More monkey cell lines, such as MA-104 and LLC-MK_2, should be tested to determine their usefulness for recovering viruses from wastewater samples. It is possible that future virus recovery systems may include immunochemical assays in combination with cell culture systems. One of the challenges of water virology is the detection of enteric viruses that are not easily recovered in cell culture systems. Moreover, it is a continuing challenge to find reliable, sensitive cell substrates that can detect the many cytopathic human pathogenic viruses in the water environment.

REFERENCES

1. **Fox, J. P.,** Human-associated viruses in water, in *Viruses in Water,* Berg, G., Bodily, H. L., Lennette, E. H., Melnick, J. L., and Metcalf, T. G., Eds., APHA, Washington, D.C., 1976, chap. 4.
2. **Goldfield, M.,** Epidemiological indicators for transmission of viruses by water, in *Viruses in Water,* Berg, G., Bodily, H. L., Lennette, E. H., Melnick, J. L., and Metcalf, T. G., Eds., APHA, Washington, D.C., 1976, chap. 7.
3. **Schmidt, N. J.,** Tissue culture in the laboratory diagnosis of viral infections, *Am. J. Clin. Pathol.,* 57, 820, 1972.
4. **Sigel, M. M., Rippe, D. F., Beasley, A. R., and Dorsey, J. M.,** Systems for detecting viruses and viral activity, in *Viruses in Water,* Berg, G., Bodily, H. L., Lennette, E. H., Melnick, J. L., and Metcalf, T.G., Eds., APHA, Washington, D.C., 1976, chap. 10.
5. **Cooney, M. K.,** Selection of cell culture systems for virus isolation, *Am. J. Med. Technol.,* 42, 36, 1976.
6. **Dunnebacke, T. H. and Zitcer, E. M.,** Preparation and cultivation of primary human amnion cells, *Cancer Res.,* 17, 1043, 1957.
7. **Cliver, D. O. and Herrmann, R. M.,** Economical tissue culture technics, *Health Lab. Sci.,* 6, 5, 1969.
8. **Dunnebacke, T. H. and Mattern, C. F. T.,** Adaptation of Coxsackie virus strain A10 to human amnion cells in tissue culture, *Proc. Soc. Exp. Biol. Med.,* 105, 553, 1960.
9. **Tobin, J. O'H.,** Notes on viruses likely to be encountered in vaccine production using monkey kidney material, *Bull. WHO,* 22, 431, 1960.
10. **Hayflick, L. and Moorehead, P. S.,** The serial cultivation of human diploid cell strains, *Exp. Cell. Res.,* 25, 585, 1961.
11. **Schmidt, N. J., Ho, H. H., and Lennette, E. H.,** Comparative sensitivity of human fetal diploid kidney cell strains and monkey cell cultures for isolation of certain human viruses, *Am. J. Clin. Pathol.,* 43, 297, 1965.
12. **Perera, H. G.,** The cytopathic effect of animal viruses, *Adv. Virus Res.,* 8, 245, 1961.
13. **Reed, L. J. and Muench, H.,** A simple method of estimating fifty percent endpoints, *Am. J. Hyg.,* 27, 493, 1938.

14. **Chang, S. L., Berg, G., Busch, K. A., Stevenson, R. S., Clarke, N. A., and Kabler, P. W.,** Application of the "most probable number" method for estimating concentrations of aminal viruses by the tissue culture technique, *Virology*, 6, 27, 1958.
15. **Pal, S. R., McQuillan, J., and Gardner, P. S.,** A comparative study of susceptibility of primary monkey kidney cells, HEp-2 cells and Hela cells to a variety of fecal viruses, *J. Hyg.*, 61, 493, 1963.
16. **Lenahan, M. F. and Wenner, H. A.,** Propagation of entero and other viruses in renal cells obtained from non-primate hosts, *J. Infect. Dis.*, 107, 203, 1960.
17. **Mingle, J. A. A. and Westwood, J. C. N.,** Maintenance of primary African green monkey kidney (pAGMK) and Vero cells at room temperature (25°C): a system for virus isolation in community practice, *Can. J. Microbiol.*, 20, 391, 1974.
18. **Lund, E.,** Influence of age and cell number of a tissue culture on its susceptibility to poliovirus, *Acta. Pathol. Microbiol. Scand.*, 51, 199, 1961.
19. **Frothingham, J. E.,** Effect of aging and rotation on human amnion cell response to polio and sindbis viruses, *Proc. Soc. Exp. Biol. Med.*, 100, 505, 1959.
20. **Melnick, J. L. and Wallis, C.,** Problems related to the use of serum and trypsin in the growth of monkey kidney cells, *Dev. Biol. Stand.*, 37, 77, 1977.
21. **Ginsburg, H. S.,** Serum and tissue inhibitors of virus, *Bacteriol. Rev.*, 24, 141, 1960.
22. **Wenner, H. A.,** Problems in working with enteroviruses, *Ann. N.Y. Acad. Sci.*, 101, 343, 1962.
23. **Mclaren, L. C., Holland, J. J., and Syverton, J. T.,** The mammalian cell-virus relationship. I. Attachment of poliovirus to cultivated cells of primate and non-primate origin, *J. Exp. Med.*, 109, 475, 1959.
24. **Green, J. A. and Baron, S.,** 5-Iododeoxyuridine potentiation of the replication *in vitro* of several unrelated RNA and DNA viruses, *Science*, 190, 1099, 1975.
25. **Staal, S. P. and Rowe, W. P.,** Enhancement of adenovirus infection in WI-38 and AGMK cells by pretreatment of cells with 5-iododeoxyuridine, *Virology*, 64, 513, 1975.
26. **Herrmann, E. C., Jr.,** The usefulness of human fibroblast cell lines for the isolation of viruses, *Am. J. Epidemiol.*, 85, 200, 1967.
27. **Barron, A. L. and Karzon, D. T.,** Studies of mutants of echovirus 6. I. Biologic and serologic characteristics, *Am. J. Epidemiol.*, 81, 323, 1965.
28. **Grayston, J. T., Loosli, C. G., Smith, M., McCarthy, M. A., and Johnston, P. B.,** Adenoviruses. I. The effect of total incubation time in HeLa cell cultures on the isolation rate, *J. Infect. Dis.*, 103, 75, 1958.
29. **Dulbecco, R.,** Production of plaques in monolayer tissue cultures by single particles of an animal virus, *Proc. Natl. Acad. Sci. U.S.A.*, 38, 747, 1952.
30. **Dulbecco, R. and Vogt, M.,** Plaque formation and isolation of pure lines with poliomyelitis viruses, *J. Exp. Med.*, 99, 167, 1954.
31. **Cooper, P. D.,** A method for producing plaques in agar suspensions of animal cells, *Virology*, 1, 397, 1955.
32. **Cooper, P. D.,** The plaque assay of animal viruses, *Adv. Virus Res.*, 8, 319, 1961.
33. **Darnell, J. E. and Sawyer, T. K.,** The basis for variation in susceptibility to poliovirus in HeLa céils, *Virology*, 11, 665, 1960.
34. **Buthala, D. A.,** Effect of carbon dioxide concentration on ability of Coxsackie A-21 to form plaques, *J. Bacteriol.*, 86, 1356, 1963.
35. **Matsumoto, M., Nakano, M., Nishikawa, F., and Tagaya, I.,** Studies on the interaction between coxsackievirus A9 and Hela cells. 1.Plaque-forming ability of coxsackievirus A9 in Hela cell cultures, *Jpn. J. Med. Sci. Biol.*, 29, 67, 1976.
36. **Hsiung, G. D. and Melnick, J. L.,** Plaque formation with poliomyelitis, coxsackie and orphan (echo) viruses in bottle cultures, *Virology*, 1, 533, 1955.
37. **Konowalchuk, J. and Speirs, J. I.,** A method for recovering enterovirus from infected samples by adsorption onto shaken suspended cells, *Can. J. Microbiol.*, 17, 1279, 1971.
38. **Youngner, J. S.,** Virus yield per cell in monkey kidney cultures inoculated with different strains of poliomyelitis virus, *J. Immunol.*, 76, 50, 1956.
39. **Wesslien , T. and Saurander, P.,** Changes in virulence of attenuated strains of poliomyelitis after passages in cultures of human embryonic tissue, *Arch. Virusforsch.*, 8, 360, 1958.
40. **Hsiung, G. D. and Melnick, J. L.,** Effect of sodium bicarbonate concentration on plaque formation of virulent and attenuated polioviruses, *J. Immunol.*, 80, 282, 1958.
41. **Hsiung, G. D.,** The use of agar overlay cultures for detection of new virus isolates, *Virology*, 9, 717, 1959.
42. **Hsiung, G. D. and Melnick, J. L.,** Morphologic characteristic of plaques produced on monkey kidney monolayer culture by enteric viruses (poliomyelitis, coxsackie and ECHO groups), *J. Immunol.*, 78, 128, 1957.
43. **Wallis, C., Melnick, J. L., and Bianchi, M.,** Factors influencing enterovirus and reovirus growth and plaque formation, *Texas Rep. Biol. Med.*, 20, 693, 1962.

44. **Baron, S. and Low, R. G.,** New maintenance medium for cell culture, *Science*, 128, 89, 1958.
45. **Takemoto, K. K.,** Plaque mutants of animal viruses, *Prog. Med. Virol.*, 8, 314, 1966.
46. **Takemori, N. and Nomura, S.,** Mutation of polioviruses with respect to size of plaque. II. Reverse mutation of minute plaque mutants, *Virology*, 12, 171, 1960.
47. **Agol, V. I. and Chumakova, M. Y.A.,** Effect of polyanions on the multiplication of two variants of poliovirus, *Acta Virol.*, 7, 97, 1963.
48. **Tytell, A. A., Torop, H. A., and McCarthy, F. J.,** Adenovirus plaque formation in Grivet monkey kidney cells, *Proc. Soc. Exp. Biol. Med.*, 109, 916, 1962.
49. **Liebhaber, H. and Takemoto, K. K.,** Alteration of plaque morphology of EMC virus with polycations, *Virology*, 14, 502, 1961.
50. **Wallis, C. and Melnick, J. L.,** Mechanism of enhancement of virus plaques by cationic polymers, *J. Virol.*, 2, 267, 1968.
51. **De Maeyer, E. and Schonne, E.,** Starch gel as an overlay for the plaque assay of animal viruses, *Virology*, 24, 13, 1964.
52. **Fenner, F. and White, D. O.,** *Medical Virology*, Academic Press, New York, 1971.
53. **Borden, E. C., Gary, G. W., Jr., and Murphy, F. A.,** Comparison of agar and agarose preparations for mengovirus plaque formation, *Appl. Microbiol.*, 20, 289, 1970.
54. **Wallis, C. and Melnick, J. L.,** Magnesium chloride enhancement of cell susceptibility to poliovirus, *Virology*, 16, 122, 1962.
55. **Wallis, C., Morales, F., Powell, J., and Melnick, J. L.,** Plaque enhancement of enteroviruses by magnesium chloride, cysteine and pancreatin, *J. Bacteriol.*, 91, 1932, 1966.
56. **Takemoto, K. K. and Liebhaber, H.,** Virus-polysaccharide interactions. II. Enhancement of plaque formation and the detection of variants of poliovirus with dextran sulfate, *Virology*, 17, 499, 1962.
57. **Feorino, P. M. and Hannon, W. H.,** Use of DEAE dextran in agar overlays to enhance size of ECHO virus plaques, *Publ. Health Rep.*, 81, 1015, 1966.
58. **Rhim, J. S. and Melnick, J. L.,** Plaque formation by reoviruses, *Virology*, 15, 80, 1961.
59. **Dubes, G. R.,** Cystine requirement for normal poliovirus action on monkey kidney tissue cultures, *Proc. Soc. Exp. Biol. Med.*, 93, 129, 1956.
60. **Darnell, J. E., Jr., Lockart, R. Z., and Sawyer, T. K.,** The effect of neutral red on plaque formation in two virus-cell systems, *Virology*, 6, 567, 1958.
61. **Wasserman, F. E.,** The inactivation of adenoviruses by ultraviolet irradiation and nitrous acid, *Virology*, 17, 335, 1962.
62. **Mosley, J. W. and Enders, J. F.,** A critique of the plaque assay technique in bottle cultures, *Proc. Soc. Exp. Biol. Med.*, 108, 406, 1961.
63. **Kaplan, A. S. and Melnick, J. L.,** Multiplication of virulent poliovirus in Capuchin monkey kidney cultures without microscopically observed cytopathogenicity, *Proc. Soc. Exp. Bio. Med.*, 90, 562, 1955.
64. **Gabrielson, M. O. and Hsiung, G. D.,** Sensitivity of agar overlay method for the recognition of enteroviruses, *Appl. Microbiol.*, 13, 967, 1965.
65. **Hatch, M. H. and Marchetti, G. E.,** A comparative study of agar overlay and standard tissue culture methods for isolation of enteroviruses, *Publ. Health Rep.*, 90, 29, 1975.
66. **Yoshii, T. and Kono, R.,** Optimal conditions for plaque assay of echoviruses in human embryonic lung fibroblast cells, *Jpn. M. Med. Sci. Biol.*, 31, 87, 1978.
67. **Nupen, E. M.,** Virus studies on the Windhoek wastewater reclamation plant (South-west Africa), *Water Res.*, 4, 661, 1970.
68. **Cooney, M. K., Hall, C. E., and Fox, J. P.,** The Seattle virus watch. III. Evaluation of isolation methods and summary of infections detected by virus isolations, *Am. J. Epidemiol.*, 96, 286, 1972.
69. **Cooney, M. K.,** Relative efficiency of cell cultures for detection of viruses, *Health Lab. Sci.*, 10, 294, 1973.
70. **Holland, J. J.,** Receptor affinities as major determinants of enterovirus tissue tropisms in humans, *Virology*, 15, 312, 1961.
71. **Righthand, V. F. and Bagshaw, J. C.,** Characterization of a regulator of host cell permissiveness to a specific enterovirus, *Virology*, 86, 148, 1978.
72. **Lonberg-Holm, K. and Philipson, L.,** Early interaction between animal viruses and cells, *Monographs in Virology*, Vol. 9, Melnick, J. L., Ed., S. Karger, Basel, 1974.
73. **Hsiung, G. D.,** Further studies on characterization and grouping of ECHO viruses, *Ann. N.Y. Acad. Sci.*, 101, 413, 1962.
74. **Schmidt, N. J., Lennette, E. H., Shon, C. W., and Dennis, J.,** The sensitivity of grivet monkey kidney cell line BS-C-1 for propagation and isolation of certain human viruses, *Am. J. Publ. Health*, 54, 1522, 1964.
75. **Hambling, M. H. and Davis, P. M.,** Susceptibility of the LLC-MK_2 line of monkey cells to human enteroviruses, *J. Hyg.*, 63, 169, 1965.

76. **Forman, M. L., Inhorn, S. L., Sheaff, E., and Cherry, J. D.,** Biological characteristics and viral spectrum of serially cultivated fetal rhesus monkey kindey cells, *Proc. Soc. Exp. Biol. Med.*, 131, 1060, 1969.
77. **Davis, P. M. and Phillpotts, R. J.,** Susceptibility of the Vero line of African green monkey kidney cells to human enteroviruses, *J. Hyg.*, 72, 23, 1974.
78. **Grabow, W. O. K. and Nupen, E. M.,** Comparison of primary vervet kidney cells with the BGM cell line for the enumeration of enteric viruses in water by means of a tube dilution technique, in *Viruses and Wastewater Treatment, Proc. Int. Symp. Viruses and Wastewater Treatment,* Goddard, M. and Butler, M., Eds., Pergamon Press, Oxford, 1981, 253.
79. **Trlifajova, V., Strizova, V., and Hejdova, E.,** The sensitivity of the human diploid strain WI-38 to different viruses, *J. Hyg. Epidemiol. Microbiol. Immunol.*, 14, 413, 1970.
80. **Rawls, W. E. and Herrmann, E. C., Jr.,** Human placenta as a source of fibroblasts for viral studies, *Am. J. Hyg.*, 80, 266, 1964.
81. **Embil, J. A. and Faulkner, R. S.,** Human diploid cell strains and their susceptibility to viruses, *Can. J. Publ. Health,* 55, 111, 1964.
82. **Wecker, I. and ter Meulen, V.,** RD cells in the laboratory diagnosis of enteroviruses, *Med. Microbiol. Immunol.*, 163, 233, 1977.
83. **Kado, G.,** Relative sensitivity of three cell substrates to the Sabin poliovirus strains, *Dev. Biol. Stand.*, 37, 261, 1977.
84. **Harris, R. E. and Pindak, F. F.,** Viral replication in human ovarian cell cultures, *Obstet. Gynecol.*, 46, 227, 1975.
85. **Marchetti, G. E. and Gelfand, H. M.,** Use of Hep-2 cells for grouping enteroviruses, *Publ. Health Rep.*, 78, 813, 1963.
86. **Hopps, H. E., Bernheim, B. C., Nisalak, A., and Tjio, J. H.,** Biological characteristics of a continuous kidney cell line derived from the African green monkey, *J. Immunol.*, 91, 416, 1963.
87. **Barron, A. L., Olshevsky, C., and Cohen, M. M.,** Characteristics of the BGM line of cells from African green monkey kidney, *Arch Virusforsch.*, 32, 389, 1970.
88. **Lee, L. H., Phillips, C. A., South, M. A., Melnick, J. L., and Yow, M. D.,** Enteric virus isolation in different cell cultures, *Bull. WHO,* 32, 657, 1965.
89. **McSwiggan, D. A. and George, R. A.,** A comparison of two cell culture systems for the primary isolation of enteric viruses, *Bull. WHO,* 43, 295, 1970.
90. **Kalter, S. S., Fuentes-Marins, R., Rodriguez, A. R., Hellman, A., Crandell, R. A., and Werthessen, N. T.,** Susceptibility of baboon *(Papio Doguera)* kidney cells to human enteroviruses, *Proc. Soc. Exp. Biol. Med.*, 111, 337, 1962.
91. **Herrmann, E. C., Jr.,** Efforts toward a more useful viral diagnostic laboratory, *Am. J. Clin. Pathol.*, 56, 681, 1971.
92. **Schmidt, N. J., Ho, H. H., and Lennette, E. H.,** Comparative sensitivity of the BGM cell line for isolation of enteric viruses, *Health Lab. Sci.*, 13, 115, 1976.
93. **Rutala, W. A., Shelton, D. F., and Arbiter, D.,** Comparative sensitivities of viruses to cell cultures and transport media, *Am. J. Clin. Pathol.*, 67, 397, 1977.
94. **Melnick, J. L. and Rennick, V.,** Infectivity titers of enterovirus as found in human stools, *J. Med. Virol.*, 5, 205, 1980.
95. **Schmidt, N. J., Ho, H. H., Riggs, J. L., and Lennette, E. H.,** Comparative sensitivity of various cell culture systems for isolation of viruses from wastewater and fecal samples, *Appl. Environ. Microbiol.*, 36, 480, 1978.
96. **Sickles, G. M., Mutterer, M., Feorino, P., and Plager, H.,** Recently classified types of coxsackie virus, group A. Behavior in tissue culture, *Proc. Soc. Exp. Biol. Med.*, 90, 529, 1955.
97. **Lehman-Grube, F. and Syverton, J. T.,** Pathogenicity for suckling mice of coxsackieviruses adapted to human amnion cells, *J. Exp. Med.*, 113, 811, 1961.
98. **Shaw, M.,** Cultivation of coxsackie virus in embryonated eggs and in chick tissue culture, *Proc. Soc. Exp. Biol. Med.*, 79, 718, 1952.
99. **Slater, E. A. and Syverton, J. T.,** The cultivation of coxsackievirus (type 4, Minnesota 1 Strain), *Proc. Soc. Exp. Biol. Med.*, 74, 509, 1950.
100. **Lenahan, M. F. and Wenner, H. A.,** Propagation of group A coxsackie viruses in primary human amnion cells. I. Cytopathic changes produced by 5 more serotypes, *Proc. Soc. Exp. Biol. Med.*, 107, 544, 1961.
101. **Wenner, H. A. and Lenahan, M. F.,** Propagation of group A coxsackie viruses in tissue cultures. II. Some interactions between virus and mammalian cells, *Yale J. Biol. Med.*, 34, 421, 1961/2.
102. **Schmidt, N. J., Ho, H. H., and Lennette, E. H.,** Propagation and isolation of group A coxackieviruses in RD cells, *J. Clin. Microbiol.*, 2, 183, 1975.
103. **Crowell, R. L. and Goldberg, B.,** Propagation and assay of group A coxsackieviruses in RD cells, *Abstr. Annu. Meet. Am. Soc. Microbiol.*, 44, 208, 1974.

104. **Sohier, R. and Gaudin, O. G.**, Monkey cell cultures in virology, *Primates Med.*, 3, 80, 1969.
105. **Hsiung, G. D. and Melnick, J. L.**, Comparative susceptibility of kidney cells from different monkey species to enteric viruses (poliomyelitis, coxsackie and ECHO groups), *J. Immunol.*, 78, 137, 1957.
106. **Dahling, D. R., Berg, G., and Berman, D.**, BGM, a continuous cell line more sensitive than primary rhesus and African green kidney cells for the recovery of viruses from water, *Health Lab. Sci.*, 11, 275, 1974.
107. **Lehman-Grube, F.**, Comparative sensitivity of mammlian cells in culture to prototype enteroviruses, *Arch. Virusforsch.*, 11, 276, 1961.
108. **Stoneback, T. A. and Crowell, R. L.**, Optimal conditions for the plaque assay and attachment of coxsackievirus A5 by RD cells, *Abstr. Annu. Meet. Am. Soc. Microbiol.*, 44, 208, 1974.
109. **Matsumoto, M., Nakano, M., and Tagaya, I.**, Studies on the interaction between coxsackievirus A9 and HeLa cells. II. Mode of growth of coxsackievirus A9 in HeLa cell cultures and the effect of sulfated polysaccharide on plaque formation, *Jpn. J. Med. Sci. Biol.*, 29, 77, 1976.
110. **Frolova, I. A. and Shirobokov, V. P.**, Comparative sensitivity of HEp-2, HeLa, MDCK and L cell cultures to group coxsackieviruses under conditions of bentonite utilization, *Mikrobiol. Zh.*, 41, 270, 1979.
111. **Sickles, G. M., Feorino, P., and Plager, H.**, Isolation and type determination of coxsackievirus, group B, in tissue culture, *Proc. Soc. Exp. Biol. Med.*, 88, 22, 1955.
112. **Crowell, R. L. and Syverton, J. T.**, The viral range *in vitro* of a malignant, human epithelial cell (strain HeLa, GEY). IV. The cytopathogenicity of C viruses, *J. Immunol.*, 74, 169, 1955.
113. **Schmidt, N. J., Midulla, M., and Lennette, E. H.**, A colorimetric measure for the typing of coxsackieviruses of group B and certain members of group A, *J. Immunol.*, 80, 454, 1958.
114. **Hollick, G. E., Reichrath, L., and Smith, T. F.**, Comparison of primary rhesus and cynomolgus monkey kidney cell cultures for viral isolation from clinical specimens, *Am. J. Clin. Pathol.*, 68, 276, 1977.
115. **Phillips, C. A.**, Enteroviruses and reoviruses, in *Manual of Clinical Microbiology*, 3rd ed., Lennette, E. H., Balows, A., Hausler, W. J., Jr., and Truant, J. P., Eds., American Society for Microbiology, Washington, D.C., 1980, 823.
116. **Barron, A. L. and Karzon, D. T.**, Cultivation of enteroviruses in hamster kidney tissue culture, *Proc. Soc. Exp. Biol. Med.*, 100, 316, 1959.
117. **Archetti, I., Weston, J., and Wenner, H. A.**, Adaption of ECHO viruses in HeLa cells; their use in complement fixation, *Proc. Soc. Exp. Biol. Med.*, 95, 265, 1957.
118. **Schmidt, N. J., Shinomoto, T. T., Dennis, J., Hagens, S. J., Fox, V. L., and Lennette, E. H.**, Colorimetric test in HeLa cell system for assay for neutralizing antibodies to ECHO viruses, *J. Lab. Clin. Med.*, 59, 687, 1962.
119. **Suto, T., Karzon, D. T., Bussell, R. H., and Barron, A. L.**, Studies of mutants of echovirus 6. II. Isolation from human alimentary tract, *Am. J. Epidemiol.*, 81, 333, 1965.
120. **Suto, T., Karzon, D. T., and Bussell, R. H.**, Studies of mutants of echovirus 6. III. Behavior in monkey and human cell culture, *Am. J. Epidemiol.*, 81, 341, 1965.
121. **Hatch, M. H. and Marchetti, G. E.**, Isolation of echoviruses with human embryonic lung fibroblast cells, *Appl. Microbiol.*, 22, 736, 1971.
122. **Podoplekin, V. D. and Idina, M. S.**, Isolation of ECHO viruses from man and their identification by the haemagglutination inhibition test, *Acta Virol.*, 7, 233, 1963.
123. **Grabow, W. O. K. and Prozesky, O. W.**, Studies on the isolation and identification of hepatitis viruses in water, *Prog. Water Technol.*, 10, 389, 1978.
123a. **Dienhardt, F. and Gust, I. D.**, Viral hepatitis, *Bull, WHO*, 60, 661, 1982.
123b. **Lemon, S. M., LeDuc, J. W., Binn, L., Escajadillo, A., and Ishak, K. G.**, Transmission of hepatitis A virus among recently captured owl monkeys, *J. Med. Virol.*, 10, 25, 1982.
124. **Provost, P. J. and Hilleman, M. R.**, Propagation of human hepatitis A virus in cell culture, *in vitro*, *Proc. Soc. Exp. Biol. Med.*, 160, 213, 1979.
125. **Frosner, G. G., Dienhardt, F., Scheid, R., Gauss-Muller, V., Holmes, N., Messelberger, V., Siegl, G., and Alexander, J. J.**, Propagation of human hepatitis A virus in a hepatoma cell line, *Infection*, 7, 1, 1979.
126. **Daemer, R. J., Feinstori, S. M., Gust, I. D., and Purcell, R. H.**, Propagation of human hepatitis A virus in African green monkey kidney cell culture: primary isolation and serial passage, *Infect. Immun.*, 32, 388, 1981.
126a. **Provost, P. J., McAleer, W. J., and Hilleman, M. R.**, In vitro cultivation of hepatitis A virus, in *Viral Hepatitis* , Szmuness, W., Alter, H. J., and Maynard, J. E., Eds., The Franklin Institute Press, Philadelphia, 1982.
126b. **Gerba, C. P., Singh, S. N., and Rose, J. B.**, Waterborne gastroenteritis and viral hepatitis, *Crit. Rev. Environ. Control*, 15(3), 213, 1985.
126c. **Simmonds, R. S., Szucs, G., Metcalf, T. G., and Melnick, J. L.**, Persistently infected cultures as a source of hepatitis A virus, *Appl. Environ. Microbiol.*, 49, 749, 1985.

127. **Yoshii, T., Natori, K., and Kono, R.,** Replication of enterovirus 70 in non-primate cell cultures, *J. Gen. Virol.*, 36, 377, 1977.
128. **Melnick, J. L. and Wenner, H. A.,** Enteroviruses, in *Diagnostic Procedures for Viral and Rickettsial Infections*, 4th ed., Lennette, E. H. and Schmidt, N. J., Eds., APHA, New York, 1969, chap. 17.
129. **Friedman, H. M. and Koropchak, C.,** Comparison of WI-38, MRC-5, and IMR-90 cell strains for isolation of viruses from clinical specimens, *J. Clin. Microbiol.*, 7, 368, 1978.
130. **Rose, H. M.,** Adenoviruses, in *Diagnostic Procedures for Viral and Rickettsial Infections*, 4th ed., Lennette, E. H. and Schmidt, N. J., Eds., APHA, New York, 1969, chap. 5.
130a. **deJong, J. C., Wigand, R., Kidd, A. H., Waddell, G., Kapsenberg, J. G., Muzeria, C. J., Wermenbol, A. G., and Firtzlaff, R. G.,** Candidate adenoviruses 40 and 41: fastidious adenoviruses from human infant stool, *J. Med. Virol.*, 11, 215, 1983.
131. **Retter, M., Middleton, P. J., Tam, J. S., and Petric, M.,** Enteric adenoviruses: detection, replication and significance, *J. Clin. Microbiol.*, 10, 574, 1979.
132. **Rowe, W. P., Huebner, R. J., Hartley, J. W., and Ward, T. G., and Parrott, R. H.,** Studies of the adenoidal-pharyngeal-conjunctival (APC) group of viruses, *Am. J. Hyg.*, 61, 197, 1955.
133. **Friedman, M. P., Lyons, M. J., and Ginsberg, H. S.,** Biochemical consequences of type 2 adenovirus and SV40 double infection of African green monkey kidney cells, *J. Virol.*, 5, 586, 1970.
134. **Hasler, P. and Wigand, R.,** The susceptibility of Vero cell cultures for human adenoviruses, *Med. Microbiol. Immunol.*, 164, 267, 1978.
135. **Tooze, J., Ed.,** *The Molecular Biology of Tumor Viruses*, Cold Spring Harbor Laboratory, Cold Spring Harbor, New York, 1973, 440.
136. **Fox, R. I. and Baum, S. G.,** Posttranscriptional block to adenovirus replication in nonpermissive monkey cells, *Virology*, 60, 45, 1974.
137. **Ginsburg, H. S.,** Identification and classification of adenoviruses, *Virology*, 18, 312, 1962.
138. **Grayston, J. T., McCarthy, M. A., Smith, M., and Loosli, C. G.,** Adenoviruses. II. The sensitivity of different tissue culture strains of cells to adenoviruses, *J. Infect. Dis.*, 103, 86, 1958.
139. **Wigand, R. and Kumel, G.,** The kinetics of adenovirus infection and spread in cell cultures infected with low multiplicity, *Arch. Virol.*, 54, 177, 1977.
139a. **Takiff, H. E., Strauss, S. E., and Garon, C. F.,** Propagation and in vitro studies of previously non-cultivatable enteral adenoviruses in 293 cells, *Lancet*, 3, 832, 1981.
140. **Kjellen, L.,** A study of adenovirus-host cell system by the plaque technique, *Virology*, 14, 234, 1961.
141. **Rouse, H. C., Bonifas, V. H., and Schlesinger, R. W.,** Dependence of adenovirus replication on arginine and inhibition of plaque formation by pleuropneumonia-like organisms, *Virology*, 20, 357, 1963.
142. **Rouse, H. C. and Schlesinger, R. W.,** An arginine-dependent step in the maturation of type 2 adenovirus, *Virology*, 33, 513, 1967.
143. **Rouse, H. C. and Schlesinger, R. W.,** The effects of arginine starvation on macromolecular synthesis in infection with type 2 adenovirus. I. Synthesis and utilization of structural proteins, *Virology*, 48, 463, 1972.
144. **Russell, W. C. and Becker, Y.,** A maturation factor for adenovirus, *Virology*, 35, 18, 1968.
145. **Everitt, E., Sundquist, B., and Philipson, L.,** Mechanism of the arginine requirement for adenovirus synthesis. I. Synthesis of structural proteins, *J. Virol.*, 8, 742, 1971.
146. **Philipson, L.,** Adenovirus assay by the fluorescent cell counting procedures, *Virology*, 15, 263, 1961.
147. **Gary, G. W., Jr., Hierholzer, J. C., and Black, R. E.,** Characteristics of noncultivable adenoviruses associated with diarrhea in infants: a new subgroup of human adenoviruses, *J. Clin. Microbiol.*, 10, 96, 1979.
148. **Goulet, N. R., Schriver, P. V., Liu, D. C., and Beiler, J. M.,** Isolation and propagation of adenoviruses in African green monkey cultures, *J. Infect. Dis.*, 115, 429, 1965.
149. **Macrae, A. D.,** Reoviruses of man, *Ann. N.Y. Acad. Sci.*, 101, 455, 1962.
150. **Rosen, L.,** Reoviruses, in *Diagnostic Procedures for Viral and Rickettsial Infections*, 4th ed., Lennette, E. H. and Schmidt, N. J., Eds., APHA, New York, 1969, chap. 9.
151. **Hsiung, G. D.,** Some distinctive biological characteristics of ECHO-10 virus, *Proc. Soc. Exp. Biol. Med.*, 99, 387, 1958.
152. **Franklin, R. M.,** Comparisons of assays for mengovirus and reovirus 3, *Proc. Soc. Exp. Biol. Med.*, 107, 651, 1961.
153. **Rhim, J. S., Kato, J. I., and Pelon, W.,** Hemagglutination by reoviruses propagated in various cell lines, *Proc. Soc. Exp. Biol. Med.*, 118, 453, 1965.
154. **Spendlove, R. S., McClain, M. E., and Lennette, E. H.,** Enhancement of reovirus infectivity by extracellular removal or alteration of the virus capsid by proteolytic enzymes, *J. Gen. Virol.*, 8, 83, 1970.
155. **Wallis, C., Smith, K. O., and Melnick, J. L.,** Reovirus activation by heating and inactivation by cooling in $MgCl_2$ solutions, *Virology*, 22, 608, 1964.
156. **Spendlove, R. S. and Schaffer, F. L.,** Enzymatic enhancement of infectivity of reovirus, *J. Bacteriol.*, 89, 587, 1965.

157. **Spendlove, R. S.,** Unique reovirus characteristics, *Prog. Med. Virol.*, 12, 161, 1970.
158. **Sargent, M. D., Long, D. G., and Borsa, J.,** Functional analysis of the interactions between reovirus particles and various proteases *in vitro, Virology*, 78, 354, 1977.
159. **Loh, P. C. and Oie, H. K.,** Role of lysine in the replication of reovirus. I. Synthesis of complete and empty virions, *J. Virol.*, 4, 890, 1969.
160. **Loh, P. C., Hashiro, G. M., and Yau, J. T.,** Effect of polycations on the early stages of reovirus infection, *Microbios*, 19, 213, 1977.
161. **Hashiro, G., Loh, P. C., and Yau, J. T.,** The preferential cytotoxicity of reovirus for certain transformed cell lines, *Arch. Virol.*, 54, 307, 1977.
162. **Wallis, C., Melnick, J. L., and Rapp, F.,** Effects of pancreatin on the growth of reovirus, *J. Bacteriol.*, 92, 155, 1966.
163. **Flewett, T. H. and Woode, G. N.,** The rotaviruses, *Arch. Virol.*, 57, 1, 1978.
164. **Banatvala, J. E.,** Viruses and diarrhea, *Trans. R. Soc. Trop. Med. Hyg.*, 73, 503, 1979.
165. **Estes, M. K. and Graham, D. Y.,** Epidemic viral gastroenteritis, *Am. J. Med.*, 66, 1001, 1979.
166. **Flewett, T. H., Thouless, M. E., Pilford, J. H., Bryden, A. S., and Candeias, J. A. N.,** More serotypes of human rotavirus, *Lancet*, 2, 632, 1978.
166a. **Wyatt, R. G., James, H. D., Pittman, A. L., Hoshino, Y., Greenberg, H. B., Kalica, A. R., Flores, J., and Kapikian, A. Z.,** Direct isolation in cell culture of human rotaviruses and their characterization into four serotypes, *J. Clin. Microbiol.*, 18, 310, 1983.
167. **Steinmann, J.,** Detection of rotavirus in sewage, *Appl. Environ. Microbiol.*, 41, 1043, 1981.
168. **Banatvala, J. E., Totterdell, B., Chrystie, I. L., and Woode, G. N.,** *In vitro* detection of human rotaviruses, *Lancet*, 2, 821, 1975.
169. **Bryden, A. S., Davies, H. A., Thouless, M. E., and Flewett, T. H.,** Diagnosis of rotavirus infection by cell culture, *J. Med. Microbiol.*, 10, 121, 1977.
170. **Schoub, B. D., Kalica, A. R., Greenberg, H. B., Bertran, D. M., Sereno, M. M., Wyatt, R. G., Chanock, R. M., and Kapikian, A. Z.,** Enhancement of antigen incorporation and infectivity of cell cultures by human rotavirus, *J. Clin. Microbiol.*, 9, 488, 1979.
171. **Wyatt, R. G., James, W. D., Bohl, E. H., Theil, K. W., Saif, L. J., Kalica, A. R., Greenberg, H. B., Kapikian, A. Z., and Chanock, R. M.,** Human rotavirus type 2: cultivation *in vitro, Science*, 207, 189, 1980.
172. **Dolin, R., Blacklow, N. R., DuPont, H., Formal, S., Buscho, R. F., Kasel, J. A., Chames, R. P., Hornick, R., and Chanock, R. M.,** Transmission of acute infectious nonbacterial gastroenteritis to volunteers by oral administration of stool filtrates, *J. Infect. Dis.*, 123, 307, 1971.
173. **Wyatt, R. G. and Kapikian, A. Z.,** Viral agents associated with acute gastroenteritis in humans, *Am. J. Clin. Nutr.*, 30, 1857, 1977.
174. **Banatvala, J. E.,** Viruses in faeces, in *Viruses and Wastewater Treatment, Proc. Int. Symp. Viruses and Wastewater Treatment*, Goddard, M. and Butler, M., Eds., Pergamon Press, Oxford, 1981, 1.
174a. **Kaplan, J. E., Gary, G. W., Baron, R. C., Singh, N., Schonberger, L. B., Feldman, R., and Greenberg, H. B.,** Epidemiology of Norwalk gastroenteritis and the role of Norwalk virus in outbreaks of acute nonbacterial gastroenteritis, *Ann. Intern. Med.*, 96, 756, 1982.
175. **Caul, E. O., Paver, W. K., and Clarke, S. K. R.,** Coronavirus particles in faeces from patients with gastroenteritis, *Lancet*, 1, 1192, 1975.
176. **Caul, E. D. and Egglestone, S. I.,** Further studies on human enteric corona-viruses, *Arch. Virol.*, 54, 107, 1977.
177. **McIntosh, K.,** Coronaviruses: a comparative review, *Curr. Top. Microbiol. Immunol.*, 63, 85, 1974.
178. **Lee, T. W. and Kurtz, J. B.,** Astroviruses detected by immunofluorescence, *Lancet*, 2, 406, 1977.
179a. **Lee, T. W. and Kurtz, J. B.,** Serial propagation of astrovirus in tissue culture with the aid of trypsin, *J. Gen. Virol.*, 57, 421, 1981.
179b. **Cubitt, W. D. and Barrett, A. D. T.,** Propagation of human candidate calicivirus in cell culture, *J. Gen. Virol.*, 65, 1123, 1984.
179c. **Khuroo, M. S.,** Study of an epidemic of non-A, non-B hepatitis. Possibility of another human hepatitis virus distinct from posttransfusion non-A, non-B type, *Am. J. Med.*, 68, 818, 1980.
179d. **Wong, D. C., Purcell, R. H., Sreenivasan, M. A., Prasad, S. R., and Pavri, K. M.,** Epidemic and endemic hepatitis in India: evidence for non-A, non-B hepatitis virus etiology, *Lancet*, 2, 876, 1980.
179e. **Pavri, K., Sreenivasan, M. A., Sehgal, A., and Prasad, S. R.,** Waterborne epidemics of non A-non B viral hepatitis in India, in *Viral Disease in Southeast Asia and the Western Pacific*, Mackenzie, J. S., Ed., Academic Press, New York, 1982.
179f. **Purcell, R. H.,** The hepatitis viruses: an overview and historical perspective, in *Viral Hepatitis*, Szmuness, W., Alter, H. J., and Maynard, J. E., Eds., The Franklin Institute Press, Philadelphia, 1982.
179g. **Sreenivasan, M. A., Arankalle, V. A., Sehgal, A., and Pavri, K. M.,** Non-A, non-B epidemic hepatitis: visualization of virus-like particles in the stool by immune electron microscopy, *J. Gen. Virol.*, 65, 1005, 1984.

180. **Keswick, B. H. and Gerba, C. P.,** Viruses in groundwater, *Environ. Sci. Technol.*, 14, 1290, 1980.
181. **Hejkal, T. W., Keswick, B., LaBelle, R. L., Gerba, C. P., Sanchez, Y., Dreesma, G., Hafkin, B., and Melnick, J. L.,** Viruses in a community water supply associated with an outbreak of gastroenteritis and infectious hepatitis, *JAWWA*, 74, 318, 1982.
182. **Markov, A. P.,** Experience in the sanitary-virological examination of waste water, *Gig. Sanit.*, 37, 94, 1972.
183. **Lydholm, B. and Nielson, A. L.,** Methods for detection of virus in wastewater, applied to samples from small scale treatment systems, *Water Res.*, 14, 169, 1908.
184. **Rolland, D. and Block, J. C.,** Simultaneous concentration of *Salmonella* and enterovirus from surface water by using micro-fiber glass filters, *Appl. Environ. Microbiol.*, 39, 659, 1980.
185. **Paul, J. R., Trask, J. D., and Culotta, C. S.,** Poliomyelitic virus in sewage, *Science*, 90, 258, 1939.
186. **Kelly, S. M.,** Detection and occurrence of coxsackie viruses in sewage, *Am. J. Publ. Health*, 43, 1532, 1953.
187. **Kelly, S. M., Clark, M. E., and Coleman, M. B.,** Demonstration of infectious agents in sewage, *Am. J. Publ. Health* 45, 1438, 1955.
188. **Kelly, S. M., Winsser, J., and Winkelstein, W., Jr.,** Poliomyelitis and other enteric viruses in sewage, *Am. J. Publ. Health*, 47, 72, 1957.
189. **Melnick, J. L., Emmons, J., Opton, E. M., and Coffey, J. H.,** Coxsackieviruses from sewage. Methodology including an evaluation of the grab sample and gauze pad collection procedures, *Am. J. Hyg.*, 59, 185, 1954.
190. **Bloom, H. H., Mack, W. N., Krueger, B. J., and Mallman, W. L.,** Identification of enteroviruses in sewage, *J. Infect. Dis.*, 105, 61, 1959.
191. **Simkova, A. and Wallnerova, Z.,** Isolation of coxsackie viruses from Danube River water, *Acta Virol.*, 17, 363, 1973.
192. **Faulkner, R. S. and Van Rooyen, C. E.,** Studies on surveiliance and survival of viruses in sewage in Nova Scotia, *Can. J. Publ. Health*, 60, 345, 1969.
193. **Farrar, L. and Hedrick, H. G.,** Recovery of enteric viruses from sewage effluent by polyelectrolyte adsorption, in *Developments in Industrial Microbiology*, Vol. 14, American Institute of Biological Sciences, Washington, D.C., 1973, 376.
194. **Kelly, S. M. and Sanderson, W. W.,** Comparison of various tissue cultures for the isolation of enteroviruses, *Am. J. Publ. Health*, 52, 455, 1962.
195. **Laveran, H., Beytout, D., Cluzel, R., Joly, B., and Herail, M.,** Demonstrating human enteric viruses in waste waters. Comparison of three methods of virus concentration, *Med. Mal. Infect.*, 4, 69, 1974.
196. **Slade, J. S.,** Enteroviruses in slow sand filtered water, *J. Inst. Water Eng. Sci.*, 32, 530, 1978.
197. **Morris, R. and Waite, W. M.,** Evaluation of procedures for recovery of viruses from water. II. Detection systems, *Water Res.*, 14, 795, 1980.
198. **Ozere, R. L., Faulkner, R., and van Rooyen, C. E.,** Enteroviruses in sewage and epidemic poliomyelitis in Eastern Canada, *Can. Med. Assn. J.*, 85, 1419, 1961.
199. **Bottiger, M.,** Isolation of viruses from sewage, with special regard to poliovirus. An evaluation of the sensitivity of three tissue culture methods, *Bull. WHO*, 56, 937, 1978.
200. **Sellwood, J. and Dadswell, J. V.,** Viruses isolated from sewage 1974—77, in *Viruses and Wastewater Treatment, Proc. Int. Symp. Viruses and Wastewater Treatment*, Goddard, M. and Butler, M., Eds., Pergamon Press, Oxford, 1981, 27.
201. **Irving, L. G. and Smith, F. A.,** One-year survey of enteroviruses, adenoviruses, and reoviruses isolated from effluent at an activated-sludge purification plant, *Appl. Environ. Microbiol.*, 41, 51, 1981.
202. **Zdrazilek, J.,** Sensitivity of human diploid embryonic lungs cells for isolation of echoviruses from sewage, *J. Hyg. Epidemiol. Microbiol. Immunol.*, 18, 2, 1974.
203. Metropolitan Sanitary District of Greater Chicago and IIT Research Institute, Viral and bacterial levels resulting from the land application of digested sewage, Report, 1979.
204. **Sekla, L., Stackiw, W., Kay, C., and Van Buckenhout, L.,** Enteric viruses in renovated water in Manitoba, *Can. J. Microbiol.*, 26, 518, 1980.
205. **Bottiger, M.,** Experiences from investigations of virus isolations from sewage over a two year period with special regard to polioviruses, *Arch. Virusforsch.*, 41, 80, 1973.
206. **Wellings, F. M., Lewis, A. L., Mountain, C. W., and Pierce, L. V.,** Demonstration: of virus in groundwater after effluent discharge into soil, *Appl. Microbiol.*, 29, 751, 1975.
207. **Loh, P. C., Fujioka, R. S., and Lau, L. S.,** Recovery, survival and dissemination of human enteric viruses in ocean waters receiving sewage in Hawaii, *Water Air Soil. Pollut.*, 12, 197, 1979.
208. **Sellwood, J.,** personal communication, 1980.

INDEX

A

B

C

D

E

F

M

N

O

P

T

U

V

W

Y

Z